COMPLEX VARIABLES

PURE AND APPLIED MATHEMATICS

A Program of Monographs, Textbooks, and Lecture Notes

MONOGRAPHS AND TEXTBOOKS IN
PURE AND APPLIED MATHEMATICS

Additional Volumes in Preparation

COMPLEX VARIABLES

An Introduction

Watson Fulks
University of Colorado
Boulder, Colorado

Marcel Dekker, Inc.　　　New York • Basel • Hong Kong

Library of Congress Cataloging-in-Publication Data

Fulks, Watson.
    Complex variables : an introduction / Watson Fulks.
        p.   cm. -- (Monographs and textbooks in pure and applied
    mathematics ; 176)
      Includes index.
      ISBN 0-8247-9079-0
      1. Functions of complex variables. I. Title. II. Series.
    QA331.7.F85 1993
    515'.9--dc20   93-26461
                                                              CIP

The publisher offers discounts on this book when ordered in bulk quantities. For more information, write to Special Sales/Professional Marketing at the address below.

This book is printed on acid-free paper.

MARCEL DEKKER, INC.
270 Madison Avenue, New York, New York 10016

Current printing (last digit):
10 9 8 7 6 5 4 3 2 1

PRINTED IN THE UNITED STATES OF AMERICA

For my granddaughter

Alexandra Elinor Fulks

Preface

This book is designed for a one semester introduction to the subject of Complex Variables. The major way in which my exposition of this material differs from other texts on the same subject is in its organization, although it also has some unique simplifying approaches to certain topics.

The most unorthodox feature is my early introduction of the integral, particularly the integral in polar coordinates. This enables me to prove preliminary forms of the Cauchy theorems for polynomials and rational functions. This in turn permits early access to the evaluation of some real integrals which are quite difficult by elementary calculus methods, and so provides an early demonstration of the computational utility of the subject.

A second important consequence of the early introduction of the integral is that it allows me to define the complex logarithm by

$$\log z = \int_1^z \frac{d\zeta}{\zeta}.$$

From this definition the ambiguity (multivaluedness) is discussed in an easy and geometrically natural way based on the paths of integration from 1 to z.

Another unusual organizational feature of my discussion is the treatment of series before the main Cauchy theorems and Taylor's theorem. This leads me to define the exponential and trigonometric functions by their power series. These functions are then seen as the natural extensions to the complex plane of the corresponding real functions. I feel that this is a much more satisfactory approach than the usual rather awkward, almost ad hoc, definition of the exponential by

$$e^z = e^x(\cos y + i \sin y).$$

An aspect of my discussions I believe to be quite useful is the treatment of analytic continuation. The emphasis is on practical methods of producing continuations and includes a good many examples and exercises. They show the utility of continuation as a tool, for example, in the evaluation of integrals.

I pay considerable attention to applications, those centering on potential theory. There are discussions of fluid flow, of heat conduction, of conformal mapping, and of evaluation of integrals. I introduce the Laplace transform and show its use in solving initial value problems for differential equations, including a Heaviside theorem. The last chapter is devoted to an elementary systematic introduction to logarithmic potential theory. This chapter contains a final section on Fourier series and their use in some problems of partial differential equations.

In the development of the material in some sections I have occasionally relegated difficult proofs to the C Exercises. These are frequently peripheral topics or are advanced topics not essential to an undergraduate course. But for those students and instructors who want to go more deeply into these subjects the exercises provide a rather complete outline with sufficient hints. A prime example is provided by the discussion of the relationship between the rectangular and the polar equations of a curve (Section 1.4). I realize, of course, that the choice of which topics to treat thoroughly in the text and which to leave to the exercises is largely a matter of taste.

Some definitions are labeled and set in italics, like the theorems. Others are given less formally as, for example,

A point z_0 is a **boundary point** of a set S if ...

The word or phrase being defined is set in bold. Also in bold are names, such as the **Cauchy–Riemann Equations**. These form the nomenclature of the course and should be learned well.

The exercises are classified as A, B, or C Exercises. This classification is meant as an approximate ordering of their difficulty. The A Exercises are straightforward applications of the textual development. Some of them may have relatively long calculations, but the method of solution should be clear. The B Exercises are conceptually deeper, and the C Exercises are intended to challenge the A students.

Finally, I want to acknowledge, with thanks, the work of the typist, Ms. Elizabeth Stimmel, who has suffered through numerous versions and revisions with calm patience and good grace.

Watson Fulks

Contents

Contents

Chapter 1

Complex Numbers and Complex Functions

1.1 Complex Numbers

One of the standard elementary properties of real numbers is that they have nonnegative squares. In fact, the square of any nonzero real number is positive. So it is clear that the equation

$$x^2 + 1 = 0 \tag{1}$$

can have no real solutions, for the left side is never less than 1 for any real x. We can study (1) formally by following the procedure used for similar equations which have real roots. We calculate: if (1) holds, then

$$x^2 = -1$$

from which

$$x = \pm\sqrt{-1}.$$

That is, by this formal procedure we get the two symbolic roots

$$x = +\sqrt{-1} \quad \text{and} \quad x = -\sqrt{-1}.$$

We can now decide to treat the symbol $\sqrt{-1}$ as a new quantity, adjoin it to the real number system, and take the rules of arithmetic of this augmented system to be identical with the rules for the real numbers except

1

that $(\sqrt{-1})^2$ is replaced by -1 whenever it occurs. This approach was in fact used a couple of centuries ago despite a certain difficulty arising from calculations like $-1 = (\sqrt{-1})(\sqrt{-1}) = \sqrt{(-1)(-1)} = \sqrt{1} = 1$. While this difficulty can be avoided, it is formally simpler to adopt the notation of Euler (pronounced "oiler"), namely $i = \sqrt{-1}$. The number i is called the **imaginary unit** and numbers of the form $a + ib$, with a and b real, are called **complex numbers**. This yields an algebraically satisfactory number system if we follow the standard rules of arithmetic except that i^2 gets replaced by (-1). It is to be expected, from our remarks concerning the equation $x^2 + 1 = 0$, that this number system is sufficient for the solution of all quadratic equations. But it is perhaps surprising that it is also sufficient for the solution of all polynomial equations. (See Theorem 3.2h, the Fundamental Theorem of Algebra.)

While this is a formally satisfactory and internally consistent number system, it still leaves one asking if we did not cheat by throwing in a previously nonsensical symbol ($\sqrt{-1}$, or equivalently, i) and proceeding to use it as though we know what it means. There is a real conceptual difficulty here which is discussed in courses in algebra. However, the simplest way to cope with this difficulty is to start over again in what appears to be an unrelated manner. If we define complex numbers as two-dimensional vectors with a peculiar multiplication rule (set up to recapture $i^2 = -1$) we get a completely logical complex number system in which all polynomial equations have solutions. The number i is then a certain distinguished vector. We now take this approach.

Definition. *The set* $\mathbb{C}$ *of* **complex numbers**, *individually denoted by* $z, \zeta, \ldots, \alpha, \beta, \ldots$, *is the set of ordered pairs of real numbers denoted by* $(x, y), (\xi, \eta), \ldots, (a, b), (c, d), \ldots$ *together with the following three rules. If* $\alpha = (a, b)$ *and* $\beta = (c, d)$, *then*

$$\alpha = \beta \quad means \quad a = c \quad and \quad b = d; \tag{A}$$

$$\alpha + \beta \quad means \quad (a + c, b + d); \tag{B}$$

$$\alpha\beta \quad or \quad \alpha \cdot \beta \quad means \quad (ac - bd, ad + bc). \tag{C}$$

The complex number system then consists of exactly the same objects as the set of two-dimensional vectors, namely, ordered pairs of real numbers. Furthermore, complex equality and addition are precisely vector equality and addition. Thus $\mathbb{C}$ can be considered as the set of 2-vectors with a peculiar multiplication law, given by (C).

We now make some developments and definitions to bring the rules of $\mathbb{C}$ into an easily accessible form. Since, by (B)

$$(a, 0) + (b, 0) = (a + b, 0)$$

and

$$(a, 0) \cdot (b, 0) = (a \cdot b, 0)$$

we see that the subset of $\mathbb{C}$ consisting of numbers of the form $(x, 0)$ is identical with the real number system $\boldsymbol{R}$. It is therefore a notational convenience to introduce the convention that

$$(x, 0) = x. \tag{D}$$

This imbeds $\boldsymbol{R}$ in $\mathbb{C}$. As a subset of $\mathbb{C}$, $\boldsymbol{R}$ is called the **real axis**.

Definition. *The complex number i is given by*

$$i = (0, 1). \tag{E}$$

The number i is called the **imaginary unit** and by (C) and (D) has the property that $i^2 = -1$. Numbers of the form $(0, y) = (y, 0)(0, 1) = yi$, for $y \neq 0$, are called **pure imaginary numbers**, or more simply, **imaginary numbers**. The set of all such numbers is called the **imaginary axis**.

The **complex conjugate** $\bar{\alpha}$ of a complex number $\alpha = (a, b)$ is given by

$$\bar{\alpha} = (a, -b) \tag{F}$$

and the **absolute value** or **modulus** $|\alpha|$ of $\alpha = (a, b)$ is given by

$$|\alpha| = \sqrt{a^2 + b^2} \,, \tag{G}$$

from which, using (C) and (F), we get

$$|\alpha|^2 = \alpha\bar{\alpha}. \tag{H}$$

We note (see Section 1.2) that if $\alpha = (a, b)$, then $|\alpha|$ is simply the distance from $(0,0)$ to the point (a, b) in the Cartesian plane.

From the commutative, associative, and distributive laws for the reals, used with (A), (B), and (C), it follows that complex numbers also obey these laws:

$$\begin{cases} \alpha + \beta = \beta + \alpha & \alpha\beta = \beta\alpha \\ (\alpha + \beta) + \gamma = \alpha + (\beta + \gamma) & (\alpha\beta)\gamma = \alpha(\beta\gamma) \\ \alpha(\beta + \gamma) = \alpha\beta + \alpha\gamma. \end{cases} \tag{I}$$

From (A), (B), (C), (D), and (E) we get

$$(a, b) = a + ib. \tag{J}$$

The two components a and b of the vector $\alpha = (a, b)$ are called, respectively, the **real part** of α and the **imaginary part** of α, and are denoted by

$$\operatorname{Re}\alpha = a, \qquad \operatorname{Im}\alpha = b. \tag{K}$$

Note that *the imaginary part of a complex number is itself a real number.* It is the coefficient of i when the number is written in the form (J).

It follows (how? See exercise A8) that

$$\alpha + 0 = \alpha, \qquad \alpha \cdot 1 = \alpha, \tag{L}$$

for all complex numbers α, and no other complex numbers than 0 and 1, respectively, have these properties. Thus, just as in R, the **additive identity** and **multiplicative identity** in $\mathbb{C}$ are 0 and 1, respectively. Also for each $\alpha = (a, b) = a + ib$ the number $-\alpha \equiv (-a, -b) = -a - ib$ is the **additive inverse** or **negative** of α.

The difference $\beta - \alpha$ can be defined by

$$\beta - \alpha = \beta + (-\alpha) \tag{M}$$

or as the unique solution z of the equation

$$\alpha + z = \beta. \tag{2}$$

Clearly $\beta + (-\alpha)$ is a solution to (2). To see that it is the only one we observe that (2) is equivalent to

$$a + x = c, \qquad b + y = d \tag{3}$$

where $z = x + iy$, $\alpha = a + ib$, and $\beta = c + id$. The system (3) has the unique solution

$$x = c - a, \qquad y = d - b$$

in the reals, which in turn is equivalent to

$$z = (c + id) - (a + ib) = \beta - \alpha.$$

As we have already mentioned, rule (C), the definition of complex multiplication, looks rather strange. However, it is simply the expression in terms of components of

$$(a + ib)(c + id) = ac + i(ad + bc) + i^2 bd$$
$$= (ac - bd) + i(ad + bc),$$

where i^2 is replaced by -1. Thus we see that it is not necessary to memorize the strange looking product law (C), for this calculation reproduces that law for us. We need only know elementary algebra, and that $i^2 = -1$.

It is in *this* feature of our construction, in *this* definition of a product, that we have added something new over the tools already available in two-dimensional vector analysis. We will see that the existence of this product has tremendously deep consequences.

The first manifestation of these consequences is that, as opposed to the dot (inner) product and the cross (vector) product defined in vector algebra, our new product admits an inverse (for nonzero factors)—that is, *complex division is possible.* To see this we seek to solve the equation

$$\alpha z = \beta \tag{4}$$

when $\alpha \neq 0$ (i.e., not both $a = 0$ and $b = 0$). Then $|\alpha| = \sqrt{a^2 + b^2} > 0$. If we multiply (4) by $\bar{\alpha}$, we get

$$\alpha \bar{\alpha} z = \bar{\alpha} \beta \quad \text{or} \quad |\alpha|^2 z = \bar{\alpha} \beta,$$

and multiplying by $1/|\alpha|^2$ we get

$$z = \bar{\alpha}\beta / |\alpha|^2$$

as the unique solution of (4). Accordingly we define

$$\frac{\beta}{\alpha} = \frac{\bar{\alpha}\beta}{|\alpha|^2}, \qquad \alpha \neq 0, \tag{N}$$

and we do not define β/α when $\alpha = 0$. In the special case where $\beta = 1$, we get the **multiplicative inverse** or **reciprocal**, $1/\alpha$, which is given by

$$\frac{1}{\alpha} = \frac{\bar{\alpha}}{|\alpha|^2} = \frac{a - ib}{a^2 + b^2} = \frac{a}{a^2 + b^2} - i\left(\frac{b}{a^2 + b^2}\right). \tag{N_1}$$

We can also solve (4) by breaking it into real and imaginary parts. We get the real system

$$ax - by = c, \qquad bx + ay = d. \tag{5}$$

The determinant of this linear system is $a^2 + b^2 = |\alpha|^2 > 0$, and Cramer's rule leads to the same solution formed above.

Let us note here early that many problems in complex variables are capable of being attacked by the two methods just now used, that is, by methods which make essential use of the complex number rules available, or by resolving into components (real and imaginary parts) and using the methods of real variables on the resulting real system. As a student you should do some of both, particularly until you get a feel for the complex approach. Resolution into components is always possible, sometimes useful, or even desirable, but *in general the complex methods are simplest.*

The formulas we have developed to express the elementary laws of complex numbers have the same form as those for the real numbers. This implies that, in particular, the rules usually listed under the heading "special products and factoring" in elementary real algebra retain that same form for complex algebra. In the following *partial list* of such formulas we use positive integer exponents as in the real case. Thus $z^2 = z \cdot z$, $z^3 = z \cdot z^2 = z \cdot z \cdot z$, etc.

$$\begin{cases} (z + \alpha)(z + \beta) = z^2 + z(\alpha + \beta) + \alpha\beta \\ \quad z^2 - \alpha^2 = (z - \alpha)(z + \alpha) \\ \quad z^3 - \alpha^3 = (z - \alpha)(z^2 + \alpha z + \alpha^2) \\ \quad z^n - \alpha^n = (z - \alpha)(z^{n-1} + \alpha z^{n-1} + \cdots + \alpha^{n-1}) \\ \quad (z + \alpha)^n = z^n + nz^{n-1}\alpha + \dfrac{n(n-1)}{1 \cdot 2} z^{n-2}\alpha^2 + \cdots + \alpha^n \end{cases} \tag{O}$$

As in the case of real variables, an expression of the form

$$P(z) = a_0 + z_1 z + a_2 z^2 + \cdots + a_n z^n = \sum_0^n a_k z^k \tag{P}$$

is called a **polynomial** in z. The a's are the **coefficients**, and if $a_n \neq 0$, the **degree** of the polynomial is n. If the coefficients are real, $P(z)$ is called a **real polynomial**, even though z, and hence $P(z)$, may take on complex values. If $P(z)$ and $Q(z)$ are polynomials, then $P(z)/Q(z)$ is called a **rational fraction** or a **rational function**, and is defined for all z except those for which $Q(z) = 0$. A rational function is called a **real rational function** if it is the ratio of two real polynomials.

Significant Topics

A Exercises

Do the computations in the first five exercises as you would in real arithmetic except for replacing i^2 by -1. The use of conjugates and absolute values may be helpful.

Let $\alpha = 3 + 2i$, $\beta = 1 - 4i$, $\gamma = \frac{1}{2} + 3i$.

1. Find

    (a) $\operatorname{Re}\alpha$ (b) $\operatorname{Re}(\alpha + \beta)$

    (c) $\operatorname{Im}(\alpha - \beta)$ (d) $\operatorname{Im}(\alpha - \gamma + \beta)$

2. Compute $(\alpha + \beta) + \gamma$ and $\alpha + (\beta + \gamma)$.
3. Compute $(\alpha\beta)\gamma$ and $\alpha(\beta\gamma)$.
4. Find   (a) $1/\alpha$,   (b) β/α.
5. Compute   (a) $\alpha^2 + \beta^2$ (b) $\alpha^2 - \beta^2$.
6. Given two complex numbers z and ζ, neither 0, is it possible that $z^2 + \zeta^2 = 0$? Try $z = 4 - i$, $\zeta = 1 + 4i$.
7. If $z = x + iy$ with x and y real, find the following in terms of x and y:

    (a) $\operatorname{Re} z^2$ (b) $\operatorname{Re} z^4$ (c) $\operatorname{Re}(1/z)$ (d) $\operatorname{Re}(1/z^2)$

    (e) $\operatorname{Im} z^2$ (f) $\operatorname{Im} z^4$ (g) $\operatorname{Im}(1/z)$ (h) $\operatorname{Im}(1/z^2)$

8. Show that if $\alpha + z = \alpha$, for any one complex α, then $z = 0$, and show that if $\alpha u = \alpha$ for any one complex $\alpha \neq 0$, then $u = 1$.

B Exercises

1. Show $\operatorname{Re}(iz) = -\operatorname{Im} z$, $\operatorname{Im}(iz) = \operatorname{Re}(z)$.

2. Show $z/z = 1$, and $\dfrac{1}{1/z} = z$ if $z \neq 0$.

3. Complete the solution of $\alpha z = \beta$, $\alpha \neq 0$, begun in the text by components, and show that it leads to the solution $z = \bar{\alpha}\beta/|\alpha|^2$ found by complex methods.

4. Show that $\alpha\beta = 0$ implies that at least one of α and β is 0.

5. Prove the cancellation laws:
   (a) $\alpha + \gamma = \beta + \gamma$ implies $\alpha = \beta$,
   (b) $\alpha\gamma = \beta\gamma$ implies $\alpha = \beta$ if $\gamma \neq 0$.

6. Show that the equation $z^2 + 1 = 0$ has exactly two solutions, namely, $\pm i$.

7. If $z = \alpha/\bar{\alpha}$, $\alpha \neq 0$, show $|z| = 1$.

8. Show

   (a) $\alpha + \bar{\alpha} = 2\operatorname{Re}\alpha$

   (b) $\alpha - \bar{\alpha} = 2i\operatorname{Im}\alpha$

   (c) $\overline{\alpha + \beta} = \bar{\alpha} + \bar{\beta}$

   (d) $\overline{\alpha\beta} = \bar{\alpha}\bar{\beta}$

   (e) $\overline{(\alpha/\beta)} = \bar{\alpha}/\bar{\beta}$

   (f) $\overline{(\bar{\alpha})} = \alpha$

   (g) $\overline{\bar{\alpha}\beta} = \alpha\bar{\beta}$

   (h) $|\alpha| = |\bar{\alpha}|$

   (i) $|\alpha\beta| = |\alpha||\beta|$

   (j) $(\bar{z})^n = \overline{(z^n)}$

9. If $P(z) = a_0 + a_1 z + a_2 z^2 + \cdots + a_n z^n$, show
   (a) $\overline{P(z)} = \bar{a}_0 + \bar{a}_1\bar{z} + \bar{a}_2\bar{a}^2 + \cdots + \bar{z}_n\bar{z}^n$
   (b) $\overline{P(\bar{z})} = \bar{a}_0 + \bar{a}_1 z + \bar{a}_2 z^2 + \cdots + \bar{a}_n z^n$
   (c) Deduce that $P(z)$ is real if and only if $\overline{P(z)} = P(\bar{z})$.

10. Show that the following laws hold for positive integer exponents.
    (a) $z^m z^n = z^{m+n}$

    (b) $(z^m)^n = z^{mn}$

    (c) $(z\zeta)^m = z^m \zeta^m$

    (d) If $z \neq 0$, $\dfrac{z^m}{z^n} = \begin{cases} z^{m-n} & \text{if } m > n, \\ 1 & \text{if } m = n, \\ 1/z^{n-m} & \text{if } m < n. \end{cases}$

11. Show that $\alpha^{2k+1} + \beta^{2k+1}$ is divisible by $\alpha + \beta$ if $\alpha + \beta \neq 0$, where k is a positive integer.

12. (Reduction to lowest terms.) Suppose that $P(z)$ and $Q(z)$ are polynomials of degree m and n, respectively, and that they have one or more common roots. Show that there are polynomials $p(z)$ and $q(z)$ of degree m' and n', respectively, with $m' < m$, $n' < n$ and $m' - n' = m - n$,

and having no common roots, for which

$$\frac{p(z)}{q(z)} = \frac{P(z)}{Q(z)}$$

for all z except the roots of $Q(z)$.

Hint: Use the result of C3 below.

13. Show that if $P(z)/Q(z)$ is an improper rational fraction, then

$$P(z)/Q(z) = p(z) + P_1(z)/Q(z)$$

where $P_1(z)/Q(z)$ is proper and $p(z)$ is a polynomial with degree equal to the difference of the degrees of $P(z)$ and $Q(z)$.

C Exercises

The Fundamental Theorem of Algebra (Theorem 3.2h) states that the equation

$$P(z) \equiv a_0 + a_1 z + a_2 z^2 + \cdots + a_n z^n = 0$$

with $n \geq 1$, and $a_n \neq 0$, has at least one solution. Prove, as outlined in the following problems, that this in turn implies the existence of exactly n roots, counting multiple roots according to their multiplicity, and further, that $P(z)$ can be completely factored as in C3 below.

1. If $P(z)$ is a polynomial of degree $n \geq 1$, then $P(z) - P(\alpha)$ is divisible by $z - \alpha$. That is, $P(z) - P(\alpha)$ contains a factor of $z - \alpha$.

   Hint: See the equations (O) in the text.

2. Deduce that if $P(\alpha) = 0$, then $P(z) = (z - \alpha)Q(z)$ where $Q(z)$ is a polynomial of degree $n - 1$.

3. Conclude from C2 that

$$P(z) = a_n(z - \alpha_1)^{p_1}(z - \alpha_2)^{p_2} \cdots (z - \alpha_k)^{p_k}$$

where a_n is the (nonzero) coefficient of z^n in $P(z)$, where the α_j's are distinct roots of $P(z)$, and where each p_j is a positive integer and

$$\sum_1^k p_j = n.$$

4. If $P(z)$ is a real polynomial and α is a root, show that $\bar{\alpha}$ is also a root.

   Hint: Use Exercise B9 above.

1.2 Geometry. Polar Coordinates

In Section 1.1 we were at some pains to point out that we can view complex numbers as vectors (with of course our new multiplication formula).
This gives an immediate geometrical interpretation of complex numbers as
points or directed segments in a plane. To be specific, we can consider
a plane with an xy-coordinate system as a geometrical realization of the
system $\mathbb{C}$ of complex numbers in the following way. We take the x-axis as
the real axis, the y-axis as the imaginary axis. The point $\alpha = a + ib$ has
for its graph the point with coordinates (a, b). Thus the real numbers (as
a subset of $\mathbb{C}$) lie on the x-axis, the pure imaginary ones on the y-axis, and
all others out in the plane between the axes.

It is also, as in vector analysis, sometimes convenient to interpret geometrically a complex number $\alpha = a + ib$ as the directed segment from the
origin to the point with coordinates (a, b), or as *any other segment having
the same length and direction.* Since complex addition coincides with vector
addition the parallelogram law is still valid. To add α and β graphically,
the vector β can be drawn as a segment starting at the point α. Then $\alpha + \beta$
is realized as the point at the terminus of that vector, or as the segment
directed from the origin to that point, as shown in the figure. By the same
token the vector $\beta - \alpha$ can be interpreted as the vector from the point α
to the point β, or as the point determined by a segment from the origin 0
having the same length and direction.

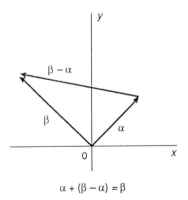

$\alpha + (\beta - \alpha) = \beta$

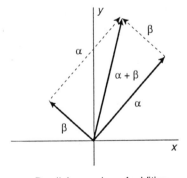

Parallelogram law of addition

It is geometrically clear that if $\alpha = a + ib$, then $|a| \leq |\alpha|$, $|b| \leq |\alpha|$, and $|\alpha| \le |a| + |b|$, for $|\alpha|$ is the hypotenuse of a right triangle with legs of length $|a|$ and $|b|$. We can give nongeometrical proofs of these statements which we now do.

1.2a Theorem. *For any complex $\alpha = a + ib$*

$$|a| = |\text{Re}\,\alpha| \le |\alpha|, \qquad |b| = |\text{Im}\,\alpha| \le |\alpha|.$$

Proof. By definition of $|\alpha|$ if $\alpha = a + ib$ we have

$$|\alpha|^2 = a^2 + b^2 \ge a^2$$

which is equivalent to the first inequality. The second one is similar. $\square$

The inequality $|\alpha| \le |a| + |b|$ is a special case of a more general one which we now prove.

1.2b Theorem. *If α and β are any two complex numbers, then*

$$|\alpha + \beta| \le |\alpha| + |\beta|,$$

and equality occurs if and only if one of α and β is a nonnegative multiple of the other.

Proof.
$$
\begin{aligned}
|\alpha + \beta|^2 &= (\alpha + \beta)(\bar{\alpha} + \bar{\beta}) \\
&= \alpha\bar{\alpha} + \alpha\bar{\beta} + \bar{\alpha}\beta + \beta\bar{\beta} \\
&= |\alpha|^2 + 2\text{Re}\,(\alpha\bar{\beta}) + |\beta|^2,
\end{aligned}
$$

the last step being by Exercise B8(a) of Section 1.1. Continuing, we get

$$
\begin{aligned}
|\alpha + \beta|^2 &\le |\alpha|^2 + 2|\alpha\bar{\beta}| + |\beta|^2 \\
&= |\alpha|^2 + 2|\alpha||\beta| + |\beta|^2 \\
&= (|\alpha| + |\beta|)^2
\end{aligned}
$$

The first step here is by Theorem 1.2a, except for the equality case which is left to the exercises (B4). Taking positive square roots completes the argument. $\square$

1.2c Corollary. $|\alpha + \beta| \ge ||\alpha| - |\beta||.$

1.2d Corollary. $|\alpha + \beta + \cdots + \gamma| \le |\alpha| + |\beta| + \cdots + |\gamma|.$

The proofs of these corollaries are left to the exercises (B1, B2).

We should point out again that if $\alpha = a + ib$, we have as a special case of Theorem 1.2b

$$|\alpha| \leq |a| + |b|,$$

for α is the sum of a and ib.

The inequality $|\alpha + \beta| \leq |\alpha| + |\beta|$ is called the **triangle inequality** because $\alpha + \beta$ is one side of a triangle whose other two sides are α and β. Geometrically then it is the statement that one side of a triangle cannot be longer than the sum of the lengths of the other two sides. *More generally the term triangle inequality refers to whichever of the last three propositions is appropriate to the context.*

Example 1. $|(z + h)^n - z^n| \leq (|z| + |h|)^n - |z|^n$.

Solution. By the binomial theorem (the last of the formulas (O) of Section 1.1) we have

$$(z + h)^n = z^n + nz^{n-1}h + \cdots + h^n$$

where we note that the binomial coefficients that occur here are *positive*. Thus

$$(z + h)^n - z^n = nz^{n-1}h + \cdots + h^n$$

from which, by the triangle inequality,

$$|(z + h)^n - z^n| \leq n|z|^{n-1}|h| + \cdots + |h|^n.$$

Since the coefficients are all positive, this last expression is of the *same structure* as the previous one with z replaced by $|z|$ and h by $|h|$. It is therefore $(|z| + |h|)^n - |z|^n$, and the solution is complete.

Example 2.

$$\left| \frac{(z + h)^n - z^n}{h} - nz^{n-1} \right| \leq \frac{(|z| + |h|)^n - |z|^n}{|h|} - n|z|^{n-1}.$$

Solution. Reasoning as above in Example 1,

$$\left| \frac{(z + h)^n - z^n}{h} - nz^{n-1} \right| = \left| \frac{1}{2} n(n - 1)z^{n-2}h + \cdots + h^{n-1} \right|$$

$$\leq \frac{1}{2} n(n - 1)|z|^{n-2}|h| + \cdots + |h|^{n-1}$$

$$= \frac{(|z| + |h|)^n - |z|^n}{|h|} - n|z|^{n-1}.$$

It is useful to introduce polar coordinates in the complex plane in the standard way as illustrated in the figure. If (x, y) are the rectangular coordinates of a point whose polar coordinates are (r, θ), then

$$x = r \cos\theta, \qquad y = r \sin\theta \tag{1}$$

where

$$r = \sqrt{x^2 + y^2},$$

is the distance from the origin ($r = |z|$ if $z = x + iy$), and θ is the angular displacement from the positive x-axis, the positive direction of rotation being counterclockwise.

The angle θ, as always in polar coordinates, is the source of some difficulty with which we will have to deal throughout this course. This difficulty arises from the lack of uniqueness of θ, for if θ_0 is one value of θ for which (1) is correct, then the formula

$$\theta_k = \theta_0 + 2k\pi$$

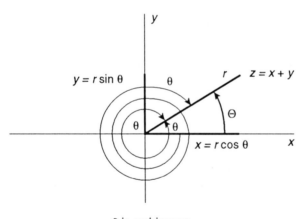

θ is ambiguous.
$\theta = \Theta + 2\,k\pi$

where k is any integer, positive or negative, gives an infinite list of other equally appropriate angles.

If $z = x + iy$, then using (1) we write

$$z = r \cos\theta + ir \sin\theta,$$

or

$$z = r(\cos\theta + i\sin\theta) \tag{2}$$

where $|z| = r$ and where θ is determined up to an additive multiple of 2π. The angle θ is called the **argument** of z, and we write

$$\theta = \arg z.$$

By this we mean that θ is one of the possible values for which (2) is valid. We observe that there is *exactly one* of the possible values of θ which satisfies

$$-\pi < \theta \leq \pi. \tag{3}$$

This value is called the **principal value** of the argument of z, and we denote it by $\operatorname{Arg} z$ and by Θ. Thus

$$\Theta = \operatorname{Arg} z$$

means the unique value for θ for which both (2) and (3) are correct.

If $z = 0 = 0 + i0$, then $r = 0$ so

$$z = 0\ (\cos\theta + i\sin\theta)$$

which holds for all values of θ, and there are no values of θ more natural than any others. Thus *we do not define* $\arg 0$.

For $z \neq 0$, the equation (2) identifies a unit vector (length 1) in the direction of z, namely

$$E(\theta) \equiv \cos\theta + i\sin\theta.$$

We will use this notation, $E(\theta)$, until Section 2.6, and so, until that section, our standard polar coordinate representation for the complex number $z = x + iy$ will be

$$z = rE(\theta); \qquad r = |z|, \quad \theta = \arg z.$$

There is one point you should keep clearly in mind. θ *is ambiguous, but $E(\theta)$ is not.* For distinct values of θ differ by multiples of 2π, and by definition of $E(\theta)$ we have $E(\theta + 2k\pi) = E(\theta)$ by the periodicity of the sine and cosine. The $2k\pi$ term embodies the ambiguity of the angle.

We consider now z and ζ given by

$$z = rE(\theta), \qquad \zeta = \rho E(\phi).$$

Their product is

$$
\begin{aligned}
z\zeta &= r\rho E(\theta)E(\phi)\\
&= r\rho(\cos\theta + i\sin\theta)(\cos\phi + i\sin\phi)\\
&= r\rho\Big[(\cos\theta\cos\phi - \sin\theta\sin\phi) + i(\sin\theta\cos\phi + \cos\theta\sin\phi)\Big]\\
&= r\rho[\cos(\theta + \phi) + i\sin(\theta + \phi)]\\
&= r\rho E(\theta + \phi).
\end{aligned}
$$

This calculation shows that *when complex numbers are multiplied*, their moduli also multiply, but *their arguments add*. We state this as a theorem.

1.2e Theorem. *If z and ζ are nonzero, then*

$$\arg(z\zeta) = \arg z + \arg \zeta. \tag{4}$$

The proof is in the previous calculation. The meaning of this equation is that for any value of $\arg z$ and any value of $\arg \zeta$, their sum is one of the values of $\arg(z\zeta)$. You should verify that (4) does not always hold if arg is replaced by Arg : take $z = \zeta = E(2\pi/3)$.

If we set $z = \zeta$ with $r = \rho = 1$, then the calculations leading up to the last theorem give

$$[E(\theta)]^2 = E(2\theta),$$

which by induction yields

$$[E(\theta)]^n = E(n\theta).$$

This is **de Moivre's theorem**, which can be written out as follows.

1.2f Theorem. $(\cos\theta + i\sin\theta)^n = \cos n\theta + i\sin n\theta.$

If we define $z^{-n} = 1/z^n$, $z^0 = 1$ for $z \neq 0$, then de Moivre's theorem holds for all integer values of n. (See Exercise B7.)

Using de Moivre's theorem one can completely discuss integer roots, and more generally, rational powers of complex numbers. We define "α is an nth root of β" to mean that $\alpha^n = \beta$. If we write $\beta \neq 0$ in polar form, $\beta = rE(\theta)$, then it is clear that *one* of the nth roots will be given by

$$\alpha = r^{1/n}E(\theta/n),$$

where $r^{1/n}$ *means the positive nth root of the positive number* r. For by the definition just given, and de Moivre's theorem

$$\alpha^n = [r^{1/n}E(\theta/n)]^n = rE\left(n \cdot \frac{\theta}{n}\right) = rE(\theta) = \beta.$$

The possibility of other roots now arises. We see immediately, for example, that if n is even and α is an nth root of β, then $-\alpha$ is another, for

$$(-\alpha)^n = (-1)^n\alpha^n = \alpha^n = \beta.$$

But what of still others? The result is the following.

1.2g Theorem. *If* $\beta \neq 0$, *and* $n \geq 1$, *then there are exactly* n *distinct* nth *roots of* β. *That is, the equation* $z^n = \beta$ *has exactly* n *distinct solutions.*

Proof. Let α be an nth root of β, and let

$$\alpha = \rho E(\phi), \qquad \beta = rE(\theta).$$

Then by de Moivre's theorem, and $\alpha^n = \beta$, we get

$$\rho^n E(n\phi) = rE(\theta)$$

from which (how? See Exercise A13.)

$$\rho^n = r, \qquad E(n\phi) = E(\theta).$$

Since ρ and r are positive numbers, we have

$$\rho = r^{1/n}$$

where $r^{1/n}$ is the positive real nth root of r. The other equation in expanded form is

$$\cos n\phi + i\sin n\phi = \cos\theta + i\sin\theta.$$

We cannot conclude $n\phi = \theta$, but rather

$$n\phi = \theta + 2k\pi, \qquad k = 0, \pm 1, \pm 2, \ldots$$

from which

$$\phi = \frac{1}{n}\theta + \left(\frac{k}{n}\right)2\pi, \qquad k = 0, \pm 1, \pm 2, \ldots \ . \tag{5}$$

Thus if α is an nth root of β, their arguments must stand in the relation (5). Any choice of k as an integer leads to an α given by

$$\alpha = r^{1/n} E\left(\frac{1}{n}\theta + \frac{k}{n}w\pi\right), \tag{6}$$

which is an nth root since

$$\alpha^n = rE(\theta + 2k\pi) = rE(\theta) = \beta.$$

Further, if k_1 and k_2 differ by a multiple of n, then (6) yields coincident values of α. To see this we write

$$k_2 = k_1 + k_0 n$$

where k_0 is an integer. Then

$$r^{1/n} E\left(\frac{1}{n}\theta + \frac{k_2}{n} 2\pi\right) = r^{1/n} E\left(\frac{1}{n}\theta + \frac{k_1 + k_0 n}{n} 2\pi\right)$$

$$= r^{1/n} E\left(\frac{1}{n}\theta + \frac{k_1}{n} 2\pi + k_0 2\pi\right)$$

$$= n^{1/n} E\left(\frac{1}{n}\theta + \frac{k_1}{n} n\pi\right).$$

Thus it is clear that choosing $k = 0, 1, \ldots, n-1$ yields n distinct α values, but any other choice of k differs from one of these by a multiple of n, and so leads to no other values of α. $\square$

We denote by $\beta^{1/n}$ the set of values (n in number) of the nth roots of β, or also any one of these roots. If $\Theta = \operatorname{Arg}\beta$, then we define the **principal value** of $\beta^{1/n}$ by

$$r^{1/n} E\left(\frac{\Theta}{n}\right), \qquad r = |\beta|.$$

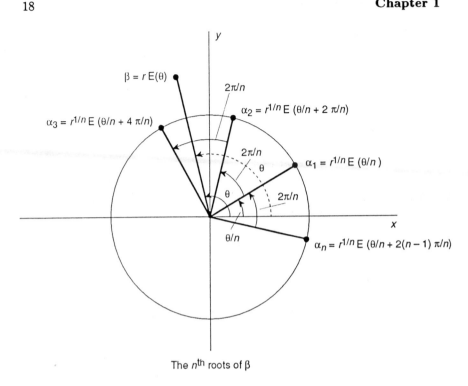

The nth roots of β

The other values of $\beta^{1/n}$, that is, the other nth roots of β, follow by consecutive rotations of $2\pi/n$, until the last rotation brings us to the principal value again.

If $\beta \neq 0$, and p/q is a rational fraction *in lowest terms*, then we define

$$\beta^{p/q} = [\beta^p]^{1/q},$$

and so $\beta^{p/q}$ has q distinct values. If p/q is *not in lowest terms* and $p/q = p_0/q_0$ in lowest terms, then we define

$$\beta^{p/q} = [\beta^{p_0}]^{1/q_0}$$

and so it has q_0 distinct values. Thus, for a rational number p/q *not in lowest terms* we distinguish between $(\beta^p)^{1/q}$ which has q values and $\beta^{p/q}$ which has $q_0 < q$ values. Thus

$$z^{4/2} = z^2$$

has one value while

$$[z^4]^{1/2} = \pm z^2$$

has two values. And $z^{3/9} = z^{1/3}$ has 3 values while $[z^3]^{1/9}$ has 9 values.

Signficant Topics	Page
Triangle inequality	11,12
Argument (arg and Arg)	14
$E(\theta) = \cos\theta + i\sin\theta$	14
$E(\theta)E(\varphi) = E(\theta + \varphi)$	15
de Moivre's Theorem	15
nth roots: $\beta^{1/n}$	16,17
Principal value	17

A Exercises

1. Find all cube roots of (a) 1, (b) $-i$, (c) -27.
2. Compute all values of

   (a) $(1+i)^{1/2}$ (b) $(-16)^{6/8}$ (c) $(16)^{6/8}$ (d) $(1+i)^{5/3}$
   (e) $(1-i)^{4/2}$ (f) $(1-i)^{1/2}$ (g) $3^{3/9}$ (h) $(3^3)^{1/9}$
   (i) $(\sqrt{3}+i)^{6/4}$ (j) $[(\sqrt{3}+i)^6]^{1/4}$

3. Describe geometrically the set of points which satisfy the following conditions.

   (a) $|z-1| = |z-i|$ (b) $|z-1| < |z-i|$
   (c) $|z-1| = 2|z-i|$ (d) $1 < |z| < 2$
   (e) $1 \le |z-1| \le 2$ (f) $\text{Re}\,(z+1) = |z|$
   (g) $\text{Re}\,(z-1) = |z|$ (h) $|z-1| + |z+1| = 3$
   (i) $|z-1| + |z+1| = 2$ (j) $\text{Arg}\,z = \pi/4$
   (k) $\pi/3 < \text{Arg}\,z \le 3\pi/4$ (ℓ) $|\text{Arg}\,z| < \pi/2$
   (m) $\text{Re}\,z > 0$ (n) $\text{Re}\,z^2 > 0$

4. Show that $\text{Re}\,z > 0$ is equivalent to $|\text{Arg}\,z| < \pi/2$. (See 3(ℓ) and (m) above.)
5. Use de Moivre's theorem to show
   (a) $\cos^3\theta - 3\cos\theta\sin^2\theta = \cos 3\theta$
   (b) Find also $\sin 3\theta$ in terms of $\sin\theta$ and $\cos\theta$.
6. Find all solutions of $z^6 - 1 = 0$.
7. Show $\max\limits_{|z|\le 1}|z^2 + 1| = 2$.
8. Write in complex notation the equation of the circle with center $2-3i$, radius 5.

9. If $|\alpha| \neq |\beta|$, then show

$$\left| \frac{\alpha}{\alpha + \beta} \right| \leq \frac{|\alpha|}{||\beta| - |\alpha||}$$

10. Show if $|\alpha| = 1$, then $1/\alpha = \bar{\alpha}$.
11. Let $z \neq 0$. Describe the location of αz relative to the location of z under each of the following conditions.

(a) $\alpha > 1$,   (b) $0 < \alpha < 1$,
(c) $\alpha < 0$,   (d) $|\alpha| = 1$,
(e) $|\alpha| = 1$, $\operatorname{Arg} \alpha = \phi$,   (f) $\alpha \neq 0$, $\operatorname{Arg} \alpha = \phi$.

12. Choose a point $z \neq 0$ in a graph of the complex plane and locate

(a) $-z$ (b) $\bar{z}$ (c) $-\bar{z}$ (d) $1/z$
(e) $-1/z$ (f) $1/\bar{z}$ (g) $-1/\bar{z}$

13. Explain why $\rho^n E(n\phi) = rE(\theta)$ gives $\rho^n = r$ and $E(n\phi) = E(\theta)$.
14. If α, β, γ are complex constants wtih $\alpha \neq 0$, show that the equation

$$\alpha z^2 + \beta z + \gamma = 0$$

has two solutions given by the quadratic formula

$$z = \frac{-\beta \pm \sqrt{\beta^2 - 4\alpha\gamma}}{2\alpha}.$$

Discuss the meaning of $\pm\sqrt{\beta^2 - 4\alpha\gamma}$ in this formula.

B Exercises

1. Prove Corollary 1.2c.
2. Prove Corollary 1.2d.
3. Prove $\arg(z/\zeta) = \arg z - \arg \zeta$ and state the meaning of the equation.
4. Show (using polar coordinates or otherwise) that $\operatorname{Re}(\alpha\bar{\beta}) = |\alpha||\beta|$ if and only if $\operatorname{Arg} \alpha = \operatorname{Arg} \beta$ and so complete the proof of the triangle inequality in the equality case.
5. Use induction starting with B4 to show that $\left| \sum_1^n \alpha_j \right| = \sum_1^n |\alpha_j|$ if and only if $\operatorname{Arg} \alpha_j$ is independent of j.

6. For $n > 1$ an integer, let $\omega = E(2\pi/n)$. Then ω is an nth root of 1. Show that if α is any one nth root of $\beta \neq 0$, then $\alpha, \alpha\omega, \alpha\omega^2, \ldots, \alpha\omega^{n-1}$ are all the nth roots of β.

7. Establish de Moivre's theorem for n negative.

8. Observe (A10 above) that if $|\alpha| = 1$, then $\bar{\alpha} = 1/\alpha$, and hence show that $|\alpha - \beta| = |1 - \bar{\alpha}\beta|$ if either $|\alpha| = 1$ or $|\beta| = 1$.

9. Show $|\alpha - \beta|^2 + |\alpha + \beta|^2 = 2(|\alpha|^2 + |\beta|^2)$.

10. If one of the values of $\beta^{p/q}$ is α, then show that one of the values of $\alpha^{q/p}$ is β.

11. Show $\displaystyle\sum_0^n z^k = 1 + z + z^2 + \cdots + z^n = \frac{1 - z^{n+1}}{1 - z}$, $z \neq 1$ and hence that

(a) $\displaystyle\sum_0^n \cos k\theta = 1 + \cos\theta + \cos 2\theta + \cos 3\theta + \cdots + \cos n\theta$

$$= \frac{1}{2} + \frac{\sin(n + \frac{1}{2})\theta}{2\sin\frac{1}{2}\theta}.$$

(b) The sum of the nth roots of any number is zero.

C Exercises

1. Show

$$n \sin\frac{2\pi}{n} < \sum_1^n \left| E\left(\frac{2k\pi}{n}\right) - E\left(\frac{2(k-1)\pi}{n}\right) \right| < 2\pi$$

and interpret geometrically.

2. Show

$$\left| \sum_1^n \alpha_k\beta_k \right|^2 \leq \left(\sum_1^n |\alpha_k|^2 \right)\left(\sum_1^n |\beta_k|^2 \right).$$

3. Let C_1 be the circle $|z| = r > 0$, and C_2 be the circle $|z| = r^{1/2}$ where $r^{1/2}$ is the positive square root of r. As $z = rE(\theta)$ moves once around C_1 from $\theta = 0$ to $\theta = 2\pi$, then show that ζ given by $\zeta = r^{1/2}E(\theta/2)$ moves from the positive x-axis to the negative x-axis on the upper half of C_2. And further, as z goes around C_1 a second time from $\theta = 2\pi$ to $\theta = 4\pi$, ζ completes one trip around C_2 by returning to the positive x-axis along the lower half of C_2.

4. Show that for $\beta \neq 0$, and m/n in lowest terms the set of values of $[\beta^m]^{1/n}$ is identical with the set of values of $[\beta^{1/n}]^m$.

5. Suppose $p(z, \zeta)$ is a polynomial in z and ζ with nonnegative coefficients. Show that $|p(z, \zeta)| \leq p(|z|, |\zeta|)$ and generalize to polynomials in n complex variables.

1.3 Functions and Sequences. Limits and Continuity

Let S (for set) be a collection of complex numbers. Geometrically we view S as a part of the complex plane. By

$$z \in S$$

we mean that z is one of the points in the collection or set S. We read this as "z is in S" or "z belongs to S" of "z is an element of S" or some other grammatically equivalent form.

If S and T are sets, then by

$$S \subset T \quad \text{or} \quad T \supset S$$

we mean that S is a **subset** of T, that each z in S is also in T. We read the first as "S is contained in T," and the second as "T contains S," or, again, some equivalent form.

Simple examples are given by the following.

Definition. *A* **neighborhood** *of a complex number α is the set of all numbers z for which*

$$|z - \alpha| < r$$

for some positive number r. A **deleted neighborhood** *of α is the set of all numbers z for which*

$$0 < |z - \alpha| < r.$$

We find it convenient to introduce a special symbol for a neighborhood of radius r of α, namely $N(\alpha, r)$. Thus $N(\alpha, r)$ is the set of points z which satisfy $|z - \alpha| < r$. That is

$$N(\alpha, r) = \{z \in \mathbb{C} : |z - \alpha| < r\}.$$

Geometrically a neighborhood of α is a disk or the inside of a circle centered at α with a radius r. The r-neighborhood of α means the disk of radius r centered at α. The geometrical meaning of a deleted neighborhood is the same, except that the point α itself is deleted from the disk. It is also called a **punctured neighborhood** or a **punctured disk**.

Let S and T be sets. Suppose that for each z in S we have a definite rule for assigning to z a number w in T. Then we say that this defines a **function** from S into T. The number w corresponding to a given z is called the **value** of the function at z. We write this as

$$w = f(z).$$

In most cases the function will be given by a formula which indicates how w is to be computed from z. The polynomials and rational functions discussed in Section 1.1 are examples of functions given by simple explicit formulas.

The set S of z's on which the function is defined is called the **domain** of the function and the set of all w's, generated by all such z's, is called the **range** of the function. Functions are also called **mappings** and are said to **map** the domain onto the range.

The formulas

$$w = \frac{1}{z(z-2)} \quad \text{and} \quad w = z^2 + 5z - 2$$

define functions of z. In the first case we have $f(z) = 1/z(z-2)$ and its domain is $\mathbb{C} - \{0\} - \{2\}$, that is, the plane punctured at 0 and 2. Its range is $\mathbb{C} - \{0\}$ (how?). In the second case $f(z) = z^2 + 5z - 2$, and here the domain and range are both $\mathbb{C}$.

If $f(z)$ is a function, by definition it will be given by a rule for determining the value $f(z)$. In most cases, this rule will be a formula, simple or complicated. Then we speak of the function $f(z)$, rather than the function f. This amounts to a bit of mathematical colloqialism since $f(z)$ is used in two senses: the function itself and the value of the function at the point z. This is standard in complex variables, and should cause no confusion. The context should make clear the sense meant in a given instance.

We examine the function $1/z$. If we set $z = x + iy$ and $w = u + iv$, the equation $w = 1/z$ becomes

$$u + iv = \frac{1}{x + iy} = \frac{x - iy}{x^2 + y^2} \quad \text{(how?)}$$

which is equivalent to

$$u = x/(x^2 + y^2), \qquad v = -y/(x^2 + y^2).$$

Thus, as in other vector cases, the single complex valued function given by the single equation

$$w = f(z)$$

is equivalent to two scalar (real) functions given by two equations of the form

$$u = \phi(x, y) \quad \text{and} \quad v = \psi(x, y).$$

Frequently we do not introduce new functional symbols (as the ϕ and ψ above) for the real and imaginary parts of a complex function, but write

$$u = u(x, y) \quad \text{and} \quad v = v(x, y).$$

So u and v are then viewed as real valued functions of the two real variables x and y. Since *knowing x and y is identical with knowing z, we can consider u and v as real valued functions of the complex variable z:*

$$u = u(z) \quad \text{and} \quad v = v(z).$$

Example 1. Break $w = z^2 + (2 + 3i)$ into its real and imaginary parts.

Solution. We write

$$\begin{aligned} w = u + iv &= z^2 + (2 + 3i) \\ &= (x^2 - y^2 + 2) + i(2xy + 3), \end{aligned}$$

so

$$\begin{aligned} u &= u(x, y) = u(z) = x^2 - y^2 + 2, \\ v &= v(x, y) = v(z) = 2xy + 3. \end{aligned}$$

In such a functional relationship we can interpret each of z and w as a point in the complex plane. It is frequently convenient to consider two separate planes, one being the z-plane in which the domain lies, and the other the w-plane in which the range lies. We then consider the function $f(z)$ as a **mapping** of each z point in the domain into a w-point of the range. When we adopt this view we tend to use mapping or map as equivalent to **function**.

But it is sometimes useful to consider both the z points and the w points in the same plane, and to describe the function by the geometry of what happens to z as it is transformed into w by the action of the function $f(z)$. Then we tend to call the function a **transformation**.

For example, the function given by

$$w = z + \alpha$$

is a **translation**, for w can be obtained from z by translating it by the constant vector α.

If α is a unit vector (that is, $|\alpha| = 1$), then

$$w = \alpha z$$

is called a **rotation**. To see the action of this function we express α and z in polar form:

$$\alpha = E(\phi) = \cos\phi + i\sin\phi$$
$$z = rE(\theta) = r(\cos\theta + i\sin\theta).$$

By multiplication we get

$$w = \alpha z = E(\phi)rE(\theta) = rE(\theta + \phi).$$

From this last equation it is clear that w is obtained from z by a rotation about the origin through the angle ϕ.

The last case for a linear equation is **dilation**:

$$w = \alpha z, \qquad \alpha > 0.$$

In this case $\arg w = \arg z$, so w lies on the same ray from the origin as z, and is further from 0 if $\alpha > 1$, closer if $\alpha < 1$. The figure illustrates the location if $\alpha > 1$.

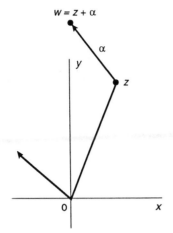

Translation
$w = z + \alpha$

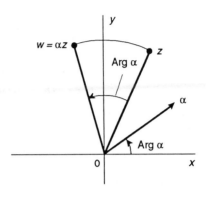

Rotation
$w = \alpha z, |\alpha| = 1$

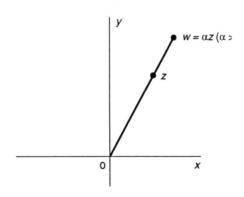

Dilation
$w = \alpha z, \; \alpha > 0$

Example 2. Let $w = 4z + 2i$. Show that the line $x - y = 1$ in the z-plane maps onto the line $u - v = 2$ in the w-plane with $z = x + iy$, $w = u + iv$.

Solution.

$$w = u + iv = 4z + 2i = 4x + (4y + 2)i$$

so that

$$u = 4x, \qquad v = 4y + 2$$

and hence

$$x = u/4, \qquad y = (v - 2)/4.$$

From $x - y = 1$ we get

$$\frac{u}{4} - \frac{v - 2}{4} = 1$$

which simplifies to $u - v = 2$.

Example 3. If z lies in the angle $A : 0 \leq \arg z \leq \phi$, where does $w = z^2$ lie?

Solution. Writing z in polar form we get

$$w = z^2 = [rE(\theta)]^2 = r^2 E(2\theta),$$

from which it is clear that the image of A under this mapping is the angle in the w-plane of opening 2ϕ as illustrated in the figure.

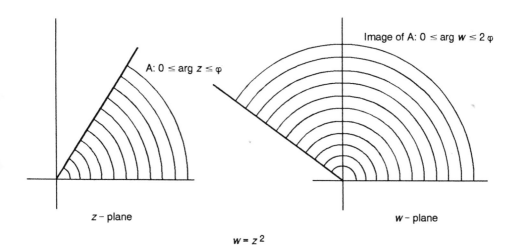

z - plane w - plane

$w = z^2$

A **complex sequence** is a sequence of complex numbers. As in the real case a sequence is defined as a function whose domain is all integers larger than or equal to some fixed integer n_0 which is usually taken to be 0 or 1. The notation here is standard. If z_n represents a typical member of the sequence, that is, the z-value corresponding to the integer n, then $\{z_n\}$ represents the whole sequence. Thus $\{2+ni\}$, with $n_0 = 0$, is the sequence whose members are

$$2, 2+i, 2+2i, \ldots, 2+ni, \ldots \ .$$

The sequential index, here written as n, may be k or j or some other letter. But i is rarely used as an index for obvious reasons.

If $\{z_n\}$ is a complex sequence, then to say that α is the limit of this sequence means, roughly, that we can make z_n as close to α as we want by taking n large. The precise definition is the following.

Definition.
$$\alpha = \lim z_n \qquad means:$$

for each $\varepsilon > 0$ there is an $N = N(\varepsilon)$ for which

$$|\alpha - z_n| < \varepsilon \quad when \quad n > N.$$

If a sequence $\{z_n\}$ has a limit α we say that it **converges**, and more precisely, that it **converges to** α. If it has no limit, we say it **diverges**.

This definition is formally identical with the definition for a real sequence. The difference comes in the meaning of the symbols, specifically, the meaning of absolute value. Geometrically the ε-condition means that z_n lies in the ε-neighborhood of α if n is large enough, that is, if $n \geq N(\varepsilon)$.

A simplified notation is

$$z_n \to \alpha \quad \text{as} \quad n \to \infty,$$

or more briefly just

$$z_n \to \alpha,$$

means

$$\alpha = \lim z_n.$$

1.3a Theorem. *Let $z_n = x_n + iy_n$, $\alpha = a + ib$. For*

$$\lim z_n = \alpha$$

it is necessary and sufficient that

$$\lim x_n = a \quad \text{and} \quad \lim y_n = b.$$

Proof. This follows from Theorems 1.2a and 1.2b. The basic inequalities are

$$\left. \begin{array}{c} |x_n - a| \\ |y_n - b| \end{array} \right\} \leq |z_n - \alpha| \leq |x_n - a| + |y_n - b|.$$

Necessity. If $z_n \to \alpha$, then given $\varepsilon > 0$ there is an N for which

$$|x_n - a| \leq |z_n - \alpha| < \varepsilon \quad \text{if} \quad n > N.$$

Thus $x_n \to a$. Similarly $y_n \to b$.

Sufficiency. If $x_n \to a$ and $y_n \to b$, then given $\varepsilon > 0$ there is an N_1 for which

$$|x_n - a| < \varepsilon/2 \quad \text{if} \quad n > N_1, \quad \text{(how?)}$$

and there is an N_2 for which

$$|y_n - b| < \varepsilon/2 \quad \text{if} \quad n > N_2. \quad \text{(how?)}$$

Then if $N = \max[N_1, N_2]$

$$|z_n - \alpha| \leq |x_n - a| + |y_n - b| < \varepsilon/2 + \varepsilon/2 = \varepsilon \quad \text{if} \quad n > N. \quad \square$$

1.3b Corollary. *If $\{z_n\}$ and $\{\zeta_n\}$ are sequences and $\zeta_n \to \alpha$, and $z_n - \zeta_n \to 0$, then $z_n \to \alpha$.*

Proof. Let $\varepsilon > 0$ be given. Then there is an N_1 for which $|\zeta_n - \alpha| < \varepsilon/2$ if $n > N_1$, and there is an N_2 for which $|z_n - \zeta_n| < \varepsilon/2$ if $n > N_2$. Then with $N = \max[N_1, N_2]$ we have for $n > N$

$$|z_n - \alpha| < |z_n - \zeta_n| + |\zeta_n - \alpha| < \varepsilon/2 + \varepsilon/2 = \varepsilon. \quad \square$$

Example 4. We know that $x^n \to 0$ if x is real and $-1 < x < 1$, that is, if $|x| < 1$. Show that $z^n \to 0$ if $|z| < 1$.

Solution. We set $|z| = r$ where $0 \leq r < 1$. Then

$$|z^n - 0| = |z^n| = |z|^n = r^n \to 0.$$

Example 5. Show $\{[(1+i)/\sqrt{2}]^n\}$ diverges.

Solution.
$$(1+i)/\sqrt{2} = \frac{1}{\sqrt{2}} + i\frac{1}{\sqrt{2}}$$

$$= \cos\frac{\pi}{4} + i\sin\frac{\pi}{4} = E\left(\frac{\pi}{4}\right)$$

By de Moivre's theorem

$$[(1+i)/\sqrt{2}]^n = E\left(n \cdot \frac{\pi}{4}\right).$$

We observe that

$$E\left((n+8)\frac{\pi}{4}\right) = E\left(n\frac{\pi}{4} + 2\pi\right) = E\left(n\frac{\pi}{4}\right),$$

so the values of this sequence rotate among the eight numbers

$$E\left(\frac{\pi}{4}\right), E\left(\frac{2\pi}{4}\right), \ldots, E\left(\frac{8\pi}{4}\right) = E(2\pi) = 1,$$

and never become and remain close to any one number.

The concept of limit extends to functions just as in the real case, and the formal definition reads the same as for the reals. Again the difference is in the meaning of absolute value.

If $f(z)$ is a complex valued function of a complex variable z, whose domain contains at least a deleted neighborhood of a point ζ, then

$$\lim_{z \to \zeta} f(z) = \alpha \quad \text{means}$$

for each $\varepsilon > 0$ there is a $\delta = \delta(\varepsilon) > 0$ for which

$$|f(z) - \alpha| < \varepsilon \quad \text{when} \quad 0 < |z - \zeta| < \delta.$$

Example 6.
$$\lim_{z \to \zeta} z^n = \zeta^n.$$

Solution. Take $0 < |z - \zeta| < 1$ as a deleted neighborhood of ζ and consider z in this neighborhood. Then

$$|z| = |z - \zeta + \zeta| \le |z - \zeta| + |\zeta| \le |\zeta| + 1 \equiv M,$$

and also $|\zeta| < |\zeta| + 1 = M$. Then by factoring,

$$|z^n - \zeta^n| = |(z - \zeta)(z^{n-1} + z^{n-2}\zeta + \cdots + z\zeta^{n-2} + \zeta^{n-1})|$$
$$\le |z - \zeta|(M^{n-1} + M^{n-1} + \cdots + M^{n-1} + M^{n-1})$$
$$= |z - \zeta|nM^{n-1} < \varepsilon$$

if $|z - \zeta| < \varepsilon/nM^{n-1} \equiv \delta(\varepsilon)$.

Example 7. Show

$$\lim_{z \to 3i}(z^2 - iz + b)/(z - 3i) = 5i.$$

Solution. We consider

$$\left| \frac{z^2 - iz + 6}{z - 3i} - 5i \right| = \left| \frac{z^2 - iz + 6 - 5iz - 15}{z - 3i} \right|$$
$$= \left| \frac{z^2 - 6iz - 9}{z - 3i} \right| = |z - 3i| < \varepsilon$$

if $|z - 3i| < \varepsilon$. That is, we take $\delta(\varepsilon) \equiv \varepsilon$.

It follows, *just as in the real case*, that the limit of a sum is the sum of the limits. The same applies to differences, to products, and also to quotients, *provided the limit of the denominator is not zero.* For the proofs of these statements, for both sequences and functions, see Exercises B1 and B2.

Again as in the real case, continuity can be defined in terms of limits, or in ε terms. The formal statement follows.

Let the domain of $f(z)$ contain a neighborhood of ζ. Then $f(z)$ is **continuous** at ζ means

$$\lim_{z \to \zeta} f(z) = f(\zeta),$$

or equivalently it means, for each $\varepsilon > 0$ there is a $\delta = \delta(\varepsilon)$ for which

$$|f(z) - f(\zeta)| < \varepsilon \quad \text{when} \quad |z - \zeta| < \delta.$$

Suppose a function is continuous at two points, say ζ_1 and ζ_2. Then to satisfy the *same* ε estimate at both points, two different δ's may be needed, say δ_1 at ζ_1 and δ_2 at ζ_2. That is,

$$|f(z) - f(\zeta_1)| < \varepsilon \quad \text{if} \quad |z - \zeta_1| < \delta_1$$

and

$$|f(z) - f(\zeta_2)| < \varepsilon \quad \text{if} \quad |z - \zeta_2| < \delta_2.$$

Then clearly, by choosing $\delta = \min(\delta_1, \delta_2)$ we can use the same δ. That is,

$$|f(z) - f(\zeta_1)| < \varepsilon \quad \text{if} \quad |z - \zeta_1| < \delta$$

and

$$|f(z) - f(\zeta_2)| < \varepsilon \quad \text{if} \quad |z - \zeta_2| < \delta$$

Thus with two points we can always achieve the same δ. Clearly with a finite number of points we can do the same sort of thing.

Suppose further that $f(z)$ is continuous on a set D containing infinitely many points. Then it still may be (also it may not, of course) that we can choose a uniform δ that will work for all points in D. In general the δ will depend on the set D as well as $\varepsilon : \delta = \delta(\varepsilon, D)$. We make the following definition.

Definition. *To say that the function $f(z)$ is* **uniformly continuous** *on a set D means that for each $\varepsilon > 0$ there is a $\delta = \delta(\varepsilon, D) > 0$ for which*

$$|f(z) - f(\zeta)| < \varepsilon \quad \text{if} \quad z \in D, \quad \zeta \in D \quad \text{and} \quad |z - \zeta| < \delta.$$

Example 8. Show that $1/z$ is continuous for $z \neq 0$, and, for each $a > 0$, it is uniformly continuous on the set $D = \{z : |z| \geq a\}$.

Solution. It suffices to prove the uniform continuity. (Why? See Exercise B13.) Suppose $|z| \geq a$ and $|\zeta| \geq a$. Then

$$\left| \frac{1}{z} - \frac{1}{\zeta} \right| = \left| \frac{z - \zeta}{z\zeta} \right| \leq \frac{|z - \zeta|}{a^2} < \varepsilon$$

if $|z - \zeta| < a^2 \varepsilon \equiv \delta$. Notice that the set D enters into δ through the number a used in defining it.

We conclude this section on limits by returning briefly to sequences to discuss sequences of functions. Suppose S is a set and that for each integer $n \geq 0$ (or more generally $n \geq n_0$) we have a complex valued function $\phi_n(z)$ defined on S. This collection is a **sequence of functions** on S which we denote by $\{\phi_n(z)\}_0^\infty$, or $\{\phi_n\}_0^\infty$, or $\{\phi_n\}$.

For each fixed z in S, $\{\phi_n(z)\}$ is a sequence of complex numbers which may or may not converge. If the sequence converges for each point in S we say the sequence is **pointwise** convergent on S. The limit will in general depend on z of course and hence is another function on S, which we denote by $\phi(z)$:

$$\lim \phi_n(z) = \phi(z) \quad \text{for each } z \text{ in } S.$$

Thus for each z in S the ε-definition of a limit must be satisfied, and so for each z in S and for each $\varepsilon > 0$ there is N for which

$$|\phi_n(z) - \phi(z)| < \varepsilon \quad \text{whenever} \quad n > N.$$

But the N in general depends on z as well as on $\varepsilon : N = N(z, \varepsilon)$.

For example, let us look at the sequence defined by $\phi_n(z) = z^n$ on the set S defined by $|z| \leq a$, where $a < 1$. Then if we denote $|z|$ by r, we have

$$|z^n - 0| = |z^n| = r^n$$

and for this to be smaller than ε, $(r^n < \varepsilon)$ we must have, taking logarithms,

$$n \ln r < \ln \varepsilon$$

or

$$n \ln \left(\frac{1}{r} \right) > \ln \left(\frac{1}{\varepsilon} \right)$$

or

$$n > \ln \left(\frac{1}{\varepsilon} \right) \Big/ \ln \left(\frac{1}{r} \right).$$

Thus we can take N to be $\ln(1/\varepsilon)/\ln(1/r)$ so that it clearly depends on ε, and on z $(|z| = r)$.

Suppose however it is possible to choose an N, depending only on ε and the set S, which works for all z in S. That is, suppose we can find an $N(S, \varepsilon)$ for which

$$|\phi_n(z) - \phi(z)| < \varepsilon \quad \text{whenever} \quad n > N(S, \varepsilon)$$

for all z in S. In this case we say $\phi_n(z)$ converges to $\phi(z)$ **uniformly** on S. The name comes from the fact that we can use the *same* or *uniform* N for all z.

We frequently write this as

$$\phi_n \to \phi \quad \text{uniformly on} \quad S.$$

Example 9. Show that $z^n \to 0$ uniformly on the set S described by $|z| \le a$ where a is some fixed number < 1.

Solution. Consider

$$|z^n - 0| = |z^n| = |z|^n = r^n \le a^n$$

for $|z| \le a$, and so

$$|z^n - 0| < \varepsilon \quad \text{if} \quad a^n < \varepsilon$$

and

$$a^n < \varepsilon \quad \text{if} \quad n > \ln\left(\frac{1}{\varepsilon}\right) \Big/ \ln\left(\frac{1}{a}\right)$$

and so we can choose $N = \ln(1/\varepsilon)/\ln(1/a)$.

Significant Topics	Page

A Exercises

1. Set $w = 3iz + 2$ and show
   (a) the line $x + 2y = 3$ maps onto $2u - v + 5 = 0$;

(b) $x^2 + y^2 = 1$ maps onto $u^2 + v^2 - 4u = 5$;

(c) the circle $u^2 + v^2 = 9$ arises from a circle in the z plane, and find its equation.

2. Let $w = z^2$ and show

   (a) $x^2 + y^2 = a^2$ maps onto $u^2 + v^2 = a^4$;

   (b) $x = a$ maps onto the parabola $v^2 + 4a^2 u = 4a^4$;

   (c) $u = c$ arises from the hyperbola $x^2 - y^2 = c$.

3. Find the limit of the following sequences.

   (a) $\dfrac{n-1}{n} + \dfrac{2i}{n}$

   (b) $\dfrac{1}{n} + \left(\dfrac{n+2}{n}\right) i$

   (c) $\dfrac{n-1}{n^2+2} + \dfrac{n+1}{n^2+3} i$

   (d) $\left(\dfrac{1}{\sqrt{3}} + \dfrac{i}{\sqrt{3}}\right)^n$

   (e) $\left(1 + \dfrac{1}{n}\right)^n + \left(1 + \dfrac{1}{n}\right)^{-n} i$

   (f) $n\left(\sqrt{1 + \dfrac{2}{n^2}} - 1\right) + \left(\dfrac{n+1}{n}\right) i$

4. Find the limits of the following sequences when the limits exist. Identify those which have no limit.

   (a) $\sin\left(2\pi n + \frac{1}{n}\right) + i \cos 2\pi n$.

   (b) $n \sin\left(2\pi n - \frac{1}{n}\right) - i\frac{1}{n} \cos n\pi$.

   (c) $n \sin\left(n\pi + \frac{1}{n}\right) - i \cos n\pi$.

   (d) $n^2 \left[\cos\left(2\pi n - \frac{1}{n}\right) - 1\right] + i\sqrt{n} \sin\left(4\pi n - \frac{1}{n}\right)$.

   (e) $[(1 + i\sqrt{3})/2]^n$.

B Exercises

1. If $\lim z_n = \alpha$ and $\lim \zeta_n = \beta$, then show

   (a) $\lim(z_n \pm \zeta_n) = \alpha \pm \beta$,

   (b) $\lim(z_n \zeta_n) = \alpha\beta$,

   (c) $\lim z_n/\zeta_n = \alpha/\beta$ if $\beta \neq 0$.

2. If $\lim_{z \to \zeta} f(z) = \alpha$ and $\lim_{z \to \zeta} g(z) = \beta$, then show

   (a) $\lim_{z \to \zeta}(f(z) \pm g(z)) = \alpha \pm \beta$,

   (b) $\lim_{z \to \zeta}[f(z)g(z)] = \alpha\beta$,

   (c) $\lim_{z \to \zeta} f(z)/g(z) = \alpha/\beta$ if $\beta \neq 0$.

3. If $f(z)$ and $g(z)$ are continuous at ζ, then show that the following are also.

(a) $f(z) \pm g(z)$ (b) $f(z)g(z)$

(c) $f(z)/g(z)$ unless $g(\zeta) = 0$.

4. Show $[(z + h)^n - z^n]/h \to nz^{n-1}$ as $h \to 0$. (See Example 2, Section 1.2)

5. Show that every polynomial is continuous in the whole plane, but no polynomial of degree > 1 is uniformly continuous there.

6. Show that $f(z)$ is continuous at ζ if and only if $\overline{f(z)}$ is continuous at ζ.

7. If $f(z) = u(z) + iv(z)$, and $\alpha = a + ib$, then show that $\lim_{z \to \zeta} f(z) = \alpha$ if and only if $\lim_{z \to \zeta} u(z) = a$ and $\lim_{z \to \zeta} v(z) = b$.

8. (a) Let $f(z) = (z^3 - 8)/(z - 2)$ for $z \neq 2$. Can one define $f(z)$ at $z = 2$ so as to be continuous there?

(b) Define each of the following so as to be continuous at $z = 2$.

(i) $\left(\frac{1}{z} - \frac{1}{2}\right)\frac{1}{z-2}$ (ii) $\left(\frac{1}{z^3} - \frac{1}{8}\right)\frac{1}{z-2}$.

9. Let A be the angle defined by $0 \leq \text{Arg}\, z \leq \phi$. Discuss the image of A under each of the mappings.

$$(a)\ w = z^3 \qquad (b)\ w = z^4 \qquad (c)\ w = z^8.$$

10. For $|z| < 1$ show that

$$\lim nz^n = \lim n^2 z^n = \lim n(n + 1)z^n = 0.$$

11. If $z \neq 1$ but $|z| = 1$, show that

$$\lim \left(\sum_1^n z^k\right)\Big/ n = 0.$$

12. In Example 8, why is it sufficient to show uniform continuity as asserted there?

C Exercises

1. Let $w = \alpha z + \beta$, $\alpha \neq 0$. Show that

(a) each line in the z-plane maps onto a line in the w-plane;

(b) each circle maps onto a circle;

(c) each parabola, ellipse, or hyperbola maps onto a parabola, ellipse, or hyperbola, respectively.

2. Let $w = 1/z$ and examine the mapping from the z-plane to the w-plane as outlined. Show that
   (a) if $|z| = 1$, then $w = \bar{z}$;
   (b) if $|z| > 1$, then $|w| < 1$, and conversely;
   (c) the ray R, described by $\arg z = \theta_0$ maps onto R' described by $\arg w = -\theta_0$, and as $z \to 0$ on R, $w \to \infty$ on R';
   (d) the circle $|z| = a$ maps onto the circle $|w| = 1/a$, and as z moves counterclockwise w moves clockwise.

3. (Continuation) Observe that if C is a circle or straight line in the z-plane, its equation can be put in the form

$$C : a(x^2 + y^2) + bx + cy + d = 0.$$

Let Γ be the image of C under the mapping $w = 1/z$. Show that the equation of Γ is

$$\Gamma : d(u^2 + v^2) + bu - cv + a = 0.$$

Deduce that if C goes through the origin, then Γ is a line; and if C does not go through the oirgin, then Γ is a circle.

4. Find the images under $w = 1/z$ of the coordinate lines $x = a$, $y = b$.

5. Prove that for α, β, γ to be the vertices of an equilateral triangle it is necessary and sufficient that

$$\frac{\beta - \alpha}{\gamma - \alpha} = \frac{\alpha - \gamma}{\beta - \gamma}.$$

1.4 Curves and Integrals

Let $I = \{t \in \mathbb{R} : a \le t \le b\}$ be a closed interval on the real axis. A **partition** Δ of I is a finite set of points in I which contains a and b. Thus, *listing the points in order of magnitude*, we have

$$\Delta = \{a_0, a_1, \dots, a_n\}$$

is a partition of I if and only if

$$a = a_0 < a_1 < \cdots < a_n = b.$$

The open intervals $I_k \equiv \{t : a_{k-1} < t < a_k\}$, $1 \le k \le n$, are called the **subintervals** of Δ.

A real valued function $f(t)$ being **piecewise continuous** on I means
(i) there is a partition Δ of I such that $f(t)$ is defined at least for $t \in I - \Delta$. On Δ itself $f(t)$ may fail to be defined or may even be ambiguously defined,
(ii) $f(t)$ is continuous on each subinterval of Δ,
(iii) $f(t)$ has one-sided limits at the points of Δ

$$f(a_k + 0) \equiv \lim_{t \to a_k+} f(t) \text{ exists for } k = 0, 1, \ldots, n-1,$$
$$f(a_k - 0) \equiv \lim_{t \to a_k-} f(t) \text{ exists for } k = 1, 2, \ldots, n.$$

A real valued function being **piecewise smooth** on an interval I means there is a partition Δ of I for which the derivative $f'(t)$ exists except (possibly) on Δ itself, and $f'(t)$ is piecewise continuous.
A complex valued function

$$f(t) \equiv g(t) + ih(t), \qquad t \in I$$

is **piecewise continuous** or **piecewise smooth** if both $g(t)$ and $h(t)$ are piecewise continuous or piecewise smooth, respectively.
It is shown in calculus courses that

$$\int_a^b f(t)dt \quad \text{or} \quad \int_I f(t)dt$$

exists if $f(t)$ real valued and is piecewise continuous on the interval $I = \{t \in \mathbb{R} : a \le t \le b\}$. Of course, the integral exists for a wider class of functions, but for simplicity we restrict our attention to piecewise continuous ones.
If $f(t) = g(t) + ih(t)$ is piecewise continuous, we define the integral of $f(t)$ on I by

$$\int_a^b f(t)dt \equiv \int_I f(t)dt \equiv \int_a^b g(t)dt + i \int_a^b h(t)dt.$$

By this we mean that the right side defines each of the two integrals on the left.
The integral of a complex valued function is also the limit of the complex Riemann sums. To see this let Δ be a partition of I. Then by the **norm** of Δ, denoted by $\|\Delta\|$, we mean

$$\|\Delta\| = \max_{1 \le k \le n} (a_k - a_{k-1})$$

and by $\Delta_k t$, the kth t-difference in Δ, we mean

$$\Delta_k t = a_k - a_{k-1}.$$

Let τ_k be any point in I_k. Then

$$\int_a^b f(t)dt = \lim_{\|\Delta\|\to 0} \sum_1^n f(\tau_k)\Delta_k t.$$

This is clear since

$$\sum_1^n f(\tau_k)\Delta_k t = \sum_1^n g(\tau_k)\Delta_k t + i \sum_1^n h(\tau_k)\Delta_k t$$

and the integrals of $g(t)$ and $h(t)$ are defined in calculus to be the limits, respectively, of the sums on the right.

The Fundamental Theorem of Calculus for complex valued functions of a real variable follows from the calculus theorem. We leave the proofs to the exercises but state the two parts of that theorem for reference.

1.4a Theorem. *If $f(t)$ is piecewise continuous on $I = \{t \in \mathbb{R} : a \le t \le b\}$, then*

$$\frac{d}{dt} \int_a^t f(\tau)d\tau = f(t)$$

at each point of continuity of $f(t)$.

1.4b Theorem. *If $F(t)$ is continuous and piecewise smooth on I, and $F'(t) = f(t)$, then*

$$\int_a^b F'(t)dt = \int_a^b f(t)dt = F(b) - F(a).$$

There is a very useful estimate for complex integrals which is a little more difficult to prove than the corresponding result for real integrals. We give one proof here and indicate another in the exercises.

1.4c Theorem. *If $f(t)$ is piecewise continuous on $I = \{t \in \mathbb{R} : a \le t \le b\}$, then*

$$\left| \int_a^b f(t)dt \right| \le \int_a^b |f(t)|dt.$$

Proof. If the integral on the left is zero, there is nothing to prove. Otherwise, by writing its value in polar form, we have

$$\int_a^b f(t)dt = r\alpha, \qquad r > 0, \ |\alpha| = 1.$$

Then since $1/\alpha = \bar{\alpha}$, we have

$$r = \int_a^b \bar{\alpha} f(t)dt = \int_a^b \text{Re}\,(\bar{\alpha} f(t))dt + i \int_a^b \text{Im}\,(\bar{\alpha} f(t))dt.$$

The last integral is zero since r is real. And

$$\text{Re}\,(\bar{\alpha} f(t)) \le |\text{Re}\,(\bar{\alpha} f(t))| \le |\bar{\alpha} f(t)| = |f(t)|,$$

and so from calculus

$$r = \int_a^b \text{Re}\,(\bar{\alpha} f(t))dt \le \int_a^b |f(t)|dt. \quad \square$$

We want to extend the notion of integral so that we can integrate over curves. So we proceed to make definitions and study curves themselves first.

Let $\phi(t)$ be a complex valued continuous function defined on $I = [a, b]$. Then $\phi(t)$ defines a **curve** C which is graphically represented by plotting in the complex plane all the points $\phi(t)$ for all t values in I. We denote this by writing

$$C : z = \phi(t), \qquad t \in I. \tag{1}$$

When the context makes it clear we sometimes do not write the $t \in I$ in (1).

Observe that the single complex equation $z = \phi(t)$ is equivalent to two real equations. Thus if $\phi(t) = \xi(t) + i\eta(t)$, then

$$x = \xi(t), \quad y = \eta(t), \quad t \in I, \tag{2}$$

is equivalent to (1). These equations (2) are called the **parametric equations** of C in elementary calculus and analytic geometry: t is called the **curve parameter** and I the **parameter interval**. In many cases t will have geometrical or physical significance. It may, for example, represent time, distance, angle, or some other geometrical or physical quantity.

Since $\phi(t)$ is a complex number for each t, we could describe it in polar form. That is, we can define a curve C by giving its equation in polar form. Thus if

$$r = r(t), \quad \theta = \theta(t) \quad t \in I$$

are continuous functions and if we define $\phi(t)$ by

$$\phi(t) = r(t)E(\theta(t))$$
$$= r(t)[\cos \theta(t) + i \sin \theta(t)], \quad t \in I,$$

then C, given by

$$z = \phi(t) = r(t)E(\theta(t)), \quad t \in I, \tag{3}$$

is a curve in polar form.

The connections between (2) and (3) are not entirely obvious. It is certainly true that given (3) we have that

$$\xi(t) = r(t) \cos \theta(t), \quad \eta(t) = r(t) \sin \theta(t)$$

are continuous if $r(t)$ and $\theta(t)$ are. So given (3) we can recapture (2). However, given (2) it is not always trivial to get (3). Certainly

$$r(t) \equiv \sqrt{\xi^2(t) + \eta^2(t)}$$

is continuous. But to define $\theta(t)$ as a continuous function of t is not easy. This is again because of the ambiguity of the argument of a complex number. We explore these difficulties in the C Exercises.

Notice that our definition of a curve C, in either rectangular or polar coordinates, carries with it an inherent notion of the orientation of C. Thus the **initial point** of C is $z_0 = \phi(a)$, and the **terminal point** is $z_1 = \phi(b)$. Then as t increases from a to b, $z = \phi(t)$ traces out C in the direction (along the curve itself) of increasing t.

Example 1. Examine

$$C : z = rE(t) = r \cos t + ir \sin t, \quad 0 \le t \le 2\pi,$$

where r is a constant.

Solution. This is equivalent to

$$x = r \cos t, \quad y = r \sin t,$$

which are standard parametric equations of a circle of radius r centered at the origin. The curve parameter here is the angle the vector z makes with the positive x-axis, for clearly $t = \arg z$ (but not always $\operatorname{Arg} z$).

A curve is **simple**, or is a **simple arc**, means that it has no points of self-intersection. Technically this means that $\phi(t) = \phi(\tau)$ only when $t = \tau$. A **simple closed** curve means that the initial point and the terminal point coincide, but there are no other points of self-intersection. This means that $\phi(t) = \phi(\tau)$ only when $t = \tau$ or when one of t and τ is a and the other is b. A simple closed curve is also called a **Jordan curve**. The circle in Example 1 is a Jordan curve.

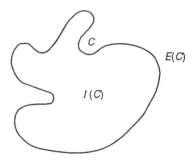

Jordan Curve Theorem: Each simple closed curve, C, has an interior, $I(C)$, and an exterior, $E(C)$.

There is a property of Jordan curves that is very easy to believe and remarkably difficult to prove in general. We will not prove this result, but we will comment further about it in Section 1.6. It is given by the following.

1.4d Theorem. (Jordan Curve Theorem) *If C is a Jordan curve, then the complex plane is composed of the three separate pieces: (i) C itself, (ii) the interior of C denoted by $I(C)$, and (iii) the exterior of C denoted by $E(C)$.*

The definition of a curve as we have given it, is much too general for our needs and purposes. For instance, there are curves (called Peano curves after the man who first constructed one) which pass through every point in a square.

Let C be a curve given by

$$C : z = \phi(t), \qquad t \in I = [a, b].$$

To say that C is **smooth** means that the derivative $\phi'(t) = d\phi/dt$ exists and is continuous in the closed interval $I = [a, b]$ and that $\phi'(t) \neq 0$ in the open interval (a, b). If C is given in rectangular form by

$$z = \xi(t) + iy(t), \qquad t \in I,$$

then clearly C is smooth if and only if $\xi'(t)$ and $y'(t)$ both exist and are continuous in I, and $\xi'^2 + \eta'^2 > 0$ in (a, b). And if C is given in polar form by

$$z = r(t)E(\theta(t)), \qquad t \in I,$$

where $\theta(t)$ is continuous in I, then C is smooth if and only if $r'(t)$ and $\theta'(t)$ exist and are continuous in I, and $r'^2 + r^2\theta'^2 > 0$ in (a, b). This is a little difficult to prove, and the proof is outlined in the C Exercises. *In specific cases we will be working with simple situations where the differentiability is clear.* Then we have

$$z' = \phi'(t) = \xi'(t) + i\eta'(t)$$

in the rectangular form, and in polar form

$$z' = \phi'(t) = r'(t)E(\theta(t)) + ir(t)E(\theta(t))\theta'(t). \quad \text{(How?)}$$

A **piecewise smooth** curve consists of a finite number of smooth pieces joined so that the terminal point of the first coincides with the initial point of the next, and so on to the last. This is equivalent to $\phi(t)$ being continuous and $\phi'(t)$ being piecewise continuous.

Let us consider the geometrical meaning of the derivative $\phi'(t)$. Let t and $t + h$ be nearby points in I. Then $\phi(t + h)$ and $\phi(t)$ will be nearby points on C, and the difference $\phi(t + h) - \phi(t)$ is the secant vector from $\phi(t)$ to $\phi(t + h)$. Then

$$[\phi(t + h) - \phi(t)]/h$$

lies along the chord and always points in the direction of increasing t. From $\phi(t)$ toward $\phi(t + h)$ if $h > 0$ and the other way if $h < 0$. The limit of this difference quotient is the derivative,

$$\phi'(t) = \lim_{h \to 0} \frac{\phi(t + h) - \phi(t)}{h}$$

which therefore represents the **tangent vector** (since $\phi'(t) \neq 0$), which points in the direction of increasing t. Then

$$\phi'(t) \big/ |\phi'(t)|$$

is the **unit tangent**, again in the direction of increasing t.

Example 2. Examine C given by

$$z = \phi(t) = \begin{cases} t + it^2 & 0 \leq t \leq 1 \\ t + i(t-2)^2 & 1 \leq t \leq 2. \end{cases}$$

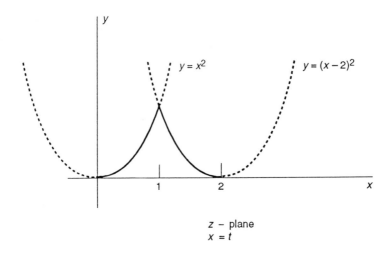

$$z - \text{plane}$$
$$x = t$$

Solution. The derivative exists in $I = [0, 2]$ except at $t = 1$.

$$z' = \begin{cases} 1 + 2it & 0 \leq t < 1 \\ 1 + 2i(t-2) & 1 < t \leq 2, \end{cases}$$

and so $\phi'(t)$ is piecewise continuous. Each of the two pieces is smooth and the curve is composed of abutting arcs of two parabolas.

The two kinds of curves of greatest interest here are straight line segments and circular arcs. If z_0 and z_1 are two distinct points in the plane, then the segment S, oriented from z_0 to z_1, can be given by

$$S : z = z_0 + t(z_1 - z_0), \qquad 0 \le t \le 1,$$

or by

$$S : z = z_0 + s(z - z_0)/L, \qquad 0 \le s \le L$$

where $L = |z_1 - z_0|$. Here s is the distance from z_0 to z along S. The connection between these two formulas is given by $t = s/L$.

A **polygonal line** is a curve made up of a finite number of straight line segments connected end to end. A polygonal line is then one kind of a piecewise smooth curve.

If z_0 is a point in the plane and $a > 0$ is a constant radius, then

$$C : z = z_0 + aE(\theta) = z_0 + a(\cos\theta + i\sin\theta), \qquad \alpha \le \theta \le \beta$$

is a circular arc centered at z_0 between the angles α and β.

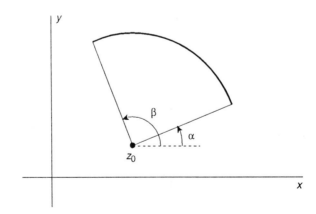

$$z = z_0 + aE(\theta)$$
$$= z_0 + a(\cos\theta + i\sin\theta)$$
$$z - z_0 = a(\cos\theta + i\sin\theta)$$
$$\text{where } \alpha \le \theta \le \beta$$

If C is a smooth or piecewise smooth curve given by

$$C : z = \phi(t) = \xi(t) + i\eta(t), \qquad t \in I = [a, b],$$

then the **length** of C is given by

$$L = L_C = \int_a^b |\phi'(t)| dt$$

This follows from the calculus formula for arc length of a curve in parametric form:

$$L = \int_a^b \sqrt{[\xi'(t)]^2 + [\eta'(t)]^2} \; dt,$$

by the same token, the arc length of C from the initial point a to an arbitrary t-value is given by

$$s = s(t) = \int_a^t |\phi'(\tau)| d\tau$$

A **contour** is a **simple closed piecewise smooth curve**. You should observe that a contour C is a Jordan curve (since it is simple and closed) and so has an interior $I(C)$ and an exterior $E(C)$. The term **contour integration** refers to integrals (which we will now define) of complex valued functions over piecewise smooth curves whether or not they are contours. A **path** is a piecewise smooth curve. A **closed path** is a piecewise smooth curve whose initial and terminal points coincide. Thus a **simple closed path is a contour**.

Let

$$C : z = \phi(t) \qquad t \in I = [a, b]$$

be a path, and let $f(z)$ be continuous (or piecewise continuous) on C. By this we mean that $f(\phi(t)) = f \circ \phi(t)$ is continuous (piecewise continuous) for $t \in I$. Then

$$\int_C f(z) dz \quad \text{means} \quad \int_a^b f(\phi(t)) \phi'(t) dt.$$

The formal substitution

$$z = \phi(t) \qquad dz = \phi'(t) dt$$

makes this definition easy to remember.

1.4e Theorem. *If $f(z)$ is piecewise continuous on the path C and if $|f(z)| \le M$ on C, and L is the length of C, then*

$$\left| \int_C f(z)dz \right| \le ML$$

Proof. This is an immediate consequence of Theorem 1.4c. For let C be given by

$$C : z = \phi(t) \qquad t \in [a, b],$$

then

$$\left| \int_C f(z)dz \right| = \left| \int_a^b f(\phi(t))\phi'(t)dt \right|$$

$$\le \int_a^b |f(\phi(t))\phi'(t)|dt$$

$$= \int_a^b |f(\phi(t))||\phi'(t)|dt \le M \int_a^b |\phi'(t)|dt$$

$$= ML. \qquad \square$$

In some cases C may be given in a nonparametric form. Under these circumstances you are expected to supply an appropriate parameterization to be used in evaluating

$$\int_C f(z)dz.$$

If C is a contour (simple closed piecewise smooth curve, or simple closed path), then the **positive direction** on C has the interior of C to the left as the tracing point $z = \phi(t)$ moves around C. Or equivalently it is the counterclockwise direction around C.

A careful study of curves would get into deep problems which are not of central concern to us in a course at the level of this book. For example, the two curves

$$C_1 : z = E(t), \qquad 0 < t < 2\pi,$$
$$C_2 : z = E(t), \qquad 0 < t < 4\pi,$$

are easily seen to have the same graph in the z-plane; C_2 merely covers the circle twice. By our definition, the arc length of C_1 is 2π and that of C_2 is 4π, for we have defined length in terms of the parameterization. While the relation between the graph of a curve and the parametric representation is in general a difficult one, we will deal only with simple cases where the connection is clear as in the preceding example.

Another difficulty arises in connection with two different parameterizations of the same curve. Suppose C is given by the two parameterizations

$$C : z = \phi(t), \qquad t \in I,$$
$$C : z = \psi(\tau), \qquad \tau \in J.$$

It is natural to expect that the arc length, the unit tangent vectors, and other geometrical properties should be the same. That these things are true are difficult to prove in general, though some simple cases are covered in the exercises.

In this connection we observe that

$$C_1 : z = E(t) = \cos t + i \sin t, \qquad -\pi \leq t \leq \pi,$$

and

$$C_2 : z = \frac{\tau^4 - 3\tau^2 + 1}{\tau^4 - \tau^2 + 1} + 2i \frac{\tau - \tau^3}{\tau^4 - \tau^2 + 1}, \qquad -1 \leq \tau \leq 1,$$

are the same geometrical curve as a little work will demonstrate. We will, as remarked above, sidestep these difficulties by restricting our attention to cases where the connections are clear.

Significant Topics

A Exercises

Compute the following integrals.

1. $\int_0^1 (2 + it)dt,$

2. $\int_0^1 (2 + ip^2)dp,$

3. $\int_0^1 (u^2 - iu^3)du,$

4. $\int_5^{-1} (6w + 3iw^2)dw,$

5. $\int_{-2}^5 f(s)ds$ where $f(s) = \begin{cases} -1 + 7is^6, & -2 \leq s < 2, \\ 2s + is^2, & 2 \leq s \leq 5, \end{cases}$

6. $\int_1^7 \phi(t)dt$ where $\phi(t) = \begin{cases} 2 + it & 1 \leq t \leq 3 \\ 3t^2 + 6it & 3 \leq t < 5 \\ 2 + it & 5 \leq t \leq 7. \end{cases}$

7. $\int_2^3 [\ln \alpha + ie^{\alpha}]d\alpha$

8. In Exercise A5 suppose $f(s)$ were modified to

$$f(s) = \begin{cases} -1 + 7is^6, & -2 \leq s \leq 2 \\ 2s + is^2, & 2 < s \leq 5. \end{cases}$$

How does this affect the value of the integral?

9. Show that $E'(\theta) = iE(\theta)$ where $' = d/d\theta$.

10. If $r(t)$ and $\theta(t)$ are both differentiable and $' = d/dt$, then show that
   (a) $[E(\theta(t))]' = iE(\theta(t))\theta'(t);$

   (b) $\begin{aligned}[r(t)E(\theta(t))]' &= r'(t)E(\theta(t)) + ir(t)E(\theta(t))\theta'(t) \\ &= E(\theta(t))[r'(t) + ir(t)\theta'(t)]; \end{aligned}$

   (c) if $\phi(t) = r(t)E(\theta(t))$, then

   $$|\phi'(t)| = \{[r'(t)]^2 + [r(t)\theta'(t)]^2\}^{1/2}.$$

11. (a) Let $\phi(t) = tE(\pi/4), -1 \leq t \leq 1$, and show that $z = \phi(t)$ is smooth.
    (b) Let $\phi(t) = t^2 E(\pi/4), -1 \leq t \leq 1$, and show that $z = \phi(t)$ is not smooth, but is piecewise smooth.

12. Given the path $C : z = (1-t)+it, 0 \le t \le 1$, sketch its graph showing its orientation. Then evaluate

$$(a) \quad \int_C x\,dz \qquad (b) \quad \int_C y\,dz \qquad (d) \quad \int_C z\,dz$$

13. Repeat Exercise A12 for $C : z = 2E(t), 0 \le t \le \pi$.
14. Repeat Exercises A12 for $C : z = 1 + E(t), 0 \le t \le \pi/2$.
15. (a) Use Theorem 1.4e to show

$$\left| \int_C (2+3x)dz \right| \le 32\pi$$

where C is the circle $|z| = 2$.
   (b) By taking the right and left halves of C separately replace 32π by 20π.

B Exercises

Let C be given by

$$z = \phi(t) = r(t)E(\theta(t)) = \xi(t) + i\eta(t), \qquad t \in I = [a, b] \qquad (*)$$

1. Show that $\phi(t)$ is continuous if and only if both $\xi(t)$ and $\eta(t)$ are continuous.
2. Show that if $r(t)$ and $\theta(t)$ are continuous so is $\phi(t)$, and hence by B1, so are $\xi(t)$ and $\eta(t)$.
3. Show that $\phi'(t)$ exists if and only if $\xi'(t)$ and $\eta'(t)$ exist. Hence show that C is smooth if and only if ξ' and η' are continuous in I and $\xi'^2 + \eta'^2 > 0$ in (a, b).
4. Show that if $r'(t)$ and $\theta'(t)$ exist, then $\phi'(t)$ exists. Deduce that if also $r'(t)$ and $\theta'(t)$ are continuous with $r'^2 + r^2\theta'^2 > 0$ in (a, b), then C is smooth.
5. Suppose $\phi(t)$ is also given by

$$\phi(t) = \rho(t)E(\psi(t)), \qquad t \in I.$$

Show
   (a) $r(t) = \rho(t)$;
   (b) $\theta(t) = \psi(t) + 2\pi k(t)$ where $k(t)$ is an integer for each t;
   (c) if $\psi(t)$ is continuous, then $k(t)$ is a constant.

6. Show that if
$$\theta(t) = \begin{cases} t, & 0 \le t < \pi \\ t + 2\pi, & \pi \le t \le 2\pi, \end{cases}$$

and $r(t) = 1 + t$, then C is a spiral. Show also that C is smooth.
Hint: reparameterize.

7. (a) Let C be the circle $|z| = 1$. Show

$$\left| \int_C \frac{dz}{12 + 5z} \right| \le \frac{2\pi}{7}$$

(b) By considering the right and left halves of C show

$$\left| \int_C \frac{dz}{12 + 5z} \right| \le \frac{20\pi}{91}$$

Hint: $|12 + 5z|$ is the distance from $5z$ to -12.

8. (Integration by parts.) (a) Suppose $f(t)$ and $g(t)$ are complex valued functions with continuous derivatives on $I = [a, b]$. Show that

$$\int_I f(t)g'(t)dt = f(b)g(b) - f(a)g(a) - \int_I f'(t)g(t)dt.$$

(b) Extend part (a) to the case where f' and g' are piecewise continuous.

C Exercises

Let C be given as in the B Exercises above.

1. Show that if $\phi'(t)$ exists, then so does $r'(t)$. Hint: $|\phi'| = \sqrt{\xi'^2 + \eta'^2}$.

2. If $\theta(t)$ is continuous, and if $\phi'(t)$ exists, show that $\theta'(t)$ exists. Hint: Form the difference quotient

$$\frac{\Delta\phi}{\Delta t} = \frac{1}{h} \left[r(t + h)E(\theta(t + h)) - r(t)E(\theta(t)) \right]$$
$$= \frac{r(t + h) - r(t)}{h} E(\theta(t + h)) + r(t) \frac{E(\theta(t + h)) - E(\theta(t))}{h}.$$

Now use the mean value theorem on the sine and cosine, and use the existence of $r'(t)$ and $\phi'(t)$ to conclude the existence of $\theta'(t)$.

3. If $\theta(t)$ is continuous and $\phi'(t)$ is continuous, show that $r'(t)$ and $\theta'(t)$ are continuous.

The following problems are concerned with converting a curve from rectangular to polar coordinates. That is, if we have C in the form

$$z = \phi(t) = \xi(t) + i\eta(t), \qquad t \in I = [a, b]$$

where ξ and η are continuous on I, we seek $r(t)$ and $\theta(t)$ *both continuous on I* so that C is also described by $z = \phi(t) = r(t) = r(t)E(\theta(t))$, $t \in I$. Clearly we must have $r(t) = \sqrt{\xi^2(t) + \eta^2(t)}$, and r is continuous on I since both ξ and η are. Furthermore, if $z(t)$ is never zero for $t \in I$, then $r(t) > 0$, and so, by the Weierstrass maximum-minimum theorem, there is a $c > 0$ for which $r(t) \geq c$ for $t \in I$. Also, by the Weierstrass theorem on the uniform continuity of continuous functions on closed bounded intervals, there is a $\delta > 0$ for which

$$|\phi(t) - \phi(\tau)| < c/\sqrt{3} \quad \text{if} \quad |t - \tau| < \delta.$$

Now let $\{a = a_0, \ldots, a_n = b\}$ be a partition of I for which $\max_k(a_k - a_{k-1}) < \delta$, and let $J_k = \{t : a_{k-1} \leq t \leq a_k\}$, $k = 1, 2, \ldots, n$.

4. (a) Define $\theta(a_0) = \theta(a)$ to be some definite value of $\arg \phi(a)$, for example $\text{Arg } \phi(a)$. Show that for each $t \in J_1$ there is a unique value of $\theta(t) = \arg \phi(t)$ for which $|\theta(t) - \theta(a_0)| < \pi/6$, and that the function $\theta(t)$ is continuous.

   (b) Show that $\theta(t)$ can be defined on J_2 so that $|\theta(t) - \theta(a_1)| < \pi/6$ and so that $\theta(t)$ is continuous on J_2 and hence on $J_1 \cup J_2$.

   (c) Show by induction that $\theta(t) = \arg \phi(t)$ can be defined on I to be continuous there.

5. Now put the pieces together from the A, B, and C exercises to show that the curve C is smooth if and only if $r(t)$ and $\theta(t)$ as defined above are differentiable on I and $r'^2 + r^2\theta'^2 > 0$ on (a, b).

1.5 Cauchy's Theorems for Polynomials and Rational Functions

We are now in a position to establish special cases of some fundamental results in the subject of complex variables. Further, the applications of these results will enable us to easily evaluate a number of real integrals which are difficult by the methods of elementary calculus.

Example 1. Evaluate

$$\int_C dz$$

where C is any path from α to β.

Solution. Let C be given by

$$z = \phi(t), \qquad a \le t \le b.$$

Then $\phi(a) = \alpha$ and $\phi(b) = \beta$, and

$$\int_C dz = \int_a^b \phi'(t)dt = \phi(t)\Big|_a^b = \phi(b) - \phi(b) = \beta - \alpha. \tag{1}$$

We comment about this calculation. We can re-express the result as

$$\int_C dz = z\Big|_\alpha^\beta = \beta - \alpha. \tag{2}$$

The reason we cannot do this first is that the *only way we have to evaluate the integral is in terms of the parameterization.* Thus (1) in fact, proves (2), but not the other way around.

Further, the evaluation of $\displaystyle\int_a^b \phi'(t)dt$ where ϕ is continuous and ϕ' is piecewise continuous (ϕ is piecewise smooth) is covered by Theorem 1.4b. Suppose, for example, there were one point of discontinuity of ϕ' say at c, $a < c < b$. Then

$$\begin{aligned}
\int_a^b \phi'(t)dt &= \int_a^c \phi'(t)dt + \int_c^b \phi'(t)dt \\
&= \phi(t)\Big|_a^c + \phi(t)\Big|_c^b \\
&= [\phi(c) - \phi(a)] + [\phi(b) - \phi(c)] \\
&= \phi(b) - \phi(a),
\end{aligned}$$

which is the same result as though ϕ' were continuous over the whole interval $[a, b]$. The same principle applies no matter how many points of discontinuity of ϕ' there are. Because ϕ itself is continuous, the contribution to the integral at these points cancel out, as it did at c just now.

The same comment applies to the other integrals here, in particular to the integral in the next theorem, a generalization of Example 1.

1.5a Theorem. *Suppose*
(i) n is an integer with $n \ne -1$
(ii) C is any path from α to β

(iii) if $n < 0$, then $\alpha \neq 0$, $\beta \neq 0$, and C does not go through the origin.
   Then

$$\int_C z^n dz = (\beta^{n+1} - \alpha^{n+1})/(n+1).$$

In particular, if C is closed, i.e., $\alpha = \beta$, then

$$\int_C z^n dz = 0.$$

Proof. We are tempted to write

$$\int_C z^n dz = \int_\alpha^\beta z^n dz = \frac{z^{n+1}}{n+1}\Big|_\alpha^\beta = (\beta^{n+1} - \alpha^{n+1})/(n+1), \qquad (3)$$

which, however, requires justification. We go back to the definition of the integral. As in the previous example we let C be

$$z = \phi(t), \qquad a \leq t \leq b$$

so that

$$\int_C z^n dz = \int_a^b (\phi(t))^n \phi'(t) dt = (\phi(t))^{n+1}/(n+1)\Big|_a^b$$
$$= ([\phi(b)]^{n+1} - [\phi(a)]^{n+1})/(n+1)$$
$$= (\beta^{n+1} - \alpha^{n+1})/(n+1)$$

which justifies the previous quick calculation (3). $\square$

If $n < 0$, why do we need the condition that $\phi(t)$ is never zero? See Exercise A5.

The following theorem, which is in reality a corollary of Theorem 1.5a, is the **Cauchy Integral Theorem** for polynomials.

1.5b Theorem. *If $p(z)$ is a polynomial and C is a closed path, then*

$$\int_C p(z) dz = 0.$$

Proof. Since $p(z)$ is a polynomial, it has the form

$$p(z) = \sum_0^n a_k z^k = a_0 + a_1 z + \cdots + a_n z^n,$$

so that

$$\int_C p(z)dz = \sum_0^n a_k \int_C z^k dz.$$

Since C is a closed curve, then the second part of Theorem 1.5a applies to each integral in the sum. Thus each integral is 0 and the proof is complete. □

Theorem 1.5a explicitly excludes the case $n = -1$. We now investigate that case, first in a rather simple situation, then in general.

Example 2. Let C be any circle centered at α. Compute

$$\int_C \frac{dz}{z - \alpha}$$

where the integral is in the positive direction.

Solution. Such a circle will have a radius; call it a. Then we parameterize C by

$$z = \alpha + aE(\theta), \qquad 0 \le \theta \le 2\pi,$$

or

$$z - \alpha = a[\cos \theta + i \sin \theta], \qquad 0 \le \theta \le 2\pi.$$

Then

$$\int_C \frac{dz}{z - \alpha} = \int_0^{2\pi} \frac{a[-\sin \theta + i \cos \theta]d\theta}{a[\cos \theta + i \sin \theta]}$$

$$= i \int_0^{2\pi} d\theta = 2\pi i$$

Notice that if C were given by the same formula, but went twice around α, that is, $0 \le \theta \le 4\pi$, then the value of the integral in Example 2 would be $4\pi i$, which is i times the change in θ over the curve.

If C is a path given in polar form

$$z = r(t)E(\theta(t)), \qquad a \le t \le b,$$

where $r(t)$ is never zero, and both $r(t)$ and $\theta(t)$ are continuous, then we will denote $\theta(b) - \theta(a)$ by $\Delta_C\theta$:

$$\Delta_C(\theta) = \theta(b) - \theta(a).$$

It should be clear that if C starts at α and ends at β, then $\Delta_C\theta$ may change from one curve C to another. Thus for example in the figure $\Delta_{C_1}\theta = \pi/2$ while $\Delta_{C_2}\theta = 5\pi/2$.

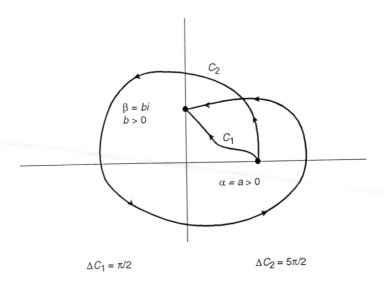

$$\Delta C_1 = \pi/2 \qquad\qquad\qquad \Delta C_2 = 5\pi/2$$

1.5c Theorem. *If C, given by*

$$z = r(t)E(\theta(t)), \qquad a \le t \le b$$

is a path from α to β and $r(t)$ is never zero, then

$$\int_C \frac{dz}{z} = \ln|\beta| - \ln|\alpha| + i\Delta_C\theta$$

where $\ln$ is the real natural logarithm.

Proof. By definition of the integral

$$\int_C \frac{dz}{z} = \int_a^b \frac{[r(t)E(\theta(t))]'\, dt}{r(t)E(\theta(t))}$$

$$= \int_a^b \frac{r'(t)E(\theta(t)) + ir(t)E(\theta(t))\theta'(t)]}{r(t)E(\theta(t))}\, dt$$

$$= \int_a^b \frac{r'(t)dt}{r(t)} + i\int_a^b \theta'(t)dt$$

$$= \ln r(t)\Big|_a^b + i\theta(t)\Big|_a^b$$

$$= \ln r(b) - \ln r(a) + i[\theta(b) - \theta(a)]$$

$$= \ln|\beta| - \ln|\alpha| + i\Delta_C\theta. \quad \square$$

Theorem 1.5a shows that certain functions have integrals whose values depend *only* on the end points no matter which paths of integration are used. For these functions we say that $\int f(z)dz$ is **independent of the path**. However, Theorem 1.5b shows that there are integrals which depend on the path. Thus, for example, if C_1 and C_2 are the paths illustrated in the figure, then by Theorem 1.5c we get

$$\int_{C_1} \frac{dz}{z} = \ln b - \ln a + i\frac{\pi}{2}$$

while

$$\int_{C_2} \frac{dz}{z} = \ln b - \ln a + i\frac{5\pi}{2}$$

even though C_1 and C_2 have the same endpoints.

1.5d Corollary. *If C is a contour with the origin in $I(C)$, then*

$$\int_C \frac{dz}{z} = 2\pi i$$

where the integral is in the positive sense.

Proof. For a contour we have $\alpha = \beta$ and $\Delta_C \theta = 2\pi$. □

1.5e Corollary. *If C is a contour with $\alpha \in I(C)$, then*

$$\int_C \frac{dz}{z - \alpha} = 2\pi i$$

where the integral is in the positive sense.

Proof. Set $z - \alpha = \zeta$ and apply Corollary 1.5d. □

1.5f Corollary. *If C is a contour with $\alpha \in E(C)$, then*

$$\int_C \frac{dz}{z - \alpha} = 0.$$

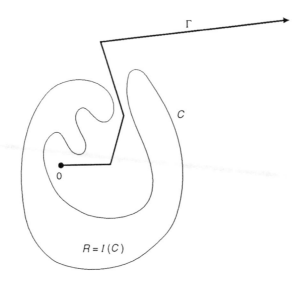

Proof. Without loss of generality, by setting $z - \alpha = \zeta$, we can assume $\alpha = 0$, so we consider $\int_C d\zeta/\zeta$. Then since $0 \in E(C)$, we can connect 0 to ∞ by a path Γ which does not intersect C. Clearly then $\Delta_C \theta = 0$ since a point tracing out C cannot cross Γ. Thus the integral is zero by Theorem 1.5c. $\square$

We come now to the second important result mentioned at the start of this section. It is called the **Cauchy Integral Formula** for polynomials.

1.5g Theorem. *If $p(z)$ is a polynomial, and C is a contour, then*

$$\frac{1}{2\pi i} \int_C \frac{p(z)}{z - \alpha}\, dz = \begin{cases} p(\alpha), & \alpha \in I(C) \\ 0, & \alpha \in E(C) \end{cases}$$

Proof. It is easy to see that $p(z) - p(\alpha)$ is divisible by $z - \alpha$. (See Exercise C1, Section 1.1.) Thus

$$\frac{p(z) - p(\alpha)}{z - \alpha} = q(z)$$

where $q(z)$ is also a polynomial, so that

$$\frac{p(z)}{z - \alpha} = p(\alpha)\frac{1}{z - \alpha} + q(z).$$

By integrating this last equation over C we get

$$\int_C \frac{p(z)}{z - \alpha} \, dz = p(\alpha) \int_C \frac{dz}{z - \alpha} + \int_C q(z) dz$$

$$= \begin{cases} p(\alpha)2\pi i + 0, & \alpha \in I(C) \\ 0 + 0, & \alpha \in E(C) \end{cases}$$

by Corollaries 1.5e and f, and the Cauchy integral theorem for polynomials. Dividing by $2\pi i$ completes the proof. $\square$

To extend this theorem to rational functions we will use the factorization of a polynomial (Exercise C3, Section 1.1) which was based on the Fundamental Theorem of Algebra. We will also need the expansion of a rational function into partial fractions. The Fundamental Theorem and the partial fraction theorem will be established in Chapters 3 and 5 respectively.

If $F(z)$ is a rational function we can express it as $F(z) = p(z) + f(z)$ where $p(z)$ is a polynomial and $f(z)$ is a proper fraction which we can take to be in lowest terms. (Exercises B12 and B13, Section 1.1.) Thus we give our attention to proper rational functions.

1.5h Theorem. *If $p(z)/q(z)$ is a proper rational function in lowest terms and*

$$q(z) = C(z - \beta_1)^{p_1}(z - \beta_2)^{p_2} \cdots (z - \beta_k)^{p_k}$$

where the β's are the distinct zeros of $q(z)$, and p_j is the multiplicity of β_j, then

$$\frac{p(z)}{q(z)} = \sum_{j=1}^{k} \sum_{n=1}^{p_j} \frac{a_{jn}}{(z - \beta_j)^n} \tag{4}$$

where the a_{jn}'s are constants.

As indicated above we will prove this in Chapter 5. The meaning of this formula is that the contribution to the sum arising from each β_j is

$$\sum_{n=1}^{p_j} \frac{a_{jn}}{(z - \beta_j)^n} = \frac{a_{j1}}{z - \beta_j} + \frac{a_{j2}}{(z - \beta_j)^2} + \cdots + \frac{a_{jp_j}}{(z - \beta_j)^{p_j}}$$

and that the sum of these contributions is then the original function. Thus, for example, we have that

$$\frac{z^4 + 2z^3 + 3z^2 + 5z - 1}{(z - 1)(z - 3)^2(z + 2i)^3}$$

has the partial fraction expansion

$$\left[\frac{a}{z - 1}\right] + \left[\frac{b_1}{z - 3} + \frac{b_2}{(z - 3)^2}\right] + \left[\frac{c_1}{z + 2i} + \frac{c_2}{(z + 2i)^2} + \frac{c_3}{(z + 2i)^3}\right].$$

The evaluation of the numerators will also be given in Chapter 4.

1.5i Lemma. *Suppose C is a contour with $\beta \in E(C)$. Then*

$$\frac{1}{2\pi i} \int_C \frac{dz}{(z - \beta)^n(z - \alpha)} = \begin{cases} 1/(\alpha - \beta)^n & \text{if} \quad \alpha \in I(C) \\ 0 & \text{if} \quad \alpha \in E(C). \end{cases}$$

Proof. By Theorem 1.5h

$$\frac{1}{(z - \alpha)(z - \beta)^n} = \frac{b_1}{z - \beta} + \frac{b_2}{(z - \beta)^2} + \cdots + \frac{b_n}{(z - \beta)^n} + \frac{a}{z - \alpha}. \qquad (5)$$

Integrating over C we have the first term integrating to 0 by Theorem 1.5f, and the other b-terms to 0 by Theorem 1.5a (with $z - \beta$ instead of z). The last term is a or 0 according as $\alpha \in I(C)$ or in $E(C)$. But from (5) we can find a: clear of fractions and let $z \to \alpha$. This gives $a = 1/(\alpha - \beta)^n$. $\square$

The next theorem is the **Cauchy Integral Formula** for rational fractions.

1.5j Theorem. *If C is a contour and $p(z)/q(z)$ is a rational function for which all the zeros of $q(z)$ are in $E(C)$, then*

$$\frac{1}{2\pi i} \int_C \frac{p(z)}{q(z)} \frac{dz}{z - \alpha} = \begin{cases} p(\alpha)/q(\alpha) & \alpha \in I(C) \\ 0, & \alpha \in E(C). \end{cases}$$

Proof. By Theorem 1.5g and Exercises C11 and C12 of Section 1.1, we need only consider the case where $p(z)/q(z)$ is a proper fraction in lowest terms. Then by Theorem 1.5h we can integrate the sum of the partial fractions. By the previous lemma, that integration yields the result. $\square$

Example 3. Evaluate

$$\int_{|z+2|=3} \frac{z^2 + 2z + 3}{(z+2)(z-3)} \, dz$$

Solution. $(z^2 + 2z + 5)/(z - 3)$ is rational and the denominator is 0 only at $z = 3$ which is outside C. Thus, by Theorem 1.5j the integral is equal to

$$2\pi i \, \frac{(-2)^2 + 2(-2) + 3}{(-2 - 3)} = -\frac{6\pi i}{5}.$$

We will now begin to use the notation

$$\int_{|z|=a} f(z)dz \quad \text{and} \quad \int_{|z-\alpha|=a} f(z)dz$$

These mean that *the integrals are to be taken in the positive sense* around the circles

$$z = aE(\theta) \quad \text{and} \quad z = \alpha + aE(\theta), \quad 0 \le \theta \le 2\pi$$

respectively.

Example 4. Evaluate

$$I = \int_{|z|=3} \frac{z^3 + 5z^2 - 4z + 6}{z - 2} \, dz.$$

Solution. Since 2 is in the interior of the circle $|z| = 3$, the Cauchy integral formula applies. We then get

$$I = 2\pi i(z^3 + 5z^2 - 4z + 6)\Big|_{z=2}$$
$$= 2\pi i(8 + 20 - 8 + 6) = 52\pi i.$$

Example 5. Evaluate

$$\int_{|z|=3} \frac{z^3 + 1}{z^2 + 4} \, dz.$$

Solution 1. We break up the integrand by partial fractions:

$$\frac{1}{z^2 + 4} = \frac{1}{(z + 2i)(z - 2i)} = \frac{1}{4i} \left[\frac{1}{z - 2i} - \frac{1}{z + 2i} \right],$$

so that

$$\int_C \frac{z^3+1}{z^2+4} = \frac{1}{4i} \int_C \frac{z^3+1}{z-2i} \, dz - \frac{1}{4i} \int_C \frac{z^3+1}{z+2i} \, dz$$

$$= \frac{1}{4i} \, 2\pi i[(2i)^3 + 1] - \frac{1}{4i} \, 2\pi i[(-2i)^3 + 1]$$

By the Cauchy integral formula with $\alpha = 2i$ in the first integral and $\alpha = -2i$ in the second. This reduces to $-8\pi i$.

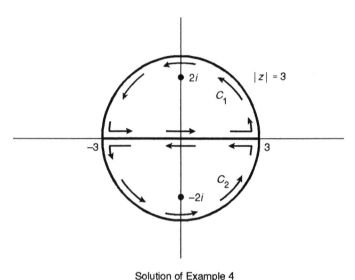

Solution of Example 4

$$\int_{|z|=3} = \int_{C_1} + \int_{C_2}$$

Solution 2. We modify the contour $C : |z| = 3$ into C_1 and C_2 as illustrated in the figure. In C_1 we integrate over the upper semicircle and long the segment from -3 to 3. In C_2 we integrate over the lower semicircle and over the same segment, but now from 3 to -3. The two integrals over the segment cancel since they are in opposite directions. Thus

$$\int_{|z|=3} \frac{z^3+1}{z^2+4} \, dz = \int_{C_1} \frac{1}{z-2i} \frac{z^3+1}{z+2i} \, dz + \int_{C_2} \frac{1}{z+2i} \frac{z^3+1}{z-2i} \, dz.$$

We evaluate the integral over C_1 by the Cauchy integral formula with $\alpha = 2i$ and $f(z) = (z^3+1)/(z+2i)$, and in the integral over C_2 we take $\alpha = -2i$

and $f(z) = (z^3 + 1)/(z - 2i)$. This gives

$$2\pi i \left(\frac{(2i)^3 + 1}{4i} \right) + 2\pi i \left(\frac{(-2i)^3 + 1}{-4i} \right),$$

again yielding $-8\pi i$ for the original integral.

Also we can now compute certain real trigonometric integrals which are difficult by the methods of elementary calculus. We illustrate this by considering

$$\int_0^{2\pi} \frac{d\theta}{3\cos\theta + 5}$$

We can connect this with contour integration through the formula for $E(\theta)$. If z is on the circle $|z| = 1$, then $z = E(\theta)$ and

$$dz = E'(\theta)d\theta = iE(\theta)d\theta = izd\theta$$

so

$$d\theta = \frac{dz}{iz}$$

and

$$E(\theta) = \cos\theta + i\sin\theta$$

and

$$\frac{1}{E(\theta)} = E(-\theta) = \cos\theta - i\sin\theta$$

so

$$\cos\theta = \frac{1}{2}\left(E(\theta) + \frac{1}{E(\theta)} \right) = \frac{1}{2}\left(z + \frac{1}{z} \right)$$

and as θ goes from 0 to 2π, z, then z goes once around $|z| = 1$ in the positive direction.

Example 6. Evaluate $I = \int_0^{2\pi} \frac{d\theta}{3\cos\theta + 5}$.

Solution. Using the substitutions indicated by the calculations in the previous paragraph we get

$$I = \int_{|z|=1} \frac{1}{\frac{3}{2}\left(z + \frac{1}{z}\right) + 5} \cdot \frac{dz}{iz}$$

$$= \frac{2}{i} \int_{|z|=1} \frac{dz}{3z^2 + 10z + 3} = \frac{2}{i} \int_{|z|=1} \frac{dz}{(3z + 1)(z + 3)}$$

$$= \frac{2}{i} \int_{|z|=1} \left[\frac{3}{8} \cdot \frac{1}{3z + 1} + \frac{1}{8} \cdot \frac{1}{z + 3}\right] dz$$

$$= \frac{1}{4i} \int_{|z|=1} \frac{dz}{z + \frac{1}{3}} + \frac{1}{4i} \int_{|z|=1} \frac{dz}{z + 3}$$

$$= \frac{1}{4i} \cdot 2\pi i + 0 = \frac{\pi}{2}$$

by Corollaries 1.5e and 1.5f.

Example 7. Evaluate

$$I = \int_0^{2\pi} \frac{d\theta}{2 + \sin\theta}$$

Solution. By the calculations just preceding Example 6 we have

$$d\theta = \frac{dz}{iz}, \qquad \sin\theta = \frac{1}{2i}\left(z - \frac{1}{z}\right)$$

and so

$$I = \int_{|z|=1} \frac{1}{2 + \frac{1}{2i}\left(z - \frac{1}{z}\right)}$$

$$= 2 \int_{|z|=1} \frac{dz}{4iz + z^2 - 1}$$

$$= 2 \int_{|z|=1} \frac{dz}{[z + (2 - \sqrt{3})i][z + (2 + \sqrt{3}i]}$$

$$= \frac{1}{i\sqrt{3}} \int_{|z|=1} \frac{dz}{z + (2 - \sqrt{3})i} - \frac{1}{i\sqrt{3}} \int_{|z|=1} \frac{dz}{z + (2 + \sqrt{3})i}$$

$$= \frac{1}{i\sqrt{3}} \cdot 2\pi i - 0 = \frac{2\pi}{\sqrt{3}}$$

If $f(\sin\theta, \cos\theta)$ is a rational function of $\sin\theta$ and $\cos\theta$, then the substitutions indicated in these last two examples reduces

$$\int_0^{2\pi} f(\sin\theta, \cos\theta) d\theta$$

to an integral of the form

$$\int_{|z|=1} F(z)dz$$

where $F(z)$ is a rational function of z. Then $F(z)$ can be expanded into partial fractions and the integral evaluated by the results of this section.

A Exercises

Evaluate:

1. $\int_C \dfrac{dz}{z^2 - 4}$ (a) $|z| = 1$; (b) $|z| = 4$; (c) $|z - 2| = 2$.

2. $\int_C \dfrac{dz}{z^2 + 9}$ (a) $|z| = 4$; (b) $|z - 3i| = 2$; (c) $|z + 3i| = 7$.

3. $\int_C \dfrac{dz}{z^2 - 1}$ (a) $|z| = \frac{1}{2}$; (b) $|z| = 5$; (c) $|z - 1| = 1$.

4. $\int_{|z|=3} \dfrac{zdz}{(z - 2)(z - 1)}$.

5. $\int_C \dfrac{dz}{z(z - 1)(z + 2)}$
   (a) $|z| = \frac{1}{2}$; (b) $|z| = \frac{3}{2}$; (c) $|z| = \frac{5}{2}$; (d) $|z - 1| = \frac{1}{2}$.

B Exercises

Evaluate:

1. $\displaystyle\int_0^{2\pi} \dfrac{d\theta}{3\sin\theta + 5}$

2. $\displaystyle\int_0^{2\pi} \frac{d\theta}{2 + \cos\theta}$

3. $\displaystyle\int_0^{\pi} \frac{d\theta}{10 + 8\cos\theta}$. Hint: If $f(\theta)$ is even, then $\displaystyle\int_0^{\pi} f(\theta)d\theta = \frac{1}{2}\int_{-\pi}^{\pi} f(\theta)d\theta$.

4. $\displaystyle\int_0^{\pi} \frac{d\theta}{1 + a^2 - 2a\cos\theta} = \frac{\pi}{1 - a^2}$, $0 < a < 1$.

5. If $\displaystyle f(z) = \sum_{-n}^{n} c_k(z - \alpha)^k$, then show $\displaystyle\int_C f(z)dz = 2\pi i c_{-1}$ for any contour C with $\alpha \in I(C)$.

6. Show $\displaystyle\int_0^{2\pi} \frac{d\theta}{1 + a\cos\theta} = \frac{2\pi}{\sqrt{1 - a^2}}$, $0 < a < 1$.

7. Show $\displaystyle\int_0^{2\pi} \frac{d\theta}{1 + a\sin\theta} = \frac{2\pi}{\sqrt{1 - a^2}}$, $0 < a < 1$. Hint: Use B6 to avoid having to use contour integration here on B7.

C Exercises

1. If C is a contour with $\alpha \in I(C)$, then show

   (a) $\displaystyle\frac{1}{2\pi i}\int_C \frac{z^2 + 2z + 3}{(z - \alpha)^2} dz = 2\alpha + 2$;

   (b) $\displaystyle\frac{1}{2\pi i}\int_C \frac{az^2 + bz + c}{(z - \alpha)^2} dz = 2a\alpha + b$.

2. Show $\displaystyle\int_0^{2\pi} \frac{d\theta}{a^2\cos^2\theta + b^2\sin^2\theta} = \frac{2\pi}{ab}$ if $a > 0$, $b > 0$.

3. Evaluate $\displaystyle\int_0^{2\pi} (\cos\theta)^{2n}\,d\theta$.

1.6 Plane Sets, Regions

In discussing point sets in the complex plane $\mathbb{C}$ (and elsewhere, but our attention here is on the plane) we need to supplement the ideas about sets which were briefly introduced in Section 1.3. Recall

$$z \in S$$

means that the point z belongs to the set S, and

$$S \subset T \quad \text{or} \quad T \supset S$$

means the set S is a subset of the set T. That is, if $z \in S$, then $z \in T$. The negations of these symbols are $\notin$ and $\not\subset$ or $\not\supset$, respectively.

If S and T are two sets, then the **union** of S and T, denoted by

$$S \cup T$$

is the set of all points in S or T or both, and the **intersection** of S and T, denoted by

$$S \cap T$$

is the set of points common to both.

The **empty set** (i.e., having no elements) is denoted by $\emptyset$. Thus the set equation

$$S \cap T = \emptyset$$

means that S and T do not meet; they have no common elements. In this case we say S and T are **disjoint**.

A set S is **bounded** if there is a constant $K > 0$ such that

$$|z| < K \quad \text{for all} \quad z \in S.$$

Otherwise it is **unbounded**.

In Section 1.3 we defined neighborhoods and deleted neighborhoods. We want to emphasize these concepts because they play a fundamental role in describing limits, and (soon to be defined) regions and derivatives. Recall that the neighborhood of z_0 of radius r is denoted by $N(z_0, r)$ and is the set of all z for which

$$|z - z_0| < r.$$

The deleted neighborhood of z_0 of radius r is the set of all z for which

$$0 < |z - z_0| < r,$$

that is, it differs from $N(z_0, r)$ by having z_0 itself deleted. Sometimes it is called a **punctured** neighborhood.

A point z_0 is an **interior point** of a set S if there is a neighborhood of z_0 in S. That is, if there is a positive r such that

$$N(z_0, r) \subset S.$$

If each member of a set is an interior point, then the set is called **open**. The simplest example of an open set is a neighborhood itself. This is not entirely tautological. To see this let $N(\alpha, a)$ be the neighborhood of radius a about α, and let $z_0 \in N(\alpha, a)$. Then if there is a neighborhood $N(z_0, r)$ for some radius $r > 0$ with $N(z_0, r) \subset N(\alpha, a)$ we can conclude that $N(\alpha, a)$ is open. It is geometrically clear that $r = [a - |z_0 - \alpha|]/2$ suffices, and the figure below illustrates the situation. The calculations are left to the exercises.

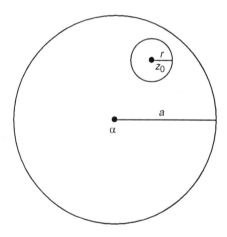

A neighborhood is open.

The **complement** of a set S, denoted by $\widetilde{S}$, is the set of all points *not* in S. For example, if $S = N(\alpha, r)$, then

$$\widetilde{S} = \{z \in \mathbb{C} : |z - \alpha| \geq r\}.$$

A set is called **closed** if its complement is open.

A point z_0 is a **boundary point** of a set S (whether or not $z_0 \in S$) if each neighborhood z_0 contains at least one point of S and one point of $\tilde{S}$. The set of all boundary points of S is called the **boundary** of S and is denoted by ∂S. The union of a set S and its boundary is called the closure of the set and denoted by $\bar{S}$. Thus

$$\bar{S} = S \cup \partial S.$$

Let S be a set and suppose that for each pair α, β of points in S there is a simple polygonal path $L \subset S$ with initial point α and terminal point β. Then we say S is a **polygonally connected** set. There are more sophisticatead definitions of connectedness, but for open sets in the plane, which is the only case where we are concerned with the concept, they are all equivalent. Hence we will merely refer to **connectedness**. Further, the restriction of simplicity on L is for the sake of clarity.

An open connected set is called a **region**. Sometimes (in other books) the word "domain" is used here. We prefer to reserve "domain" to mean the set of definition of a function.

A point ζ is a **limit point** of a set S if there is sequence $\{\alpha_n\}$ of points in S with each $\alpha_n \neq \zeta$ for which $\alpha_n \to \zeta$. A limit point of S may or may not belong to S. A boundary point of a region is always a limit point. A necessary and sufficient condition that a set S be closed is that it contain all its limit points. These statements are discussed in the exercises.

We now state a theorem which is usually proved in advanced calculus. It is intuitively clear from geometry that the result is expected.

1.6a Theorem. *Let S and T be two closed sets one of which is bounded, and suppose $S \cap T = \emptyset$, that is, S and T have no points in common. Then there is a positive minimal distance between them; that is, there is a number $a > 0$ such that $|z - \zeta| \geq a$ for every pair of points z, ζ with $z \in S$ and $\zeta \in T$. Furthermore there is a $z_0 \in S$ and $\zeta_0 \in T$ for which $|z_0 - \zeta_0| = a$.*

We restate the Jordan Curve Theorem (1.4d) in a slightly different form.

1.6b Theorem. *If C is a Jordan curve, then the complement of C consists of two disjoint regions, one of which is bounded and the other not bounded. C is their common boundary.*

The proof is easy and obvious for certain simple curves such as a circle or a square, but is too complicated to give in total generality. We will accept it and use it, though most instances in which we use it are sufficiently simple that its truth will be clear.

If C is a Jordan curve the bounded region of the complement is called the **interior** of C and denoted by $I(C)$. The unbounded region is called the **exterior** of C and denoted by $E(C)$. A region R is a **Jordan region** if there is a Jordan curve C for which $R = I(C)$.

A region R is called **simply connected** if for every Jordan curve $C \subset R$ we also have $I(C) \subset R$. A neighborhood is simply connected. A Jordan region is simply connected. An **annulus**, a region of the form

$$R = \{z \in \mathbb{C} : a < |z - \alpha| < b, \qquad b > a \geq 0\},$$

is not simply connected since the circle $C = \{z \in \mathbb{C} : |z - \alpha| = (b - a)/2\}$ is in R but $I(C) \not\subset R$ since $\alpha \in I(C)$ and $\alpha \notin R$. A region which is not simply connected is called **multiply connected**. Some of these ideas are pursued in the problems.

A region R is said to be **star-shaped** with respect to a point α if for any $z \in r$ we have the segment from α to z also in R. The point α is called a **radiating center** of R. When we say a region is star shaped, or is a **star region**, we mean there is a radiating center with respect to which it is star shaped. The interior of a circle, an ellipse or a square is star shaped with respect to any of its points. The complex plane with the negative real axis deleted is star shaped with respect to any point on the nonnegative real axis, but no others. A region which is star shaped with respect to all its points is called **convex**.

In the real number system we usually distinguish two "infinities" corresponding to the two ends of the number line: $+\infty$ and $-\infty$. In the complex number system it is customary, for easons that will become clear as we proceed, to consider a single "point at ∞". By an ε-neighborhood of ∞ we mean the *exterior* of a circle of radius $1/\varepsilon : |z| > 1/\varepsilon$.

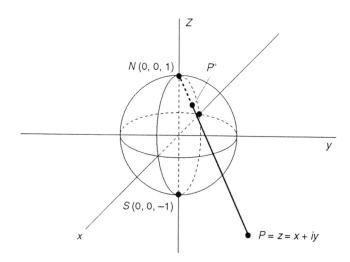

To see in a geometrical way what is involved we use **stereographic projection**. We construct a sphere of radius 1 centered at the origin of the z-plane and take a Z-axis perpendicular to the z-plane. We take the point $N(0,0,1)$ as a center of projection, and call it the "north pole." Then a point P in the plane corresponds to a point P' on the sphere if P and P' are on the same straight line through N. It is clear that points on the unit circle correspond to themselves, points outside to the upper hemisphere, and points inside to the lower, with the origin itself corresponding to the "south pole" $S(0,0,-1)$.

This correspondence gives a one-to-one map of the plane onto the sphere punctured at N; there is no point in the plane corresponding to N. We do note, however, that when P is in the plane and far from the origin, in whatever direction, then P' is close to N. A neighborhood of ∞ in the plane corresponds to a neighborhood of N on the surface of the sphere. Thus we adjoin the symbol ∞ to the complex number system and visualize it as the image, under stereographic projection, of the north pole of the sphere in the limit sense as described above; finite points "near" ∞ correspond to points on the sphere near N.

A Exercises

1. Complete the proof that $N(\alpha, a)$ is open by showing for the given value of r that $N(z_0, r) \subset N(\alpha, a)$. That is, show that if $|z - z_0| < r$, then $|z - \alpha| < a$.

2. Show that $\partial N(\alpha, a) = \{z : |z - \alpha| = a\}$.

3. Show that $N(\alpha, a)$ is connected and hence is a region.

4. Show that the set $R = \{z = x + iy : 0 < x < 1, \ 0 < y < 1\}$ is a region.

5. Show that $R = \{z = x + iy : 4x^2 + 9y^2 < 36\}$ is a region, that $\partial R = \{z = x + iy : 4x^2 + 9y^2 = 36\}$. What are $\bar{R}$ and $\tilde{R}$?

6. For any set S show that $\partial S = \partial \tilde{S}$. Give an example to show that there are sets for which $\partial S \neq \partial \bar{S}$.

7. If R is an open set and $\alpha \in \partial R$, then show that α is a limit point of R.

8. If R is an open set and $\alpha \in \partial R$, then show that α is a limit point of $\tilde{R}$.

9. Let $S = \{z = x + iy : 0 \le x < 1, \ 0 \le y < 1\}$. Find all limit points of S which are in ∂S.

10. Show that $R = \{z = x + iy : 0 < x < 1\}$ is an unbounded region. Find ∂R.

11. Show that $R = \{z : \operatorname{Re} z > 0\}$ is an unbounded region. Find ∂R.

12. Let $R = \{z : |z| < 1\} \cap \{0 < \arg z < 2\pi\}$. Show that
    (a) R is not a Jordan region,

$f(\zeta)$ is differentiable at β, $g(z)$ it is differentiable at α, and $g(\alpha) = \beta$, then $f(g(z))$ is differentiable at α and the formula for the derivative is $f'(\beta)g'(\alpha)$. These formulas are all discussed in the exercises.

The notation is taken over from real calculus essentially unchanged. Thus if z is a point where $f(z)$ is differentiable, we write

$$f'(z), \quad \frac{d}{dz} f(z), \quad \frac{df(z)}{dz}$$

or more briefly

$$f' \quad \text{or} \quad \frac{df}{dz}$$

for the derivative. If we have $w = f(z)$, we sometimes write

$$w' \quad \text{or} \quad \frac{dw}{dz}$$

for the derivative.

Example 1. Compute $f'(z)$ if $f(z) = 3z^2 + 2z$.

Solution. Let z be any point, and $h = \Delta z$. The difference quotient is

$$\frac{\Delta f}{\Delta z} = \frac{3(z+h)^2 + 2(z+h) - 3z^2 - 2z}{h}$$
$$= \frac{6zh + 3h^2 + 2h}{2} = 6z + 2 + 3h$$
$$\to 6z + 2 \quad \text{as} \quad h \to 0$$

so

$$f'(z) = 6z + 2.$$

It is clear that the formula for the complex derivative of $3z^2 + 2z$ in $\mathbb{C}$ is the same as the formula for the real derivative of $3x^2 + 2x$ in $\mathbb{R}$. The reason of course is that the algebraic rules for $\mathbb{C}$ are identical with those for $\mathbb{R}$.

2.1a Theorem. $(z^n)' = \dfrac{d}{dz} z^n = nz^{n-1}$.

Proof. From Example 2, Section 1.2 we have

$$\left| \frac{(z+h)^n - z^n}{h} - nz^{n-1} \right| \le \frac{1}{2} n(n-1)|z|^{n-2}|h| + \cdots + |h|^{n-1}$$
$$\to 0 \quad \text{as} \quad h \to 0$$

since each term on the right contains a positive integral power of $|h|$. $\square$

2.1b Corollary. *Any polynomial is differentiable.*

It is clear from Theorem 2.1a that the formula for differentiating a polynomial is the same as in the real case: If

$$P(z) = \sum_{0}^{n} a_k z^k,$$

then

$$P'(z) = \sum_{0}^{n} k a_k z^{k-1} = \sum_{1}^{n} k a_k z^{k-1}.$$

2.1c Corollary. *Any rational function is differentiable except where the denominator is zero.*

Example 2. Compute the derivative of $(z - 2)/(z^2 + 1)$.

Solution.
$$\frac{(z^2 + 1)(1) - (z - 2)(2z)}{(z^2 + 1)^2} = \frac{1 + 4z - z^2}{(z^2 + 1)^2}.$$

This is valid except where $z^2 + 1 = 0$: for all z except $z = \pm i$.

Let us suppose that $f(z)$ is defined in a region R and differentiable at a point $z \in R$. We set $\Delta z = h + ik$, where $h = \operatorname{Re} \Delta z$ and $k = \operatorname{Im} \Delta z$. then with $f(z) = u(x, y) + iv(x, y)$ we form the difference quotient at z:

$$\frac{\Delta f}{\Delta z} = \frac{f(z + \Delta z) - f(z)}{\Delta z}$$

$$= \frac{[u(x + h, y + k) - u(x, y)] + i[v(x + h, y + k) - v(x, y)]}{h + ik}.$$

We first restrict Δz to be real (and $= h$):

$$\frac{\Delta f}{\Delta z} = \frac{u(x + h, y) - u(x, y)}{h} + i\frac{v(x + h, y) - v(x, y)}{h}.$$

As $h \to 0$ the left side becomes $f'(z)$ and each of the difference quotients on the right becomes a partial derivative with respect to x. There results

$$f'(z) = u_x(x, y) + iv_x(x, y), \tag{4}$$

where the subscripts indicate partial derivatives. Similarly if we restrict Δz to be pure imaginary (and $= ik$), we get

$$\frac{\Delta f}{\Delta z} = \frac{1}{i} \frac{u(x, y+k) - u(x,y)}{k} + \frac{v(x, y+k) - v(x,y)}{k}$$

from which we get as $\Delta z = ik \to 0$

$$f'(z) = v_y(x,y) - iu_y(x,y), \tag{5}$$

since $1/i = -i$.

2.1d Theorem. (Cauchy–Riemann Equations) If $f(z) = u(x,y) + iv(x,y)$ is differentiable at a point $z = x + iy$, then

$$u_x(x,y) = v_y(x,y), \qquad u_y(x,y) = -v_x(x,y).$$

Proof. We equate the real and imaginary parts of the right sides of (4) and (5) since both formulas represent the same number, namely $f'(z)$. □

Example 3. Verify the Cauchy–Riemann equations for $f(z) = z^2$.

Solution. $z^2 = x^2 - y^2 + i2xy$, so

$$u = x^2 - y^2, \qquad v = 2xy.$$

Thus

$$u_x = 2x = v_y,$$
$$u_y = -2y = -v_x.$$

It is true that there are complex functions which have derivatives at isolated points and no where else (Exercise C2). However, we are primarily interested in those which have a derivative in a region.

Definition. $f(z)$ is **analytic in a region** R means that it is differentiable at each point of R, and $f(z)$ is **analytic at a point** α means there is a region R, with $\alpha \in R$, in which $f(z)$ is analytic. To say that $f(z)$ is **entire**, or is an **entire function**, means that it is analytic in the whole plane. (In the older literature the word integral is used as we use entire, but we will not use integral in that sense.)

Some surprising results follow from analyticity. Clearly $f(z)$ must be continuous in any region where $f(z)$ is analytic, since differentiability implies continuity (Exercise A3). However, the real surprise is that $f'(z)$ is itself analytic, as we shall see.

If $f(z)$ is the derivative of another function $F(z)$ in a region R, then $F(z)$ is called an **anti-derivative** of $f(z)$, an (**indefinite**) **integral** of $f(z)$, or finally a **primitive** of $f(z)$. Each of these merely means that $F'(z) = f(z)$ in R. Clearly if $F(z)$ is an integral of $f(z)$, then so is $F(z)+K$ for any constant K. We shall soon see that there are no others.

2.1e Theorem. *If $F(z)$ is an indefinite integral of a continuous function $f(z)$ in a region R, α and β are points in R, and C is a path in R from α to β, then*

$$\int_C f(z)dz = F(\beta) - F(\alpha).$$

Proof. Let C be given by

$$C : z = \phi(t) \qquad t \in I = [a, b],$$

with $\phi(a) = \alpha$, $\phi(b) = \beta$. Then

$$\int_C f(z)dz = \int_a^b f(\phi(t))\phi'(t)dt$$

$$= \int_a^b F'(\phi(t))\phi'(t)dt = \int_a^b \frac{d}{dt} F(\phi(t))dt$$

$$= F(\phi(b)) - F(\phi(a)) = F(\beta) - F(\alpha). \ \square$$

2.1f Corollary. *If $F(z)$ and $G(z)$ are both indefinite integrals of $f(z)$, then $G(z) = F(z) + K$ where K is a constant.*

Proof. Let $H(z) = G(z) - F(z)$. Then $H'(z) = 0$ in R. Let C be any path in R from a point α to an arbitrary z. Then by Theorem 2.1e

$$0 = \int_C 0 \cdot d\zeta = \int_C H'(\zeta)d\zeta = H(z) - H(\alpha) = H(z) - K,$$

where K is $H(\alpha)$. Hence $H(z) = K$ or $G(z) = F(z) + K$. $\square$

2.1g Corollary. *If $F(z)$ is an indefinite integral of a continuous function $f(z)$ in a region R, then*

$$\int_C f(z)dz = 0$$

for each closed path C in R.

Proof. By Theorem 2.1e with $\alpha = \beta$. $\square$

2.1h Theorem. *If $f(z)$ is continuous in a region R, if $\int_C f(z)dz = 0$ for each closed path C in R, and if we define $F(z)$ in R by*

$$F(z) = \int_\alpha^z f(\zeta)d\zeta$$

over a path C in R from α to z, then $F(z)$ is analytic in R and $F'(z) = f(z)$.

Proof. We first observe that

$$\int_\alpha^z f(\zeta)d\zeta$$

is independent of the path in R (why?) and hence, with α fixed, defines a function of z alone, which we denote by $F(z)$. Let us now fix our attention on one z in R. For h sufficiently small $z + h$ is also in R (why?) and we can take the integral defining $F(z + h)$ to be on a path from α to z, and then on the segment from z to $z + h$. So

$$F(z + h) - F(z) = \int_z^{z+h} f(\zeta)d\zeta$$

where the integral is over the segment from z to $z + h$. Then

$$\frac{\Delta f}{\Delta z} - f(z) = \frac{F(z + h) - F(z)}{h} - f(z)$$

$$= \frac{1}{h}\int_z^{z+h} f(\zeta)d\zeta - f(z) = \frac{1}{h}\int_z^{z+h} [f(\zeta) - f(z)]d\zeta.$$

Since $f(z)$ is continuous in R, given $\varepsilon > 0$, there is an $a > 0$ for which

$$|f(\zeta) - f(z)| < \varepsilon \quad \text{if} \quad |z - \zeta| < a.$$

So for $|h| < a$, we have, by Theorem 1.4d,

$$\left|\frac{\Delta f}{\Delta z} - f(z)\right| \leq \frac{1}{|h|}\, \varepsilon \cdot |h| = \varepsilon \quad \text{if} \quad |\Delta z| = |h| < a.$$

This proves the theorem. $\square$

There is a near-converse of the theorem about the Cauchy–Riemann equations which we now give.

2.1i Theorem. *Suppose* $u = u(x, y)$ *and* $v = v(x, y)$ *have continuous partial derivatives in a region R and satisfy the Cauchy–Riemann equations there. If we define* $f(z) = u(x, y) + iv(x, y)$, *(with* $z = x + iy$*), then* $f(z)$ *is analytic in R, and* $f'(z) = u_x + iv_x = v_y - iu_y$.

Proof. What we must show is that $f'(z)$ exists at each point in R. So let $\alpha = a + ib$, be an arbitrary but fixed point in R. Then it suffices to show that $f'(\alpha)$ exists. This we do. We let $\Delta z = h + ik$. For $|\Delta z|$ sufficiently small $\alpha + \Delta z$ is in R and

$$\frac{\Delta f}{\Delta z} = \frac{f(\alpha + \Delta z) - f(\alpha)}{\Delta z}$$

$$= \frac{[u(a + h, b + k) - u(a, b)] + i[v(a + h, b + k) - v(a, b)]}{\Delta z}$$

$$= \frac{\Delta u + i\Delta v}{\Delta z}.$$

Since u and v have continuous partial derivatives we have

$$\Delta u = u_x(a, b)h + u_y(a, b)k + \eta_1 |\Delta z|,$$
$$\Delta v = v_x(a, b)h + v_y(a, b)k + \eta_2 |\Delta z|,$$

where η_1 and $\eta_2 \to 0$ as $|\Delta z| \to 0$. Then substituting these formulas into the expression for $\Delta f / \Delta z$, collecting terms, and using the Cauchy–Riemann equations we get

$$\frac{\Delta f}{\Delta z} = u_x(a, b) + iv_x(a, b) + \eta |\Delta z| / \Delta z$$

where $\eta = \eta_1 + i\eta_2$. But $|\Delta z|/\Delta z$ has absolute value 1 and $\eta \to 0$ as $\Delta z \to 0$, so

$$f'(\alpha) = \lim_{\Delta z \to 0} \frac{\Delta f}{\Delta z} = u_x(a, b) + iv_x(a, b). \;\; \Box$$

This theorem can be used to show analyticity of a function, but it should not be overworked. We have a few examples and exercises on it, but generally it is simpler to use complex methods. (Recall the remarks in Section 1.1.)

Example 4. Show that $f(z) = e^x E(y)$ is an entire function.

Solution. $e^z E(y) = e^x \cos y + i e^x \sin y$ so

$$u = e^x \cos y, \qquad v = e^x \sin y$$

from which

$$u_x = e^x \cos y = v_y$$
$$u_y = -e^x \sin y = -v_x$$

and these derivatives are continuous everywhere. Thus $f(z)$ is entire by Theorem 2.1i. Also by that theorem

$$f'(z) = u_x + i v_x = e^x \cos y + i e^x \sin y = f(z).$$

Thus $f(z)$ is not only entire, but also satisfies the complex differential equation and initial condition:

$$f'(z) = f(z), \qquad f(0) = 1$$

It is useful to define a notion of analyticity at ∞. We will say that $f(z)$ is **analytic at** ∞ means that (1) $f(z)$ is analytic in a neighborhood of ∞, that is, for $|z| > N$ for some number N, and (2) $f\left(\dfrac{1}{\zeta}\right)$, after simplification, is analytic at the origin.

Example 5. Show that $f(z) = (1 + z^2)/(z^2 - 1)$ is analytic at ∞.

Solution. $f(z)$ is analytic for $|z| > 1$, and

$$f\left(\frac{1}{\zeta}\right) = \frac{1 + \frac{1}{\zeta^2}}{\frac{1}{\zeta^2} - 1} = \frac{\zeta^2 + 1}{1 - \zeta^2}$$

which is analytic at the origin.

Example 6. Show that $f(z) = 2 + 3z$ is not analytic at ∞.

Solution. $f(z)$ is analytic for all z, (all finite z, that is), but

$$f\left(\frac{1}{\zeta}\right) = 2 + \frac{3}{\zeta} = (2\zeta + 3)/\zeta$$

is *not* analytic at $\zeta = 0$.

As we shall later show, if $f(z) = u + iv$ is analytic, then its second and higher derivatives all exist and are continuous. Anticipating this result, we look at some of its implications for the real and imaginary parts of $f(z)$. The first observation is that the second and all higher partial derivatives of u and v exist and are continuous. We differentiate the Cauchy–Riemann equations:

$$u_x = v_y; \qquad u_y = -v_x$$
$$u_{xx} = (v_y)_x; \quad u_{yy} = -(v_x)_y;$$
$$= v_{yx}; \qquad\quad = -v_{xy}.$$

But since v_{xy} and v_{yx} are continuous, they are equal. So adding u_{xx} and u_{yy} we get

$$u_{xx} + u_{yy} = 0.$$

Similarly

$$v_{xx} + v_{yy} = 0.$$

The differential equation

$$\phi_{xx} + \phi_{yy} = 0, \tag{3}$$

satisfied by $\phi = u$ and by $\phi = v$ in a region of analyticity of $f(z) = u + iv$, is called **Laplace's equation.** Its solutions are called **harmonic functions.** Our calculations have proved the following.

2.1j Theorem. *If $f(z) = u + iv$ is analytic in a region R, and if u and v have continuous second derivatives there, then both u and v are harmonic in R.*

Laplace's equation (3) is almost certainly the most studied of all partial differential equations because it arises in so many different applied situations. The differential operator

$$\frac{\partial^2}{\partial x^2} + \frac{\partial^2}{\partial y^2}$$

is called the (two-dimensional) **Laplacian.** Other notations for the Laplacian are ∇^2 $\left(\text{since } \left(\mathbf{i}\,\dfrac{\partial}{\partial x} + \mathbf{j}\,\dfrac{\partial}{\partial y}\right) \bullet \left(\mathbf{i}\,\dfrac{\partial}{\partial x} + \mathbf{j}\,\dfrac{\partial}{\partial y}\right) = \dfrac{\partial^2}{\partial x^2} + \dfrac{\partial^2}{\partial y^2}\right)$ and Δ. We will use Δ so that the symbol Δu, in this context, means $u_{xx} + u_{yy}$, and Laplace's equation is then $\Delta u = 0$.

If u is the real part of an analytic function in a region R, it is then harmonic there. The function v is called the **harmonic conjugate** of u. If

u is known but not v, then v can be found (theoretically) by integrating the Cauchy–Riemann equations. It is also true that if u is harmonic in a simply connected region R, then there is a function $f(z)$ which is analytic in R for which $u = \mathrm{Re}\,(f(z))$ (see Theorem 3.2i). To find $f(z)$ is equivalent to finding v which, as noted above, can be found through the Cauchy–Riemann equations.

Example 7. Verify that $u = \frac{1}{2}\,\ln(x^2 + y^2)$ is harmonic for $x > 0$, find a conjugate harmonic function v, and then construct an analytic function $f(z) = u + iv$.

Solution. We compute

$$u_x = \frac{x}{x^2 + y^2}; \qquad\qquad u_y = \frac{y}{x^2 + y^2}2$$

$$x_{xx} = \frac{(x^2 + y^2) - x(2x)}{(x^2 + y^2)^2}; \qquad u_{yy} = \frac{(x^{2+y}) - y(2y)}{(x^2 + y^2)^2}$$

$$= \frac{y^2 - x^2}{(x^2 + y^2)^2}; \qquad\qquad = \frac{x^2 - y^2}{(x^2 + y^2)^2}.$$

Thus $u_{xx} + u_{yy} = 0$, and u is indeed harmonic. In fact, we have shown that u is harmonic except at the origin. The reason we take $x > 0$ will be clear when we compute v.

The conjugate harmonic function v must satisfy

$$v_x = -u_y = -\frac{y}{x^2 + y^2}$$

and

$$v_y = u_x = \frac{x}{x^2 + y^2}$$

Then, since v_x is the derivative of v with respect to x with y held fixed, we can recapture v by integrating v_x with respect to x with y held fixed. Thus

$$v = -\int \frac{y\,dx}{x^2 + y^2} = \int \frac{\left(-\frac{y}{x^2}\right)dx}{1 + \left(\frac{y}{x}\right)^2}$$

$$= \arctan\frac{y}{x} + c(y)$$

where the "constant of integration" may depend on y. Similarly

$$v = \int \frac{x\,dy}{x^2 + y^2} = \int \frac{\frac{dy}{x}}{1 + \left(\frac{y}{x}\right)^2}$$

$$= \arctan\frac{y}{x} + k(x)$$

where this time the "constant" may depend on x. But for those two formulas to both be correct, we must have $k(x) = c(y)$ and hence the "constant" must indeed be constant, independent of both x and y. Thus

$$v = \arctan \frac{y}{x} + C$$

and we find

$$f(z) = \frac{1}{2} \ln(x^2 + y^2) + i \arctan \frac{y}{x} + iC.$$

If we introduce polar coordinates we have that

$$f(z) = \ln r + i\theta + iC$$

(where $z = rE(\theta)$) is analytic in the right half-plane $|\theta| < \dfrac{\pi}{2}$.

Where did we use $x > 0$? (See Exercise A7).

<table>
<tr><td>Significant Topics</td><td>Page</td></tr>
</table>

A Exercises

1. (a) If $f(z)$ and $g(z)$ are differentiable at a point z, show that

$$[f(z) \pm g(z)]' = f'(z) \pm g'(z),$$
$$[f(z)g(z)]' = f(z)g'(z) + f'(z)g(z),$$
$$\left[\frac{f(z)}{g(z)}\right]' = \frac{g(z)f'(z) - f(z)g'(z)}{[g(z)]^2}, \quad g(z) \neq 0.$$

(b) If $f(z)$ and $g(z)$ are analytic in a region so are $f(z)\pm g(z)$, $f(z)g(z)$, and also $f(z)/g(z)$ except where $g(z) = 0$.

2. Compute the derivatives of

(a) $z^3 + 3z^2 - 2z + 1$

(b) $(z^2 - 1)/(z^2 + 1)$

(c) $(z^2 - 1)(z^2 - 2z)$

(d) $\dfrac{z - 1}{z + 1}$

3. Show that if $f'(\alpha)$ exists, then $f(z)$ is continuous at $z = \alpha$.

4. Show that each polynomial is an entire function, that is, if $P(z)$ is a polynomial, it is analytic everywhere.

5. Show that each rational function is analytic in the whole plane except at zeros of the denominator.

6. If $f(z)$ is rational in the form $p(z)/q(z)$ where $\deg q \geq \deg p$, then $f(z)$ is analytic at ∞.

7. Where is the condition $x > 0$ used in the solution of Example 7?

8. Show that if $f(z) = u + iv$ is twice differentiable (that is, $f'(z)$ is differentiable) in a region R, then $f''(z) = u_{xx} + iv_{xx} = -u_{yy} + iv_{yy} = v_{xy} - iu_{xy}$.

B Exercises

1. Show that if $f(z)$ is analytic in a region R, and either (a) u, (b) v, (c) $|f(z)|$ is a constant, then $f(z)$ itself must be constant.

2. If $f(z) = u + iv$ is entire, and if $u = x^3 - 3xy^2$, find v from the Cauchy–Riemann equations, and express $f(z)$ as a polynomial in z, which is unique up to a pure imaginary constant.

3. Prove that $f(z) = \cos x \cosh y - i \sin x \sinh y$ is entire.

4. If $f(z) = u + iv$ is entire, and $u = \sin x \cosh y$, then find v.

5. Show that the following functions are nowhere differentiable.

(a) $f(z) = \bar{z}$ (b) $f(z) = x$ (c) $f(z) = xy$

6. If $f(z) = x^3yz/(x^6 + y^2)$, for $z = x + iy \neq 0$ and $f(0) = 0$, show that $\Delta f/\Delta z \to 0$ at the origin where $\Delta z \to 0$ along any radius, but that $f'(0)$ does not exist.

7. (**Chain Rule**) Show that if $f(\zeta)$ is differentiable at β, $g(z)$ is differentiable at α and $g(\alpha) = \beta$, then $f(g(z))$ is differentiable at α and its derivative is $f'(\beta)g'(\alpha)$. Further show that if $f(\zeta)$ is analytic at β and $g(z)$ is analytic at α, then $f(g(z))$ is analytic at α.

8. Show that $(z^2 + z - 1)/(z^3 + 4z^2 - 3z + 1)$ is analytic at ∞.

9. Show that $2z^2 + 3z + 5$ is *not* analytic at ∞.

10. In Example 7, we showed that $u = \frac{1}{2} \ln(x^2 + y^2)$ is harmonic for $x^2 + y^2 > 0$, so, in particular, in the upper half plane $y > 0$. Find v there so that $u + iv$ is analytic for $y > 0$.

11. If $p(z)$ and $q(z)$ are both polynomials of degree n, and if $p^{(k)}(\alpha) = q^{(k)}(\alpha)$, $k = 0, 1, \ldots, n$, for some point α, then $p(z) \equiv q(z)$. Hint: This is the same as showing that if $h(z)$ is a polynomial of degree $\leq n$, and if $h^{(k)}(\alpha) = 0$, $k = 0, 1, \ldots, n$, then $h(z) \equiv 0$.

12. Let $p(z)$ be a polynomial of degree n.

    (a) For α given, show $p(z) = \sum_0^n b_k(z - \alpha)^k$ where $b_k = p^{(k)}(\alpha)/k!$.

    Hint: Differentiate k times and set $z = \alpha$. Use B11 above.

    (b) Deduce that if C is a contour and $\alpha \in I(C)$, then

    $$\frac{1}{2\pi i} \int_C \frac{p(z)}{(z - \alpha)^{k+1}}\, dz = p^{(k)}(\alpha)/k!$$

13. If $f(z) = u + iv$ is analytic in a region R, show that uv is harmonic in R (a) by computing $\Delta(uv)$; (b) without computing $\Delta(uv)$.

C Exercises

1. If $f(z)$ is analytic in R and $f(z)$ is never zero, show that $v = \ln|f(z)|$ satisfies the differential equation $v_{xx} + v_{yy} = 0$ in R. Here ln is the real natural logarithm.

2. Define $f(z)$ to be z^2 if $z = x + iy$ where *both* x and y are rational, and to be zero otherwise. Show that $f(z)$ is continuous *only* at $z = 0$ and is differentiable there.

3. Show that any polynomial of degree larger than 0 is not analytic at ∞.

4. (a) If $f(z)$ is analytic in a region R, show

    $$\Delta|f(z)|^2 = 4|f'(z)|^2.$$

    (b) If both $f(z)$ and $g(z)$ are analytic in R and

    $$|f(z)|^2 + |g(z)|^2 = C,$$

    where C is constant, show that both $f(z)$ and $g(z)$ are constant.

2.2 Harmonic Functions in Physical Problems

In this section we observe how harmonic functions arise in certain physical situations. Our derivations here are quite heuristic (quick-and-dirty). We leave it to the physical scientists to make a more careful development of the arguments involved in a more complete development.

This is to lay a foundation for the study of the mathematical properties of harmonic functions. This, as we noted in Section 2.1, is important in many applied areas. This study is called **potential theory**, aspects of which will appear intermittently throughout the book, and is discussed at some length in the last chapter.

A. Heat Conduction

We consider a heat conducting lamina L (flat plate) occupying a region R of the xy-plane. We assume the lamina to be insulated on its surfaces (top and bottom) so that heat can enter or leave only through the edge (= boundary of R). We will denote the temperature at a point (x, y) at time t by $u(x, y, t)$, and the heat content of L, or a portion of L, by H measured in, say, calories. We will work from the following experimental facts.

(i) Heat flows from high to low temperature.

(ii) The rate of heat flow across a line in R is jointly proportional to the length of the line and the derivative of u in the direction orthogonal to the line.

(iii) The change in the amount of heat (ΔH) when heat flows in or out of an area is jointly proportional to the mass of the area and the change in temperature (Δu).

We apply these facts to a small rectangle A as illustrated. We denote the mass density of L by ρ so that the mass of A is

$$\Delta m = \rho \Delta x \Delta y.$$

Then by (iii) the change ΔH in heat in A in a time interval Δt is given by

$$\Delta H = c \Delta m \Delta u = c \rho \Delta x \Delta y \Delta u$$

where Δu is the corresponding temperature change at a mean point in A. The constant of proportionality c is called the **specific heat** of the material of L. Then the average rate of change of heat in A over the time interval Δt is

$$\frac{\Delta H}{\Delta t} = c\rho\Delta x\Delta y\,\frac{\Delta u}{\Delta t}. \tag{1}$$

Further, by (ii), the rate of heat flow into A from the left is

$$-k\Delta y\frac{\partial u}{\partial x}\Big|_x \tag{2}$$

where the derivative is evaluated at a mean point between y and $y + \Delta y$, and from the right is

$$k\Delta y\frac{\partial u}{\partial x}\Big|_{x+\Delta x}$$

where the proportionality constant k is called the **thermal conductivity** of the material of L. Similarly the rates in the y direction are

$$-k\Delta x\frac{\partial u}{\partial y}\Big|_y \quad \text{and} \quad k\Delta x\frac{\partial u}{\partial y}\Big|_{y+\Delta y}$$

respectively. (Explain the algebraic signs in these expressions. See Exercise A4.) Then, by addition,

$$\frac{dH}{dt} = k\left\{\Delta y\left[\frac{\partial u}{\partial x}\Big|_{x+\Delta x} - \frac{\partial u}{\partial x}\Big|_x\right] + \Delta x\left[\frac{\partial u}{\partial y}\Big|_{y+\Delta y} - \frac{\partial u}{\partial y}\Big|_y\right]\right\}$$

or, more briefly

$$\frac{dH}{dt} = k\left[\Delta y\Delta(\partial u/\partial x) + \Delta x\Delta(\partial u/\partial y)\right]. \tag{3}$$

Since $\Delta H/\Delta t \to dH/dt$ as $\Delta t \to 0$ we have approximately

$$c\rho\Delta x\Delta y\frac{\Delta u}{\Delta t} = k\left[\Delta y\Delta(\partial u/\partial x) - \Delta x\Delta(\partial u/\partial y)\right].$$

We divide by $k\Delta x\Delta y$, and let all of $\Delta t, \Delta x, \Delta y \to 0$, and get

$$\left(\frac{c\rho}{k}\right)\frac{\partial u}{\partial t} = \frac{\partial^2 u}{\partial x^2} + \frac{\partial^2 u}{\partial y^2},$$

and by a change of scale we can write this as

$$\frac{\partial u}{\partial t} = \frac{\partial^2 u}{\partial x^2} + \frac{\partial^2 u}{\partial y^2} \tag{4}$$

where all derivatives are evaluated at (x, y, t). This is the (normalized) two-dimensional **equation of heat conduction**, or more briefly, the **heat equation**.

In a similar way, if $u(x, y, t)$ represents not temperature but concentration of a substance diffusing into porous material occupying R, we come to the same equation. Also if u is a probability density function so that $u \Delta x \Delta y$ represents the probability that a particle moving randomly in R will be in A at time t, as for example in Brownian motion, we again come to the same equation. We will not give the derivations for these cases.

We are particularly interrested in **steady state** heat conduction. This is the case where the temperature at each point in R is constant in time, though it is in general variable from point to point. Thus $\partial u / \partial t = 0$ so equation (4) becomes

$$\frac{\partial^2 u}{\partial x^2} + \frac{\partial^2 u}{\partial y^2} = 0 \qquad (5)$$

and therefore *the temperature function in this case is a harmonic function.*

From the interpretation of harmonic functions as representing steady state heat flow we can give a heuristic physical "proof" of the **maximum principle** for Laplace's equation. We will later give a mathematical proof.

2.2a Theorem. *If $u(x, y)$ is harmonic in a region R and is not a constant, then it cannot achieve a maximum value at any point of R.*

Proof. We give a heuristic proof by contradiction. Suppose there were a point (x_0, y_0) in R for which

$$u(x_0, y_0) \geq u(x, y)$$

for all (x, y) in R. Since u is *not a constant*, there are portions of R in which

$$u(x_0, y_0) > u(x, y).$$

Then, since $u(x, y)$ represents a temperature function, heat would flow away from (x_0, y_0) lowering the temperature there. Thus the temperature *would change with time* and we would not be in the steady state. This contradiction completes the argument. $\square$

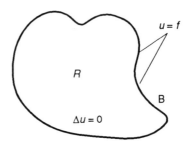

The Dirichlet Problem

Now let R be a region with boundary B and let $f(\zeta) = f(\xi, \eta)$ be a continuous function given on B. We can ask if there is a function $u(x, y)$, continuous in the closed set $\bar{R} = R \cup B$, which satisfies Laplace's equation in R and coincides with f on B. This is called the **Dirichlet problem** for Laplace's equation for R. (We will sometimes relax the condition of continuity on f.) We do not yet attempt to solve the Dirichlet problem, but we can, on the basis of Theorem 2.2a establish its uniqueness property. We begin with a lemma which is another, and weaker, form of the maximum principle.

2.2b Lemma. *If R is a bounded region and u continuous in $\bar{R} = R \cup B$ and harmonic in R itself, then for each point (x, y) in R we have*

$$u(x, y) \leq \max_B u(\xi, \eta),$$

and if there is a point in R at which equality holds, then u is constant in R.

Proof. If u is constant, then we have equality for all points (x, y) in R. If u is not constant, then it cannot achieve a maximum in R. But u is continuous in the closed set $R \cup B$, and so, by a theorem of real calculus, it achieves a maximum at some point of $R \cup B$. Thus the maximum must be achieved on B since it cannot be in R. $\square$

2.2c Theorem. *If R is a bounded region, then the Dirichlet problem for R as described above can have no more than one solution u.*

Proof. Suppose u_1 and u_2 are both solutions. Then set

$$w = u_1 - u_2$$

and observe that w is harmonic in R, continuous in $R \cup B$, and $w = 0$ on B. By the lemma, $w(x, y) \leq 0$ in R. But the *same argument* applies to $-w$, so $-w(x, y) \leq 0$ in R. Thus we get

$$0 \leq w(x, y) \leq 0 \quad \text{in} \quad R$$

and so $u_1(x, y) \equiv u_2(x, y)$ in R. $\square$

We can see by examples that the boundedness of R is essential here. Some of the exercises are concerned with this fact.

B. Potentials

Newton's law of gravitation states that two material particles attract each other with a force which is proportional to the product of their masses divided by the square of the distance between them. This, of course, is the "inverse square law." We consider a particle of mass m at $Q(\xi, \eta, \zeta)$ and a unit particle at $P(x, y, z)$. If $\boldsymbol{\rho}$ is the vector from Q to P, and ρ is its magnitude, then the force on the unit particle due to the attraction of m at Q is

$$\mathbf{F} = -\left(\frac{Gm}{\rho^2}\right) \cdot \frac{\boldsymbol{\rho}}{\rho} = -Gm\frac{\boldsymbol{\rho}}{\rho^3} \tag{6}$$

where G is the constant of proportionality and is called the universal gravitation constant. By choosing our units properly we can assume that $G = 1$. In terms of components, then, we have

$$\mathbf{F} = \frac{-m\boldsymbol{\rho}}{\rho^3} = m\frac{(x - \xi)\mathbf{i} + (y - \eta)\mathbf{j} + (z - \zeta)\mathbf{k}}{[(x - \xi)^2 + (y - \eta)^2 + (z - \zeta)^2]^{3/2}}. \tag{7}$$

We further assume that if we have several particles of mass m_i at Q_i, then the force on a unit particle is

$$\mathbf{F} = -\sum m_j \frac{\boldsymbol{\rho}_j}{\rho_j^3} \tag{8}$$

and, still further, that if we have a continuous distribution of matter over a set S, then the force on a unit particle generated by the mass distribution is

$$\mathbf{F} = -\int_S \mu \frac{\boldsymbol{\rho}}{\rho^3} \, dS \tag{9}$$

where $\boldsymbol{\rho} = (\xi - x)\mathbf{i} + (\eta - y)\mathbf{j} + (\zeta - z)\mathbf{k}$, (ξ, η, ζ) is the point of integration, and μ is the density. Then dS is an element of length, surface area, or volume according as the mass distribution is curvilinear, on a surface, or throughout a volume.

Coulomb's law for electrostatics has the same form as Newton's except that particles of opposite charge attract while particles of the same charge repel each other, and the masses in Newton's law are replaced by the charges in Coulomb's law. So the force, in the two cases, has the same mathematical structure, except for the algebraic sign for the case of repulsion.

A **potential** of a force field $\mathbf{F}$ is a real valued function φ for which

$$\nabla\varphi = \mathbf{F},$$

so clearly, if φ is a potential so is $\varphi + C$. In practice C is usually chosen to normalize some convenient physical condition. The name, potential, ultimately derives from the notion of potential energy.

One can integrate $\mathbf{F}$ as given by (7) to get that

$$\varphi = \frac{m}{\rho}. \tag{10}$$

is a potential, normalized so that $\varphi = 0$ for $\rho = \infty$. Similarly

$$\varphi = \sum \frac{m_j}{\rho_j}$$

and

$$\varphi = \int_S \frac{\mu}{\rho}\, dS$$

are potentials for (8) and (9), respectively.

Let us consider (10) with $(\xi, \eta, \zeta) = 0$ and $m = 1$. Then

$$\varphi = \frac{1}{\rho} = \frac{1}{(x^2 + y^2 + z^2)^{1/2}}$$

$$\nabla\varphi = \frac{(x\mathbf{i} + y\mathbf{j} + z\mathbf{k})}{(x^2 + y^2 + z^2)^{3/2}}$$

$$\nabla \cdot \nabla\varphi = \varphi_{xx} + \varphi_{yy} + \varphi_{zz}$$

$$= \frac{y^2 + z^2 - 2x^2}{\rho^5} + \frac{x^2 + z^2 - 2y^2}{\rho^5} + \frac{x^2 + y^2 - 2z^2}{\rho^5}$$

$$= 0.$$

So φ satisfies Laplace's equation in three dimensions. Any solution of Laplace's equation in three dimensions is called a (three-dimensional) harmonic function, and if such a function is independent of z, it is a two-dimensional harmonic function, and thus is the real part of an analytic function.

Let us consider the force generated on a unit mass at the point $(x, y, 0)$ by a linear mass distribution with linear density μ along the z-axis which we view as a material wire. An element of the wire at z of length dz would attract the unit mass with a certain force. The symmetrically located element at $-z$ would attract it with a force of equal magnitude. The vertical components of these forces would cancel and the resultant would be a force whose magnitude would be twice the projection onto the xy-plane:

$$|d\mathbf{F}| = 2\mu dz \cdot \frac{\sqrt{x^2 + y^2}}{(x^2 + y^2 + z^2)^{3/2}},$$

and the whole wire would then apply a force

$$|\mathbf{F}| = 2\mu\sqrt{x^2 + y^2} \int_0^\infty \frac{dz}{(x^2 + y^2 + z^2)^{3/2}}$$

to the unit particle at $(x, y, 0)$. If we set $\mathbf{r} = x\mathbf{i} + y\mathbf{j}$, and $|\mathbf{r}| = r$, we get

$$|\mathbf{F}| = 2\mu r \int_0^\infty \frac{dz}{(r^2 + z^2)^{3/2}}$$
$$= 2\mu r \left[\frac{1}{r^2} \frac{z}{(r^2 + z^2)^{1/2}} \right]_0^\infty$$
$$= \frac{2\mu}{r}.$$

If we project everything onto the xy-plane, then this result can be interpreted as a particle of mass 2μ attracting a unit particle by an inverse first power law. To point the force in the right direction, that is, toward the origin, we multiply by $-\mathbf{r}/r$ to get

$$\mathbf{F} = -2\mu \frac{\mathbf{r}}{r^2}$$
$$= -\frac{2\mu(x\mathbf{i} + y\mathbf{j})}{x^2 + y^2}$$

and it is clear that

$$\varphi = -2\mu \ln r = 2\mu \ln \left(\frac{1}{r} \right)$$

is a potential in two dimensions for this force. And we compute

$$\nabla \varphi = \frac{-2\mu(x\mathbf{i} + y\mathbf{j})}{(x^2 + y^2)}$$

and

$$\nabla \bullet \nabla \varphi = \varphi_{xx} + \varphi_{yy}$$

$$= -2\mu \left[\frac{y^2 - x^2}{r^4} + \frac{x^2 - y^2}{r^4} \right] = 0.$$

and so our potential function is a two-dimensional harmonic function. The theory of Laplace's equation in three dimensions is called **Newtonian potential theory**, and in two is called **logarithmic potential theory**. The last chapter is devoted to logarithmic potential theory.

C. Fluid Flow

We consider the steady (i.e., time independent) flow of an ideal fluid over a horizontal plane which we take as the xy-plane. In particular we assume that the flow is frictionless (zero viscosity), irrotational, and incompressible in a region R. And we take the velocity vector at the point (x, y) to be

$$\mathbf{V} = p(x, y)\mathbf{i} + q(x, y)\mathbf{j}$$

and assume that p and q are continuously differentiable in R.

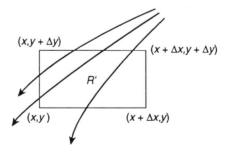

Let R' be a subregion of R. Since the flow is incompressible, the rate of flow into R' must equal the rate of flow out. Thus the net rate of flow into R' must be zero. We take R' to be a small rectangle as illustrated.

Then the net rate of flow into R is

$$0 = \int_y^{y+\Delta y} p(x + \Delta x, \eta)d\eta - \int_y^{y+\Delta y} p(x, \eta)d\eta$$

$$+ \int_x^{x+\Delta x} q(\xi, y + \Delta y)d\xi - \int_x^{x+\Delta x} q(\xi, y)d\xi$$

$$= \int_y^{y+\Delta y} \left[p(x + \Delta x, \eta) - p(x, \eta) \right] d\eta$$

$$+ \int_x^{x+\Delta x} \left[q(\xi, y + \Delta y) - q(\xi, y) \right] d\xi.$$

We apply the mean value theorem of differential calculus to the integrand in each of the last two integrals to get

$$\Delta x \int_y^{y+\Delta y} p_x(\xi', \eta)d\eta + \Delta y \int_x^{x+\Delta x} q_y(\xi, \eta')d\xi = 0$$

where ξ' and η' are mean points. Then applying the mean value theorem for integrals we get

$$p_x(\bar{\xi}, \bar{\eta})\Delta x \Delta y + q_y(\bar{\bar{\xi}}, \bar{\bar{\eta}})\Delta x \Delta y = 0.$$

We divide by $\Delta x \Delta y$ and let $\Delta x \to 0$ and $\Delta y \to 0$. Since p_x and q_y are continuous, we get

$$p_x(x, y) + q_y(x, y) = 0, \tag{11}$$

or, equivalently

$$\nabla \bullet \mathbf{V} = 0. \tag{12}$$

The meaning of the statement that the flow is irrotational is the following. For any contour C, with C and $I(C)$ in the field of flow we have

$$\int_C \mathbf{V} \bullet \mathbf{T} ds = 0$$

where $\mathbf{T}$ is the unit tangent and ds is the element of arc. This integral measures the net tangential velocity component and its being zero means there is no tendency to rotate and form an eddy. For our simple rectangle this gives

$$\int_y^{y+\Delta y} \left[q(x + \Delta x, \eta) - q(x, \eta) \right] d\eta - \int_x^{x+\Delta x} \left[p(\xi, y + \Delta y) - p(\xi, y) \right] d\xi = 0.$$

Arguing as before we get

$$q_x(x,y) - p_y(x,y) = 0. \tag{13}$$

Now, going to complex rather than vector notation we write

$$v(z) = p(x,y) + iq(x,y)$$

and call this the **complex velocity**. Then (11) and (13) verify that

$$\overline{v(z)} = p(x,y) - iq(x,y)$$

is analytic in R, for they are just exactly the Cauchy–Riemann equations for $\overline{v(z)}$.

We show in Section 3.1 that if a function such as $\overline{v(z)}$ is analytic in a simply connected region (our proof is actually given only for star-shaped regions), then it has an indefinite integral $f(z)$ given by

$$f(z) = \int_{z_0}^{z} \overline{v(\zeta)}d\zeta$$

where the integral is over any path from z_0 to z in R. $f(z)$ is called the **complex potential** of the flow. If we write

$$f(z) = \varphi(z) + i\psi(z)$$

we get

$$v = \bar{f}' = \varphi_x - i\psi_x = \psi_y + i\varphi_y = p + iq,$$

so

$$\varphi_x = p, \qquad \varphi_y = q,$$

from which, in vector notation,

$$\nabla\varphi = p\mathbf{i} + q\mathbf{j}.$$

By analogy with the potential of a force field, φ is called the **velocity potential** of the flow. The curves $\varphi = c$, for constant c's are called the **equipotential lines**, and since $\mathbf{V} = \nabla\varphi$ is orthogonal to $\varphi = c$, the lines of flow are given by the orthogonal trajectories of the system $\varphi = c$. We observe

$$\nabla\psi = \psi_x\mathbf{i} + \psi_y\mathbf{j} = -q\mathbf{i} + p\mathbf{j}$$

from which

$$\nabla\varphi \cdot \nabla\psi = (p\mathbf{i} + q\mathbf{j}) \cdot (-q\mathbf{i} + p\mathbf{j})$$
$$= 0.$$

Thus the orthogonal trajectories (lines of flow) are given by $\psi = c$.

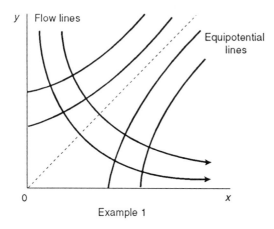

Example 1

Example 1. Describe the flow in the first quadrant whose complex potential is given by $f(z) = z^2$.

Solution. $f(z) = x^2 - y^2 + 2ixy$. So

$$\varphi = x^2 - y^2, \qquad \psi = 2xy$$

Thus the equipotential lines are the hyperbolas $x^2 - y^2 = c$ and the lines of flow are the hyperbolas $2xy = c$. The velocity vector is

$$\mathbf{V} = \nabla\varphi = 2x\mathbf{i} - 2y\mathbf{j}.$$

A Exercises

1. Show that the following polynomials satisfy Laplace's equation.
   (a) x; (b) y; (c) xy; (d) $x^3 - 3xy^2$; (e) $3x^2y - y^3$.
2. By inspection find a harmonic function in the rectangle $\{0 \le x \le 2;\ 0 \le y \le 1\}$ with boundary values as indicated in the figure.

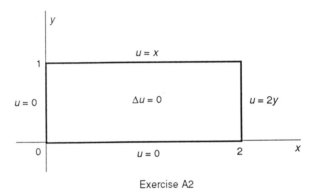

Exercise A2

3. Let a, b, μ be positive numbers. A unit mass is located at $-b$ on the x-axis, and a wire of linear density μ lies along the positive x-axis from 0 to a. Find the force exerted by the wire on the unit mass. Find the limit of the force as $a \to \infty$.
4. Explain the algebraic signs in the expressions for heat flow across the sides of the rectangle in part A.

B Exercises

1. Show that $r^2 \cos 2\theta$ and $r^2 \sin 2\theta$ are harmonic in the plane.
2. Find a non-zero harmonic function in the angle $0 < \theta < \pi/2$ which is zero on the boundary $\theta = 0$ and $\theta = \pi/2$.
3. Show that $\sin x \sinh y$ is harmonic in the semi-infinite strip $\{0 < x < \pi,\ 0 < y < \infty\}$, and is zero on the boundary.
4. Do B2 and B3 contradict the uniqueness theorem?

5. Suppose a wire with linear density μ lies along the positive z-axis. It attracts a unit particle at $(x, y, 0)$. Show that the force on the particle is

$$\mathbf{F} = \mu\left[-x\mathbf{i} - y\mathbf{j} + \sqrt{x^2 + y^2}\,\mathbf{k}\right]\Big/(x^2 + y^2).$$

6. Show that if $u(x, y)$ is harmonic in a region R, and is not a constant, then u cannot achieve a minimum value at any point of R.

2.3 Infinite Series

An **infinite series** is an indicated sum of the form

$$\alpha_0 + \alpha_1 + \alpha_2 + \cdots + \alpha_n + \cdots \tag{1}$$

or

$$\sum_0^\infty \alpha_n \quad \text{or} \quad \sum_0^\infty \alpha_k. \tag{2}$$

The α's are its **terms** and the numbers

$$S_n = \sum_0^n \alpha_k = \alpha_0 + \alpha_1 + \cdots + \alpha_n \tag{3}$$

are its **partial sums**. Though we have shown the series starting with the zeroth term, it can start with a term of any index. For example

$$\sum_{-3}^\infty \alpha_k = \alpha_{-3} + \alpha_{-2} + \cdots + \alpha_n + \cdots,$$

$$\sum_5^\infty \alpha_k = \alpha_5 + \alpha_6 + \cdots + \alpha_n + \cdots$$

are admissible series and can be brought into the form (1) or (2) by a change of index. When we can do so unambiguously, we sometimes write simply

$$\sum \alpha_k \quad \text{or} \quad \sum \alpha_n$$

without explicitly writing the limits on the $\sum$.

If the limit S of the partial sums exists ($S = \lim S_n$), we call it the **sum** of the series and write

$$S = \sum_0^\infty \alpha_n. \tag{4}$$

If the series has a sum, that is, the sequence $\{S_n\}$ has a limit, we say the series **converges**; otherwise it **diverges**. And if $\sum |\alpha_n|$ converges, we say $\sum \alpha_n$ **converges absolutely**.

In dealing with infinite series we are dealing with two sequences namely $\{\alpha_n\}$ and $\{S_n\}$, the **terms** and the **partial sums**. One of the simplest properties of convergent series is that $\alpha_n \to 0$. In practice this amounts to a divergence test since some, but not all, divergent series have this property. Hence if $\alpha_n \not\to 0$, then $\sum \alpha_n$ diverges, but if $\alpha_n \to 0$, $\sum \alpha_n$ may or may not converge. The result is the following.

2.3a Theorem. *If $\sum \alpha_n$ converges, then $\alpha_n \to 0$.*

Proof. $\alpha_n = \sum_0^n \alpha_k - \sum_0^{n-1} \alpha_k = S_n - S_{n-1} \to S - S = 0.$ $\square$

A very simple but useful consequence of Theorem 2.3a is the boundedness of the terms of a convergent series. Since the convergence of $\sum \alpha_n$ implies $\alpha_n \to 0$, there is a N such that $|\alpha_n| < 1$ for $n > N$. Then taking

$$M = \max \left[|\alpha_0|, |\alpha_1|, \dots, |\alpha_N|, 1 \right]$$

we have the following result.

2.3b Theorem. *If $\sum \alpha_n$ converges, there is a constant M such that for all n*

$$|\alpha_n| \leq M.$$

The single most important series (certainly for this course, and probably for all of complex variable theory) is the **geometric series**:

$$G(z) = 1 + z + z^2 + \cdots + z^n + \cdots = \sum_0^\infty z^n. \tag{5}$$

Two related series are those generated from $G(z)$ by formal termwise differentiation:

$$G'(z) = 1 + 2z + 3z^2 + \cdots + nz^{n-1} + (n+1)z^n + \cdots$$
$$= \sum_0^\infty (n+1)z^n = \sum_1^\infty nz^{n-1}, \tag{6}$$

$$G''(z) = 2 + 6z + 4 \cdot 3z^2 + \cdots + n(n-1)z^{n-2} + \cdots$$
$$= \sum_0^\infty (n+2)(n+1)z^n = \sum_2^\infty n(n-1)z^{n-2}. \tag{7}$$

The following theorem evaluates the sums of these series and justifies the prime notation.

2.3c Theorem. *For $|z| \geq 1$, the series (5), (6), and (7) diverge. For $|z| < 1$ they have sums given by*

$$G(z) = \frac{1}{1-z}, \quad G'(z) = \frac{1}{(1-z)^2}, \quad G''(z) = \frac{2}{(1-z)^3}.$$

Proof. In the proof we need to know

$$r^n \to 0, \quad nr^n \to 0, \quad \text{and} \quad n^2 r^n \to 0 \quad \text{as} \quad n \to \infty \tag{8}$$

for $0 \leq r < 1$. These are proved in calculus courses and are assigned here as an exercise (B1).

Clearly all these series diverge for $|z| \geq 1$ since then $|z^n| = |z|^n \geq 1$ and so by Theorem 2.3a all three series diverge.

For $|z| < 1$ we denote the nth partial sum of $G(z)$ by $G_n(z)$ and write

$$\begin{cases} G_n(z) = \displaystyle\sum_0^n z^k = 1 + z + z^2 + \cdots + z^n \\[2mm] \quad\;\; = \dfrac{1 - z^{n+1}}{1-z} = \dfrac{1}{1-z} - \dfrac{z^{n+1}}{1-z} \end{cases} \tag{9}$$

Then, with $|z| = r < 1$, we have

$$\left| G_n(z) - \frac{1}{1-z} \right| \leq \frac{r^{n+1}}{|1-z|} \to 0$$

by (8) above. This establishes the sum of $G(z)$. By termwise differentiation of (9) we get partial sums of $G'(z)$ and $G''(z)$ and using the other parts of (8) we evaluate their sums. The remaining details are left to the exercises (A2). $\square$

We now recall (with at least sketches of the proofs) some of the convergence tests for real series studied in calculus.

I. Theorem. *If there is a $K > 0$ for which $a_k \geq 0$ for $k > K$, then $\sum a_k$ converges if and only if the partial sums are bounded.*

The sequence $\{S_n\}$ is non-decreasing for $n > K$, and it is a fundamental property of the real numbers that such a sequence has a limit if and only if it is bounded.

II. Theorem (Comparison Test). *Suppose there is an integer $K > 0$ for which $0 \leq a_k \leq b_k$ for $k > K$. Then*
(i) $\sum a_k$ converges if $\sum b_k$ does;
(ii) $\sum b_k$ diverges if $\sum a_k$ does.

Proof. Let $\{S_n\}$ be the partial sums of $\sum a_k$ and $\{S'_n\}$ be the partial sums of $\sum b_k$. Then

$$S_n - S_K = \sum_{K+1}^{n} a_k \leq \sum_{K+1}^{n} b_k = S'_n - S'_K,$$

from which Theorem I implies Theorem II. (How?) $\square$

A consequence of the comparison test is that absolute convergence implies convergence.

III. Theorem. *If $\sum |a_k|$ converges, so does $\sum a_k$.*

Proof. We define p_k and n_k by

$$p_k = \begin{cases} a_k & \text{if} \quad a_k \geq 0 \\ 0 & \text{if} \quad a_k < 0 \end{cases} \quad ; \quad n_k = \begin{cases} 0 & \text{if} \quad a_k \geq 0, \\ -a_k & \text{if} \quad a_k < 0, \end{cases}$$

so that $a_k = p_k - n_k$ and $|a_k| = p_k + n_k$. Then

$$0 \leq p_k \leq |a_k| \quad \text{and} \quad 0 \leq n_k \leq |a_k|$$

from which both $\sum p_k$ and $\sum n_k$ converge by Theorem II. Then

$$\sum a_k = \sum (p_k - n_k) = \sum p_k - \sum n_k$$

converges since the limit of a sum is the sum of the limits. $\square$

We now return to complex series. We first want to extend Theorem III to complex series.

2.3d Theorem. *If $\sum |\alpha_k|$ converges, so does $\sum \alpha_k$.*

Proof. Set $\alpha_k = a_k + ib_k$. Then

$$|a_k| \leq |\alpha_k| \quad \text{and} \quad |b_k| \leq |\alpha_k|$$

hence $\sum a_k$ and $\sum b_k$ converge by Theorem II. Then $\sum \alpha_k$ converges by Theorem 1.3a. $\square$

The following tests are more general and easier to prove than the elementary tests for real series usually given in calculus courses. The usual limit forms are then consequences of these.

2.3e Theorem. (Cauchy's root test) *(a) If there is a positive $\lambda < 1$, and an integer N for which*

$$\sqrt[n]{|\alpha_n|} \le \lambda, \qquad n \ge N,$$

then $\sum \alpha_n$ converges. (b) If there are infinitely many distinct values of n for which

$$\sqrt[n]{|\alpha_n|} \ge 1,$$

then $\sum \alpha_n$ diverges.

Proof. In part (b) we have $|\alpha_n| \ge 1$ for infinitely many values of n so $\alpha_n \nrightarrow 0$, and $\sum \alpha_n$ diverges by Theorem 2.3a.

For part (a) we have

$$|\alpha_n| < \lambda^n, \qquad n \ge N$$

and $\sum |\alpha_n|$ converges by comparison with $G(\lambda)$. Thus $\sum \alpha_n$ converges by Theorem 2.3d. $\square$

2.3f Corollary. *If $\lim \sqrt[n]{|\alpha_n|} = a$ exists, then*
(i) $\sum \alpha_n$ converges if $a < 1$,
(ii) $\sum \alpha_n$ diverges if $a > 1$ (including $a = \infty$).

Proof. (i) For $\lambda = (1 + a)/2 < 1$, there is an N for which $\sqrt[n]{|\alpha_n|} \le \lambda$ if $n \ge N$ (how?). So Theorem 2.3e implies convergence.
(ii) Take $\lambda = (1+a)/2$, then $1 < \lambda < a$ if a finite, and take $\lambda = 2$ if $a = \infty$. There is an N such that $\sqrt[n]{|\alpha_n|} \ge \lambda$ for $n \ge N$. Again Theorem 2.3e applies showing divergence here. $\square$

2.3g Theorem. (Ratio Test) *If there is an N for which $\alpha_n \ne 0$ for $n > N$, and a positive $\lambda < 1$ for which*

$$|\alpha_{n+1}/\alpha_n| \le \lambda \qquad n > N$$

then $\sum \alpha_n$ *converges. If there is an* N *for which*

$$|\alpha_{n+1}/\alpha_n| \geq 1, \quad for \quad n > N,$$

then $\sum \alpha_n$ *diverges.*

Proof. For the first part $|\alpha_{N+1}| \leq \lambda |\alpha_n|$, and $|\alpha_{N+2}| \leq \lambda |\alpha_{N+1}| \leq \lambda^2 |\alpha_N|$ and by induction $|\alpha_{N+k}| \leq \lambda^k |\alpha_N|$. By ssetting $N + k = n$ this becomes

$$|\alpha_n| \leq \frac{|\alpha_N|}{\lambda^N} \lambda^n, \quad n \geq N,$$

and $\sum \alpha_n$ converges by comparison with $(|\alpha_n|/\lambda^N)G(\lambda)$.

For the second part $|\alpha_{N+1}| \geq |\alpha_N| > 0$, and by induction

$$|\alpha_{N+k}| \geq |\alpha_N| > 0$$

so $\lim \alpha_n \neq 0$ and the series diverges by Theorem 2.3a. $\square$

2.3h Corollary. *If* $\lim |\alpha_{n+1}/\alpha_n| = a$ *exists, then*
(i) $\sum \alpha_n$ *converges if* $a < 1$,
(ii) $\sum \alpha_n$ *diverges if* $a > 1$ *(including* $a = \infty$*).*

The proof is analogous to that of Corollary 2.3e and is left as an exercise. These last two corollaries are called the **limit forms** of the **root test** and **ratio test** respectively. They give no information in the case $a = 1$, for there are series (for each test) for which $a = 1$ which converge, and others which diverge (see Exercises A2, A3).

If $w_k(z)$ is a sequence of complex valued functions defined on a set D, then

$$\sum_0^\infty w_k(z)$$

defines an infinite series of functions. For each fixed $z \in D$, this series is a series of complex numbers, and if the series converges for each $z \in D$, we say that $\sum_0^\infty w_k(z)$ converges **pointwise** in D. If we denote the sum by $f(z)$ and partial sums by $S_n(z)$, then pointwise convergence means

$$\lim S_n(z) = f(z) \quad \text{for each} \quad z \in D.$$

which, in turn, means that for each $\varepsilon > 0$ there is $N = N(z, \varepsilon)$ for which

$$|f(z) - S_n(z)| < \varepsilon \quad \text{whenever} \quad n > N. \tag{1}$$

If it is possible to choose the N to depend on ε and the set D, but to be independent of the point z at which convergence is considered (that is, N is uniform — the same — for all $z \in D$) so that (1) is *valid for all* $z \in D$ *and the same* N, then we say the series **converges uniformly** on D. The concept of uniform convergence plays a very important role in analytic function theory. A useful criterion for uniform convergence is the following.

2.3i Theorem. (Weierstrass M-test) *If $\sum w_k(z)$ is a series of complex functions on a set D, and if there is a convergent series of positive constants $\sum M_k$ for which $|w_k(z)| \leq M_k$, then $\sum w_k(z)$ converges uniformly on D.*

Proof. $\sum |w_k(z)|$ converges for each $z \in D$ by comparison with $\sum M_k$. So $\sum w_k(z)$ converges for each $z \in D$ by Theorem 2.3d. We denote the sum by $f(z) : f(z) = \sum w_k(z)$. By the definition of convergence, given $\varepsilon > 0$ there is an N for which $\sum\limits_{N+1}^{\infty} M_k < \varepsilon$. Then for $n > N$ we get

$$|f(z) - S_n(z)| = \left| f(z) - \sum_0^n w_k(z) \right| = \left| \sum_{n+1}^{\infty} w_k(z) \right|$$

$$\leq \sum_{n+1}^{\infty} |w_k(z)| \leq \sum_{n+1}^{\infty} M_k < \varepsilon$$

uniformly for all $z \in D$. $\square$

Example. Show that $\sum\limits_1^{\infty} z^n / n^2$ converges uniformly on $D : |z| \leq 1$.

Solution. For $z \in D$, that is, for $|z| \leq 1$

$$\left| \frac{z^n}{n^2} \right| \leq \frac{1}{n^2}.$$

So if we choose $M_n = 1/n^2$, then by elementary calculus $\sum 1/n^2$ converges, and by the M-test our series converges uniformly on the disk $|z| \leq 1$.

A Exercises

1. Establish the summations for $G'(z)$ and $G''(z)$ given in Theorem 2.3c.
2. The series $\sum 1/n$ diverges (it is the harmonic series) and $\sum 1/n^2$ converges. Show that for each of these series

$$\lim |\alpha_{n+1}/\alpha_n| = 1.$$

3. (See Exercise B2) It is known that $\sqrt[n]{n} \to 1$. Use this to show that for both series in A2 above

$$\lim \sqrt[n]{|\alpha_n|} = 1.$$

4. If $a > b > 0$, and $a > 1$ test $\sum 1/(a^n - b^n)$ for convergence.
5. Test $\sum \alpha_n$ for convergence where α_n is

   (a) $\dfrac{n!}{n^n}$ (b) $\dfrac{n!}{(1000)^n}$ (c) $\dfrac{n^3(n+1)^n}{(3n)^n}$

6. Show that $G(z)$ (see Theorem 2.3c) converges uniformly for $|z| \le a$ where $a < 1$.
7. Show that $\sum_{1}^{\infty}(z^n + 1)/n^2$ converges uniformly for $|z| \le 1$.

B Exercises

1. Show that if $0 \le r < 1$, then $r^n \to 0$, $nr^n \to 0$, and $n^2 r^n \to 0$ as $n \to \infty$.

   Hint: Set $\rho = 1/r > 1$ and apply l'Hospital's rule to x^p/ρ^x where p is a non-negative integer.

2. Show $\sqrt[n]{n} \to 1$ as $n \to \infty$. Hint: Set $y = x^{1/x}$, take logarithms, apply l'Hospital's rule and show $\ln y \to 0$ as $x \to \infty$.

3. If $\sum |\alpha_n|$ converges, show that $\sum \alpha_n^2$ converges.

4. Show that $\displaystyle\sum_0^\infty \left(\frac{z}{1+z}\right)^n$ converges for $x = \operatorname{Re} z > -\frac{1}{2}$, and find its sum.

5. Prove Corollary 2.3h.

C Exercise

1. Show that if $\sum |\alpha_n|^2$ and $\sum |\beta_n|^2$ converge, then so does $\sum \alpha_n \beta_n$. Hint: First assume α_n, β_n real and positive, and start with $(a_n - b_n)^2 \geq 0$.

2.4 Power Series: Domination, Radius of Convergence

The kind of series of functions in which we are most interested at present are series of the form

$$\sum_0^\infty c_k z^k \tag{1}$$

where the c's are (complex) constants called the **coefficients**. Series of the form $\sum_0^\infty c_k(z - \alpha)^k$ can be subsumed under (1) by the substitution (translation) $\zeta = z - \alpha$. The series (1) is said to be **centered** at 0, the other at α. We restrict our attention for the present to series centered at 0. Such series (either form) are called **power series**. We will frequently write $\sum c_k z^k$ for (1), not indicating the limits of summation.

To say that a power series $\sum a_n z^n$ is **dominated** by another $\sum b_n z^n$, which we write as

$$\sum a_n z^n \ll \sum b_n z^n \quad \text{or} \quad \sum b_n z^n \gg \sum a_n z^n, \tag{2}$$

means

$$|a_n| \leq |b_n|, \qquad n = 0, 1, 2, 3, \ldots, \tag{3}$$

and we are mostly interested in the case where $b_n \geq 0$. The concept of one power series dominating another is, as you might expect, useful in formulating certain general comparison tests and other results for power series.

We examine the domain of convergence of power series. Clearly any power series (1) converges for $z = 0$, for then all partial sums are a_0. In fact some converge only at $z = 0$, while some others converge for all values of z.

Example 1. Show $\sum n! z^n$ converges only for $z = 0$.

Solution. Let $z \neq 0$ be fixed. Then $|(n+1)! z^{n+1} / n! z^n| = (n+1)|z| > 1$ when $n > (1/|z|) - 1$, so the series diverges by Theorem 2.3g.

Example 2. Show $\sum z^n / n!$ converges for all z.

Solution. We need only consider $z \neq 0$. Then

$$\left| \frac{z^{n+1}}{(n+1)!} \middle/ \frac{z^n}{n!} \right| = \frac{|z|}{n+1} < \frac{1}{2} \quad \text{when} \quad n > 2|z| - 1,$$

and the series converges by Theorem 2.3g.

In order to investigate the convergence behavior of other series we will use a fundamental property of the real numbers called thee **least upper bound property**. It is the following: Suppose S is a nonempty set of real numbers which is bounded above, that is, there is a constant M such that $r \leq M$ for all $r \in S$. Then there is a real number ρ satisfying the following conditions:

(a) If $r \in S$, then $r \leq \rho$. (Thus ρ is an upper bound for S.)

(b) For each number $m < \rho$ there is an $r \in S$ for which $m < r \leq \rho$. (Thus there is no smaller upper bound.)

The number ρ is called the **least upper bound** or the **supremum** of S, and is denoted by

$$\rho = \sup S.$$

If S has a maximal element, then that must be ρ, but, for example, if S is the interval $\{r : 0 \leq r < 1\}$, then S has no maximal element but its supremum is 1.

We begin with the following preliminary results.

2.4a Lemma. *If $\sum c_k z^k$ converges at a point z_0 with $|z_0| = a > 0$, then there is a number $M > 0$ for which $|c_k| < M/a^k$, or equivalently*

$$\sum c_k z^k \ll M G(z/a)$$

where G represents the geometric series of powers of z/a.

Proof. The equivalence of the two statements is clear. And by Theorem 2.3b, there is an M for which $|c_k|a^k = |c_k z_0^k| \leq M$. □

2.4b Lemma. *(i) If $\sum c_k z^k$ converges for z_0, with $|z_0| = a$, then it converges absolutely for any z with $|z| < a$. (ii) If it diverges for z_1 with $|z_1| = b$, then it diverges for any z with $|z| > b$.*

Proof. (i) Choose z with $|z| < a$. Then by Lemma 2.4a, there is an M for which

$$|c_k z^k| \leq (M/z^k)|z^k| = M(r/a)^k.$$

Thus $\sum |c_k z^k|$ converges by comparison with $MG(r/a) = M \sum (r/a)^k$.

(ii) Choose z with $|z| > b$. If $\sum c_k z^k$ converges, then $\sum c_k z_1^k$ would converge by part (i). □

Now we define the **radius of convergence** ρ of a series $\sum c_k z^k$ as follows.
(i) $\rho = 0$ if $\sum c_k z^k$ converges only for $z = 0$.
(ii) $\rho = \infty$ if $\sum c_k z^k$ converges for all $z \in \mathbb{C}$.
(iii) If $\sum c_k z^k$ converges for some non-zero value of z and diverges for another, we take S to be the set of all real numbers r for which there is a z with $|z| = r$ and the series converges for this z. Then we define

$$\rho = \sup S.$$

2.4c Theorem. *If $\sum c_k z^k$ has a radius of convergence ρ, then the series converges absolutely for any z with $|z| < \rho$ and diverges for any z with $|z| > \rho$.*

Proof. If $\rho = 0$ or if $\rho = \infty$, there is nothing to prove. If $0 < \rho < \infty$, then choose z with $|z| < \rho$. By the definition of the least upper bound there is a number r, $|z| < r < \rho$ for which there is a z_0 with $|z_0| = r$ so that $\sum c_k z_0^k$ converges. By Lemma 2.4b, $\sum c_k z^k$ converges absolutely.

If $|z| > \rho$, then $\sum c_k z^k$ diverges by the definition of least upper bound. □

Thus a power series (1) always has associated with it a number ρ (or $\rho = \infty$) its radius of convergence. It converges in the disk $D = \{z : |z| < \rho\}$ and diverges for $|z| > \rho$. In case $\rho = 0$ the disk is degenerate (in fact empty) and for $\rho = \infty$ the divergence set is empty. The disk D is called the **disk of convergence**. The boundary of the disk, the circle $C = \{z : |z| = \rho\}$, is called the **circle of convergence**. The question of convergence on C is in general a difficult one, and we will see by examples that many possibilities occur. (See Exercise A4).

The general problem of determining the radius of convergence is a difficult one. This is discussed in the C Exercises. We now give some simple tests for ρ which are based on the limit forms of the ratio and root tests.

2.4d Theorem. *If* $\lim |c_{n+1}/c_n| = a$ *exists (and possibly* $= \infty$*), then the radius of convergence* ρ *of* $\sum c_k z^k$ *is given by* $\rho = 1/a$ *with the convention that* $1/0 = \infty$, $1/\infty = 0$.

Proof. We apply the limit form of the ratio test (Corollary 2.3h) to $\sum c_k z^k$ for arbitrary but fixed $z \neq 0$. Then

$$\left| c_{n+1} z^{n+1} \middle/ c_n z^n \right| = \left| c_{n+1} \middle/ c_n \right| |z| \to a|z|.$$

If $a = \infty$, the series diverges for any $z \neq 0$, and so $\rho = 0$. If $a = 0$, the series converges for all z, and so $\rho = \infty$. If $0 < a < \infty$, the series converges for $a|z| < 1$ and diverges for $a|z| > 1$; that is, $\rho = 1/a$. $\square$

2.4e Theorem. *Suppose* $a = \lim \sqrt[n]{|c_n|}$ *exists. Then the radius of convergence* ρ *of* $\sum c_n z^n$ *is given by* $\rho = 1/a$*, where again* $1/0 = \infty$ *and* $1/\infty = 0$.

The proof, based on Corollary 2.3f is given as Exercise B6.

Example 3. Find the radius of convergence of $\sum z^k/k^2$.

Solution. $a = \lim(n+1)^2/n^2 = \lim \left(1 + \dfrac{1}{n}\right)^2 = 1$, and so $\rho = 1/a = 1$.

Example 4. Find the radius of convergence of $\sum z^k/k^k$.

Solution. $a = \lim \sqrt[n]{1/n^n} = \lim 1/n = 0$, so $\rho = \infty$.

Example 5. Find the radius of convergence of $\sum [(n+1)/n]^{n^2} z^n$.

Solution. $a = \lim \sqrt[n]{[(n+1)/n]^{n^2}} = \lim \left(1 + \dfrac{1}{n}\right)^n = e$, where $e \approx 2.71828\ldots$ is the base of nnatural logarithms. Hence $\rho = 1/e \approx 0.367\ldots$

2.4f Theorem. *If $\sum \beta_k z^k$ has a radius of convergence ρ and $\sum \alpha_k z^k \ll \sum \beta_k z^k$, then $\sum \alpha_k z^k$ has a radius of convergence $\rho' \geq \rho$.*

Proof. If $\rho = 0$ or ∞, there is nothing to prove. Otherwise it suffices to show that $\sum \alpha_k z^k$ converges for any z with $|z| = r < \rho$. For such a z

$$|\alpha_k z^k| = |\alpha_k| r^k \leq |\beta_k| r^k,$$

and so $\sum |\alpha_k z^k|$ converges by comparison with $\sum |\beta_k| r^k$ which converges since $r < \rho$. $\square$

It is sometimes desirable to perform the arithmetic operations on infinite series. If $\sum a_k z^k$ and $\sum b_k z^k$ have radii of convergence ρ_1 and ρ_2, then it is elementary that $\sum a_k z^k \pm \sum b_k z^k = \sum (z_k \pm b_k) z^k$ within the smaller disk of convergence of the two series. (How?)

We want to consider the product $f(z) \cdot g(z)$ where $f(z) = \sum a_n z^n$ with radius of convergence ρ_1 and $g(z) = \sum b_n z^n$ with radius of convergence ρ_2. Then let ρ be the smaller of ρ_1, ρ_2. We would expect that for $|z| < \rho$, we could multiply termwise and collect powers of z. The formal result of this is

$$p(x) = \sum c_n z^n$$

where

$$c_n = \sum_{j+k=n} a_j b_k = \sum_{j=0}^{n} a_j b_{n-j}$$

But it is not a priori clear that $p(z)$ has a nonzero radius of convergence, and if so it is not clear that $p(z)$ is then the product of $f(z)$ and $g(z)$.

To facilitate the proof of these facts we set up the following notation:

$$f_n(z) = \sum_{0}^{n} a_k z^k, \quad g_n(z) = \sum_{0}^{n} b_k z^k$$

$$p_n(z) = \sum_{0}^{n} c_k z^k$$

$$C_n = \sum_{j=0}^{n} |a_j| |b_{n-j}| \quad \text{(hence } C_n \geq |c_n|)$$

$$P_n(z) = \sum_{0}^{n} C_n z^k, \quad P(z) = \sum_{0}^{\infty} C_n z^k$$

2.4g Theorem. *In the notation above,* $P(z)$, *and hence* $p(z)$, *has a radius of convergence* $\geq \rho$, *and for* $|z| < \rho$

$$p(z) = f(z)g(z).$$

Proof. To prove that $P(z)$ has aa radius of convergence $\geq \rho$ it suffices to show that the series for $P(z)$ converges for any $|z| < \rho$. So we choose z, with $|z| = r < \rho$ and choose R so that $r < R < \rho$. Then both $\sum a_k z^k$ and $\sum b_k z^k$ converge at $z = R$ and so by Lemma 2.4a there is an $M > 0$ for which $|a_k| < M/R^k$ and $|b_k| \lessgtr M/R^k$. Then

$$C_k = \sum_{j=0}^{k} |a_j||b_{k-j}| < \sum_{0}^{k} \frac{M}{R^j} \frac{M}{R^{k-j}} = M^2(k+1)/R^k.$$

Thus

$$p(z) \ll P(z) \ll M^2 G'(z/R)$$

and since $G'(z/R)$ converges for $|z| = r$ so do $p(z)$ and $P(z)$. This proves that the radius of convergence of $p(z)$ is $\geq \rho$.

We observe that the product $f_n(z)g_n(z)$ is the sum of a finite number of terms of the form

$$(a_j z^j)(b_k z^k) = a_j b_k z^{k+j}.$$

In fact it is the sum of all such terms for which $0 \leq j \leq n$ and $0 \leq k \leq n$. Similarly $p_m(z)$ is the sum of all such terms for which $j + k \leq m$. Hence if we choose $m \geq 2n$ and consider the difference

$$\Delta \equiv p_m(z) - f_n(z)g_n(z),$$

then Δ is again a finite sum of such terms. (Why do we want $m \geq 2n$?) If we let $T_{m,n}$ be a symbol for the set of indices of the terms in Δ, then

$$\Delta = \sum_{T_{m,n}} a_j b_k z^{j+k}$$

and hence

$$|\Delta| \leq \sum_{T_{m,n}} |a_j||b_k| r^{k+j} \leq \sum_{n < j+k \leq m} |a_j||b_k| r^{j+k}$$

$$= \sum_{k=n+1}^{m} \left(\sum_{j=0}^{k} |\alpha_j||b_{k-j}| \right) r^k = \sum_{n+1}^{m} C_k r^k a$$

$$= P_m(r) - P_n(r).$$

Then as $m \to \infty$ we get

$$|p(z) - f_n(z)g_n(z)| \leq P(r) - P_n(r),$$

and as $n \to \infty$ we get

$$|p(z) - f(z)g(z)| \leq 0,$$

from which $p(z) = f(z)g(z)$.

The remaining arithmetic operation is division. We state the following theorem and delay the proof to the last paragraph of Section 3.3. See also Exercise C2 here.

2.4h Theorem. *If $f(z) = \sum a_k z^k$ with $a_0 = 1$, with a nonzero radius of convergence, then $1/f(z)$ can be written as a power series with a nonzero radius of convergence:*

$$1/f(z) = \sum b_k z^k$$

where the b's are determined recursively by

$$b_0 = 1,$$

$$b_n = -\sum_1^n a_k b_{n-k}.$$

Significant Topics	Page

A Exercises

1. Find the radius of convergence of $\sum a_n z^n$ where a_n is given below.

(a) $(\frac{1}{5})n^4(n^3 + 1)$ (b) $(2n)!/n!$

(c) $n!/(2n)!$ (d) $n!/n^n$

(e) $n^{2n}/(n!)^2$ (f) $[1 + (-1)^n 3]^n$

(g) $(a^n + b^n)$, $a > b > 0$ (h) $1/n^p$, $p > 0$

(i) n^p, $p > 0$

2. Discuss the following as in A1 above.

$$\text{(a) } \frac{3 \cdot 5 \cdot 7 \cdots (2n+1)}{4 \cdot 7 \cdot 10 \cdots (3n+4)} \qquad \text{(b) } \frac{1}{3^n} \left[\frac{1 \cdot 3 \cdots (2n+1)}{2 \cdot 4 \cdots (2n+2)} \right]^3$$

$$\text{(c) } \left(1 + \frac{1}{n^2} \right)^{n^3}$$

3. Why do $\sum c_n z^n$ and $\sum c_n z^{n+1}$ have the same radius of convergence?

4. Show that

   (a) $\sum_0^\infty z^n$ had radius of convergence $\rho = 1$, and converges nowhere on $|z| = 1$.

   (b) $\sum_1^\infty z^n/n$ has radius of convergence $\rho = 1$, and converges for $z = -1$, but diverges for $z = +1$.

   (c) $\sum_1^\infty z^n/n^2$ has radius of convergence $\rho = 1$ and converges everywhere on $|z| = 1$.

5. If $a_n = 1 + \dfrac{1}{n!}$ and $b_n = 1 - \dfrac{1}{n!}$ determine the radius of convergence of

   $$\text{(a) } \sum a_n z^n \qquad \text{(b) } \sum b_n z^n \qquad \text{(c) } \sum (a_n - b_n) z^n.$$

B Exercises

1. If $M \geq |a_n| \geq m > 0$ where m and M are constants, show that the radius of convergence of $\sum a_n z^n$ is 1.

2. Generalize the previous problem as follows:

   (a) If there are constants $M > 0, p$ for which $|a_n| \leq M n^p$, then $\sum a_n z^n$ has radius of convergence $\rho \geq 1$.

   (b) If there are constants $m > 0, p$ for which $|a_n| \geq m n^p$, then $\rho \leq 1$.

3. Find the radius of convergence of each series given below. Note that $a_n = 0$ for infinitely many values of n.

   $$\text{(a) } \sum 2n z^{2n} \qquad \text{(b) } \sum n! z^{2n+1} \qquad \text{(c) } \sum z^{2n+1}/n!$$
   $$\text{(d) } \sum z^{n!} \qquad\qquad \text{(e) } \sum n^2 z^{n^n}$$

4. Consider the series

   $$1 + (3z) + (2z)^2 + (3z)^3 + \cdots + (2z)^{2k} + (3z)^{2k+1} + \cdots$$

Show that this is

$$G((2z)^2) + 3zG((3z)^2)$$

where G is the geometric series. Deduce that the radius of convergence is $1/3$.

5. If $\sum c_k z^k$ has radius of convergence $\rho > 0$, then show that for any r, $0 < r < \rho$, it converges uniformly in $\{z : |z| \leq r\}$.

6. Prove Theorem 2.4e.

7. If $\sum \alpha_n z^n$ and $\sum \beta_n z^n$ have finite radii of convergence $\rho_1 > 0$ and $\rho_2 > 0$ respectively, say what you can about the radius of convergence of

(a) $\sum (\alpha_n \pm \beta_n) z^n$ (b) $\sum \alpha_n \beta_n z^n$ (c) $\sum (\alpha_n / \beta_n) z^n$.

C Exercises

1. Using Cauchy's root test (Theorem 2.3e) establish the Cauchy–Hadamard formula for the radius of convergence:

$$\rho = 1/\varlimsup \sqrt[n]{|a_n|}$$

with the convention that $1/0 = \infty$, $1/\infty = 0$.

2. Suppose, in Theorem 2.4h, that $1/f(z)$ has a power series as asserted. Then $(\sum a_k z)(\sum b_k z) = 1$. Multiply the series and derive the recursion formula for the b's given in the theorem.

2.5 Analyticity of Power Series

A power series $\sum c_k z^k$ with a nonzero radius of convergence defines a function in its disk D of convergence (which is the whole plane if $\rho = \infty$). And on any smaller concentric disk it converges uniformly (Exercise B5, Section 2.4) and absolutely (Theorem 2.4c). Furthermore, for each n, the approximating sum S_n is a polynomial:

$$S_n(z) = \sum_0^n c_k z^k.$$

Thus each partial sum is analytic in D. One could then hope that the sum of the series would be analytic in D. This turns out to be the case.

We begin with the preliminary result that termwise differentiation does not alter the radius of convergence.

2.5a Lemma. *Let the radii of convergence of*

$$f(z) = \sum c_k z^k \quad and \quad g(z) = \sum k c_k z^{k-1}$$

be ρ and ρ' respectively. Then $\rho = \rho'$.

Proof. Multiplying $g(z)$ by z does not alter its radius of convergence (why?), so

$$z g(z) = \sum k c_k z^k$$

has radius of convergence ρ'. But

$$|k c_k| \geq |c_k| \qquad k \geq 1$$

so $\rho' \leq \rho$ by Theorem 2.3f.

Next we show that $\rho' \geq \rho$ by showing that $g(z)$ converges for any z with $|z| < \rho$. Choose such a z, say $|z| = r < \rho$. Then choose a with $r < aa < \rho$, (e.g., $a = (r + \rho)/2$ if $\rho < \infty$, or $a = r + 1$ if $\rho = \infty$). Then $\sum c_n a^n$ converges and so, by Lemma 2.3a, there is a constant M for which

$$|c_k| \leq M \mid a^k.$$

We compute

$$|k c_k z^{k-1}| \leq k \frac{M}{a^k} r^{k-1} = \frac{M}{a} k \left(\frac{r}{a}\right)^{k-1}$$

Now $0 < r/a < 1$ so $\sum k c_k z^{k-1}$ is dominated by $(M/a) G'(r/a)$. $\square$

2.5b Theorem. *If $f(z) = \sum c_n z^n$ has radius of convergence $\rho > 0$, then it is analyic for $|z| < \rho$ and*

$$f'(z) = \sum n c_n z^{n-1}, \qquad |z| < \rho. \tag{1}$$

Proof. It suffices to show, for arbitrary fixed z with $|z| = r < \rho$, that $f'(z)$ exists and is given by (1). With such a z, choose a between r and ρ (e.g., take $a = (r + \rho)/2$). Then take h so that $|h| < a - r$. This gives

$$|z + h| \leq |z| + |h| < r + a - r = a < \rho$$

so that $z + h$ is also in the open disk of convergence. We now set

$$g(z) = \sum nc_n z^{n-1}$$

and consider

$$D \equiv \frac{\Delta f}{\Delta z} - g(z) = \frac{f(z + h) - f(z)}{h} - g(z)$$

$$= \sum c_n \left[\frac{(z + h)^n - z^n}{h} - nz^{n-1} \right]$$

By Example 2, Section 1.2, with $|z| = r$, $|h| = k$, we get

$$\left| \frac{(z + h)^n - z^n}{h} - nz^{n-1} \right| \leq \frac{(r + k)^n - r^n}{k} - nr$$

and, by Lemma 2.3a, there is a constant M for which $|c_n| \leq M/a^n$. Then

$$|D| \leq M \sum \frac{1}{a^n} \left[\frac{(r + h)^n - r^n}{k} - mr^{n-1} \right]$$

$$= M \left\{ \frac{1}{k} \left[\sum \left(\frac{r + k}{a} \right)^n - \sum \left(\frac{r}{a} \right)^n \right] - \frac{1}{a} \sum n \left(\frac{r}{a} \right)^{n-1} \right\}$$

$$= M \left\{ \frac{1}{k} \left[\frac{1}{1 - (r + k)/a} - \frac{1}{1 - r/a} \right] - \frac{1}{a} \frac{1}{(1 - r/a)^2} \right\}$$

the sums beign by Theorem 2.3c since $r < r + k < a$. Then

$$|D| \leq M \left[\frac{1}{k} \frac{a}{a - r - k} - \frac{1}{k} \frac{a}{a - r} - \frac{a}{(a - r)^2} \right]$$

$$= M \frac{ak}{(a - r)^2 (a - r - k)} \to 0 \quad \text{as} \quad |h| = k \to 0. \quad \square$$

By applying this theorem to $f'(z)$ we see that $f''(z)$ exists and can be computed termwise. By induction the same applies to a derivative of any order. This observation yields the following result.

2.5c Corollary. *If $f(z) = \sum c_n z^n$ has a nonzero radius of convergence ρ, then for $|z| < \rho$, $f(z)$ is analytic, has derivatives of all orders, and*

$$f^{(k)}(z) = \sum_{n=k}^{\infty} \frac{n!}{(n - k)!} c_n z^{n-k}.$$

2.5d Corollary. *If $f(z) = \sum c_n z^n$ has a nonzero radius of convergence ρ, then*

$$c_k = f^{(k)}(0)/k!.$$

Proof. Set $z = 0$ in the formula of the preceding corollary. $\square$

This last result is important in two ways: it not only gives a formula for the coefficients but gives also a uniqueness theorem.

2.5e Theorem. *If $f(z) = \sum c_n z^n = \sum b_n z^n$, then $c_n = b_n$ for $n = 0, 1, 2, \ldots$.*

Proof. By Corollary 2.5d we have

$$c_n = f^{(n)}(0)/n! = b_n. \quad \square$$

2.5f Corollary. *If $f(z) \equiv \sum c_n z^n$ has a nonzero radius of convergence ρ, then, for any complex number c,*

$$F(z) \equiv c + \sum_0^\infty c_n z^{n+1}/(n+1)$$

has radius of convergence ρ and

$$F'(z) = f(z), \qquad |z| < \rho.$$

Proof. That the radius of convergence is ρ is left to the exercises (A6). Then $F'(z) = f(z)$ by Theorem 2..5b. $\square$

2.5g Theorem. *If $\sum c_k z^k$ has radius of convergence $\rho > 0$, and if C is any path (piecewise smooth curve) from α to β lying in the disk of convergence, then*

$$\int_C f(z)dz = \sum_0^\infty c_k \beta^k/(k+1) - \sum_0^\infty c_k \alpha^k/(k+1),$$

and if C is a closed path $(\alpha = \beta)$, then

$$\int_C f(z)dz = 0.$$

The proof is clear from Corollary 2.5f.

Since power series are analytic, we can seek analytic solutions to differential equations in the form of power series. Let us consider the following problem to find a solution to a first order differential equation which achieves a given value at the origin. Such a problem is called an **initial value problem (IVP)**.

We seek a function $w(z)$, analytic in at least a neighborhood of the origin which satisfies the conditions

$$IVP : w'(z) = aw(z), \qquad w(0) = 1, \tag{2}$$

We try a solution in the form

$$w = \sum_0^\infty c_k z^k = c_0 + c_1 z + c_2 z^2 + \cdots \tag{3}$$

Then

$$w' = \sum_1^\infty k c_k z^{k-1} = \sum_0^\infty (k+1) c_{k+1} z^k,$$

so that if such a solution exists we must have

$$\sum_0^\infty (k+1) c_{k+1} z^k = a \sum_0^\infty c_k z^k = \sum_0^\infty a c_k z^k$$

holding within some circle of convergence. By Theorem 2.5e we then have

$$(k+1) c_{k+1} = a c_k, \qquad k = 0, 1, 2, \ldots . \tag{4}$$

This last equation is called an **iteration** or **recursion formula**. Since the initial condition $w(0) = 1$ implies that $c_0 = 1$, we can use the recursion formula (4) to successively determine all the coefficients in (3). Thus taking $k = 0$ in (4) we get

$$c_1 = a c_0 = a$$

Then $k = 1$ gives

$$2 c_2 = a c_1 = a^2$$

or

$$c_2 = a^2/2 = a^2/2!.$$

By induction we get in general

$$c_k = a^k/k!.$$

Thus we have determined a possible solution to be

$$w = \sum_0^\infty (a^k/k!)z^k = \sum_0^\infty (az)^k/k!$$

It is easy to see that this series has an infinite radius of convergence. Thus it can be differentiated as indicated and so satisfies the IVP. In the next section we sum the series.

Significant Topics   **Page**

A Exercises

1. Show that the function

$$f(z) = 1 + \sum_{n=1}^\infty \frac{\alpha(\alpha-1)\cdots(\alpha-n+1)}{n!} z^n$$

is analytic for $|z| < 1$, and that

$$(1+z)f'(z) = \alpha f(z), \qquad f(0) = 1$$

2. If $f(z) = 1 + \dfrac{z^3}{3!} + \dfrac{z^6}{6!} + \cdots + \dfrac{z^{3n}}{(3n)!} + \cdots$, show that $f'''(z) = f(z)$, $f(0) = 1$, $f'(0) = f''(0) = 0$.

3. If $f(z) = 1 + z + \dfrac{z^2}{2!} + \cdots + \dfrac{z^n}{n!} + \cdots$, show that $f'(z) = f(z)$, $f(0) = 1$.

4. If $f(z) = 1 + z + \dfrac{z^2}{2!} + \cdots + \dfrac{z^n}{n!} + \cdots$, then show $a + \displaystyle\int_0^z f(t)\,dt = f(z)$.

5. a If $f(z) = 1 - z + z^2 - z^3 \pm \cdots$, $|z| < 1$, then show $\int_0^z f(t)\,dt = z - \dfrac{z^2}{2} + \dfrac{z^3}{3} - \dfrac{z^4}{4} - \cdots$, $|z| < 1$.

6. Complete the proof of Corollary 2.5f by showing the radius of convergence of F is ρ.
7. Find a power series solution to the IVP $w' = 2zw$, $w(0) = 1$.
8. Find a power series solution to the IVP
   (a) $w'' = w$, $w(0) = 1$, $w'(0) = 0$.
   (b) $w''' = w$, $w(0) = 1$, $w'(0) = w''(0) = 0$.
   (c) $w^{(n)} = w$, $w(0) = 1$, $w'(0) = \cdots = w^{(n-1)}(0) = 0$.
   where $w^{(n)}$ means $(d/dz)^n w$.

2.6 Exponential, Trigonometric, and Hyperbolic Functions

From elementary calculus we recall that

$$e^x = 1 + x + \frac{x^2}{2!} + \frac{x^3}{3!} + \cdots + \frac{x^n}{n!} + \cdots = \sum_0^\infty \frac{x^n}{n!}.$$

Accordingly we *define* e^z by

$$e^z \equiv 1 + z + \frac{z^2}{2!} + \frac{z^3}{3!} + \cdots + \frac{z^n}{n!} + \cdots = \sum_0^\infty \frac{z^n}{n!} \tag{1}$$

for all complex z. since the radius of convergence ρ is ∞, we have e^z defined for all $z \in \mathbb{C}$. And, clearly, when $z = x$, we have $e^z = e^x$. That is, our definition of the complex exponential is compatible with the real exponential we already know. Thus we see that the solution to the initial value problem we solved at the end of the previous section is e^{az}.

Similarly, we define the sine and cosine by their series:

$$\sin z \equiv \sum_0^\infty \frac{z^{2n+1}(-1)^n}{(2n+1)!} = z - \frac{z^3}{3!} + - \cdots + \frac{(-1)^n z^{2n+1}}{(2n+1)!} + \cdots; \tag{2}$$

$$\cos z \equiv \sum_0^\infty \frac{(-1)^n z^{2n}}{(2n)!} = 1 - \frac{z^2}{2!} + \frac{z^3}{3!} - + \cdots + \frac{(-1)^n z^{2n}}{(2n)!} + \cdots. \tag{3}$$

We define the other complex trigonometric functions in terms of the sine and cosine:

$$\begin{aligned} \tan z &= \sin z / \cos z, & \cot z &= 1/\tan z \\ \sec z &= 1/\cos z, & \csc z &= 1/\sin z. \end{aligned} \tag{4}$$

It is clear that

$$\frac{d}{dz}\sin z = 1 - \frac{z^2}{2!} \pm \cdots + \frac{(-1)^n z^{2n}}{(2n)!} + \cdots$$

$$= \cos z$$

and similarly

$$\frac{d}{dz}\cos z = -\sin z,$$

so that all the trigonometric differentiation formulas are exactly the same as in elementary calculus.

The hyperbolic functions are defined in terms of the exponential:

$$\sinh z \equiv (e^z - e^{-z})/2 \qquad (5)$$

$$\cosh z \equiv (e^z + z^{-z})/2 \qquad (6)$$

and the others by

$$\tanh z \equiv \sinh z / \cosh z, \quad \coth z \equiv 1/\tanh z,$$
$$\text{sech } z \equiv 1/\cosh z, \qquad \text{csch } z \equiv 1/\sinh z. \qquad (7)$$

These formulas say what we mean by these functions, both trigonometric and hyperbolic, which by these definitions coincide with the corresponding real functions when z is real. What is not at all clear is the behavior of these functions in the complex plane off the real axis. We examine some of their properties now.

We begin with the exponential. One of the fundamental properties of the real exponential is that it satisfies $e^x e^y = e^{x+y}$ and we would hope that the complex exponential would satisfy the same law.

2.6a Theorem. $e^z e^\zeta = e^{z+\zeta}$.

Proof. By Theorem 2.4g (how does this theorem apply?) we get

$$e^z e^\zeta = \left(\sum \frac{z^k}{k!}\right)\left(\sum \frac{\zeta^k}{k!}\right) = \sum_{k=0}^{\infty}\left(\sum_{j=0}^{k} \frac{z^j}{j!}\frac{\zeta^{k-j}}{(k-j)!}\right)$$

$$= \sum_{k=0}^{\infty}\frac{1}{k!}\left(\sum_{j=0}^{k}\frac{k!}{j!(k-j)!}z^j\zeta^{k-j}\right) = \sum_{k=0}^{\infty}\frac{(z+\zeta)^k}{k!},$$

the last equality by the binomial theorem. But this last sum is just exactly $e^{z+\zeta}$. $\square$

2.6b Corollary. e^z *is never* 0.

Proof. Set $\zeta = -z$ in the previous theorem:

$$e^z e^{-z} = 1.$$

If there were a z for which $e^z = 0$, the product $e^z e^{-z}$ would be 0 and not 1. $\square$

2.6c Corollary. $e^{-z} = 1/e^z$.

Proof. Divide $e^z e^{-z} = 1$ by e^z. $\square$

We can now connect our function E defined by $E(\theta) = \cos\theta + i\sin\theta$ with the exponential.

2.6d Theorem. $E(y) = e^{iy}$ for real y.

Proof. By definition of e^z with $z = iy$

$$e^{iy} = \sum \frac{(iy)^n}{n!} = \sum_{n \text{ even}} \frac{(iy)^n}{n!} + \sum_{n \text{ odd}} \frac{(iy)^n}{n!}$$

$$= \sum_0^\infty \frac{(iy)^{2k}}{(2k)!} + \sum_0^\infty \frac{(iy)^{2k+1}}{(2k+1)!}.$$

But $i^{2k} = (i^2)^k = (-1)^k$, so

$$e^{iy} = \sum_0^\infty \frac{(-1)^k y^{2k}}{(2k)!} + i\sum_0^\infty \frac{(-1)^k y^{2k+1}}{(2k+1)!}$$

$$= \cos y + i\sin y = E(y). \quad \square$$

2.6e Corollary. $e^z = e^{x+iy} = e^x \cos y + i e^x \sin y$.

The proof of this separation of e^z into its real and imaginary parts is clear and left as an exercise (A2).

From the periodicity of the real trigonometric functions and their connections with the complex exponential, it is by now not surprising that it too is periodic. To be precise, when we say that an entire function has **period** p we mean that $f(z + p) = f(z)$ for each $z \in \mathbb{C}$.

2.6f Theorem. *(a) e^z has period $2\pi i$.*
(b) If p is a period of e^z, then $p = 2k\pi i$ for some integer k.

Proof. (a) We calculate

$$e^{z+2\pi i} = e^z e^{2\pi i} = e^z[\cos 2\pi + i \sin 2\pi] = e^z.$$

(b) Suppose p is a period of e^z. Then

$$e^{z+p} = e^z,$$

so, if $p = a + ib$,

$$e^p = e^{a+ib} = 1,$$

from which

$$|e^p| = e^a = 1$$

which implies $a = 0$. (Why?) Thus

$$e^{ib} = \cos b + i \sin b = 1,$$

so that

$$\cos b = 1, \qquad \sin b = 0$$

from which $b = 2k\pi$ for some integer k. $\square$

We have seen connections between the complex exponential and the real trigonometric functions. These persist with the complex trigonometric functions.

2.6g Theorem. (Euler Formulas)

$$e^{iz} = \cos z + i \sin z, \qquad e^{-iz} = \cos z - i \sin z,$$
$$\cos z = (e^{iz} + e^{-iz})/2, \qquad \sin z = (e^{iz} - e^{-iz})/2i.$$

Proof. The formula for e^{-iz} follows from that for e^{iz} by noting that by definition ((2) and (3)) $\cos z$ is even and $\sin z$ is odd. The formulas for $\cos z$ and $\sin z$ follow from the first two by adding and dividing by 2, and by subtracting and dividing by $2i$. It remains to establish the formula for e^{iz}.

We use the definition (1) with z replaced by iz. The calculations are then formally identical with the proof of Theorem 2.6d. $\square$

2.6h Theorem.

$$\sin iz = i \sinh z, \qquad \cos iz = \cosh z$$
$$\sinh iz = i \sin z, \qquad \cosh iz = \cos z$$

Proof. We prove the first formula. The others are similar:

$$\sin iz = (e^{i(iz)} - e^{-i(iz)})/2i = (e^{-z} - e^{z})/2i$$
$$= -i(e^{-z} - e^{z})/2 = i(e^{z} - e^{-z})/2 = i \sinh z. \quad \square$$

It is also true that *all* trigonometric identities, including the differentiations, which are valid for the real functions extend in exactly the same form to the complex functions.

2.6i Theorem. (Addition formula for the sine)

$$\sin(z + \zeta) = \sin z \cos \zeta + \cos z \sin \zeta.$$

Proof. By the Euler formulas

$$\sin z \cos \zeta = \frac{1}{4i} (e^{iz} - e^{-iz})(e^{i\zeta} + e^{-i\zeta})$$
$$= \frac{1}{4i} [e^{i(z+\zeta)} + e^{-i(z-\zeta)} - e^{i(z-\zeta)} - e^{-i(z+\zeta)}]$$

and

$$\cos z \sin \zeta = \frac{1}{4i} [e^{i(z+\zeta)} - e^{-i(z-\zeta)} + e^{i(z-\zeta)} - e^{-i(z+\zeta)}]$$

so

$$\sin z \cos \zeta + \cos z \sin \zeta = \frac{1}{2i} [e^{i(z+\zeta)} - e^{-i(z+\zeta)}]$$
$$= \sin(z + \zeta). \quad \square$$

The remaining identities follow easily from this one. Thus differentiating both sides with respect to ζ yields the addition formula for the cosine, namely,

$$\cos(z + \zeta) = \cos z \cos \zeta - \sin z \sin \zeta,$$

and the addition formulas for sinh and cosh follow from these by Theorem 2.6h.

Further, if we set $z = \zeta$ in the addition formula for the sine and $z = -\zeta$ in that for the cosine, we get

$$\sin 2z = 2 \sin z \cos z,$$

$$\cos^2 z + \sin^2 z = 1,$$

and all the others follow similarly.

We can also use Theorem 2.6i to separate $\sin z$ into its real and imaginary parts. With $z = x + iy$ we compute

$$\sin z = \sin(x + iy) = \sin x \cos iy + \cos x \sin iy$$

or

$$\sin z = \sin x \cosh y + i \cos x \sinh y \qquad (8)$$

(See Exercises B3,4; Section 2.1.) Similarly

$$\cos z = \cos x \cosh y - i \sin x \sinh y. \qquad (9)$$

One of the most emphasized and very heavily used properties of the real sine and cosine functions is that they are bounded by 1 in absolute value. This is *not* true of the sine and cosine in the complex plane. In fact each of these achieves every value in $\mathbb{C}$. We will not prove this here, but will examine an illustrative special case. The general case will be discussed in the next section.

Example 1. Find all solutions of the equation

$$\sin z = 2.$$

Solution. This is (by (8)) equivalent to

$$\sin x \cosh y = 2 \quad \text{and} \quad \cos x \sinh y = 0.$$

Then either

$$\cos x = 0 \quad \text{or} \quad \sinh y = 0.$$

But if $\sinh y = 0$, then $y = 0$ from which $\cosh y = 1$ which implies

$$|\sin x \cosh y| = |\sin x| \le 1 < 2,$$

and so $\sin x \cosh y \neq 2$. Thus we must have $\sinh y \neq 0$, and so must have $\cos x = 0$. This implies

$$x = (2n+1)\pi/2, \qquad n = 0, \pm 1, \pm 2, \ldots .$$

Further, since $\sin x \cosh y = 2$, and $\cosh y \geq 1 > 0$, we must have $\sin x > 0$, which eliminates the odd values of n and gives

$$x = (4k+1)\pi/2, \qquad k = 0, \pm 1, \pm 2, \ldots .$$

Thus $\sin x = 1$ and we have left to solve

$$\cosh y = 2.$$

Since the cosh is even, we have two solutions

$$y = \pm \cosh^{-1} 2$$

so that finally

$$z = (4k+1)\pi/2 \pm i \cosh^{-1} 2, \qquad k = 0, \pm 1, \pm 2, \ldots$$

represents all solutions of our equation.

Example 2. Solve the differential equation

$$w'' + aw' + bw = 0.$$

Solution. We have seen that the equation $w' = aw$ has a solution e^{az}, so we try $w(z) = e^{\lambda z}$ where λ is an unknown constant. For $e^{\lambda z}$ to be a solution we must have

$$\lambda^2 e^{\lambda z} + a\lambda e^{\lambda z} + be^{\lambda z} = 0,$$

and since $e^{\lambda z}$ is never zero, λ must be chosen so that

$$\lambda^2 + a\lambda + b = 0.$$

Thus we get two values of λ, namely

$$\lambda_1 = \frac{-a + \sqrt{a^2 - 4b}}{2}, \qquad \lambda_2 = \frac{-a - \sqrt{a^2 - 4b}}{2}$$

With these two values of λ we have both $w_1(z) = e^{\lambda_1 z}$ and $w_2(z) = e^{\lambda_2 z}$ are solutions. So also is

$$w(z) = C_1 e^{\lambda_1 z} + C_2 e^{\lambda_2 z}$$

for each pair of constants C_1, C_2.

Significant Topics

A Exercises

1. Show that $e^z = e^\alpha \sum_0^\infty \dfrac{(z-\alpha)^k}{k!}$. Hint: $e^z = e^\alpha e^{z-\alpha}$.

2. Prove Corollary 2.6e.

3. Separate $\sinh z$ and $\cosh z$ into real and imaginary parts; that is, write each in the form $u + iv$.

4. Show that

$$\lim_{y \to \infty} \frac{\cos z}{e^y} = \frac{1}{2} e^{-iz}; \qquad \lim_{y \to -\infty} \frac{\cos z}{e^{-y}} = \frac{1}{2} e^{iz}.$$

5. Replace $\cos z$ by $\sin z$ in A4 and compute the limits.

6. (a) Show that $\tan z$ is analytic everywhere except $z = \left(n + \frac{1}{2}\right)\pi$, $n = 0$, $\pm 1, \pm 2, \ldots$.

   (b) Show that $\cot z$ is analytic everywhere except $z = n\pi$, $n = 0$, $\pm 1, \pm 2, \ldots$.

   (c) Where do $\tanh z$ and $\coth z$ fail to be analytic?

7. (a) Show that $\cot z$ has the form

$$\cot z = \frac{1}{z} + g(z), \qquad 0 < |z| < R$$

where $g(z)$ is analytic for $|z| < R$.

(b) How big can R be chosen?

8. In Example 1, the answer involved $\pm \cosh^{-1} 2$. Show that $\cosh^{-1} 2 = \ln(2 + \sqrt{3})$ and that $-\cosh^{-1} 2 = \ln(2 - \sqrt{3})$.

9. If a is a positive constant, sketch the graph of

$$\left.\begin{array}{ll} \text{(a)} & z = ae^{i\theta}, \\[2mm] \text{(b)} & z = \alpha + ae^{i\theta}, \end{array}\right\} \quad 0 \le \theta \le 2\pi.$$

10. Compute the limit as $z \to 0$ of

$$\text{(a)} \ \frac{e^z - 1}{z}, \qquad \text{(b)} \ \frac{\sin z}{z}, \qquad \text{(c)} \ \frac{1 - \cos z}{z}, \qquad \text{(d)} \ \frac{1 - \cos z}{z^2}.$$

11. Show that

$$\text{(a)} \ \frac{d}{dz} \tan z = \sec^2 z, \qquad \frac{d}{dz} \cot z = -\csc^2 z.$$

12. Find all solutions of
   (a) $e^z = 1 - i\sqrt{3}$,   (b) $\cos z = 2$

13. Discuss the periods of
   (a) $\sinh z$ (b) $\tanh z$

14. Show that $\tan z = i$ and $\tan z = -i$ have no solutions.

15. Show $|e^z| = e^x \le e^{|x|} \le e^{|z|}$.

16. (a) Evaluate $\displaystyle\int_\alpha^\beta e^z dz$ where the integral is over any path from α to β.

   (b) Similarly evaluate $\displaystyle\int_\alpha^\beta \sin z dz$ and $\displaystyle\int_\alpha^\beta \cos z dz$.

B Exercises

1. Show

$$\text{(a)} \ |\sin z|^2 = \sin^2 x + \sinh^2 y = \cosh^2 y - \cos^2 x$$

$$= \frac{1}{2}(\cosh 2y - \cos 2x)$$

$$\text{(b)} \ |\cos z|^2 = \cos^2 x + \sinh^2 y = \cosh^2 y - \sin^2 x$$

$$= \frac{1}{2}(\cosh 2y + \cos 2x)$$

   (c) $|\sinh z|^2 = \sinh^2 x + \sin^2 y.$
   (d) $|\cosh z|^2 = \sinh^2 x + \cos^2 y.$

2. Show, for any integer n, positive, negative or zero, that

$$e^z/z^n \to \infty \quad \text{as} \quad z \to \infty \quad \text{in} \quad |\arg z| \le \frac{\pi}{2} - \varepsilon$$

$$e^z/z^n \to 0 \quad \text{as} \quad z \to \infty \quad \text{in} \quad |\arg -z| \le \frac{\pi}{2} - \varepsilon$$

for each ε, $0 < \varepsilon < \frac{\pi}{2}$.

3. Show that $\tan z = \dfrac{\sin 2x + i \sinh 2y}{\cos 2x + \cosh 2y}$.

4. (a) Solve the differential equation

$$w''(z) = w(z).$$

   (b) Solve the IVP

$$w''(z) = w(z), \quad w(0) = 1, \quad w'(0) = 0.$$

   (c) Show

$$|e^z - 1| \le e^{|z|} - 1$$

$$|e^z - z - 1| \le e^{|z|} - |z| - 1$$

$$\left| \frac{e^z - 1}{z} \right| \le e^{|z|}.$$

5. (a) Prove if $m > n$

$$\sum_{k=n}^{m} z^k = \frac{z^n - z^{m+1}}{1 - z} \quad \text{if} \quad z \ne 1.$$

   (b) Set $z = e^{i\theta}$, $n = 0$ in (a) and get

$$\sum_0^m e^{ik\theta} = \frac{1 - e^{i(m+1)\theta}}{1 - e^{i\theta}}.$$

   Deduce

$$\frac{1}{2} + \sum_0^m \cos k\theta = \frac{\sin(m + \frac{1}{2})\theta}{2\sin(\frac{1}{2}\theta)}$$

   and

$$\sum_1^m \sin k\theta = \frac{\sin(\frac{1}{2}m\theta)\sin\frac{1}{2}(m+1)\theta}{\sin\frac{1}{2}\theta}.$$

   (c) Set $z = e^{2i\theta}$, $n = 0$ in (a), then multiply by $e^{i\theta}$ and deduce

$$\sum_0^{n-1} \cos(2k + 1)\theta = \frac{\sin 2n\theta}{2\sin\theta}.$$

6. Solve $w'' - 2aw' + a^2w = 0$. By the method of Example 2 we get only $w = e^{az}$. Find a second solution by setting $w(z) = u(z)e^{az}$ and solving for $u(z)$.

C Exercise

1. (a) Solve the differential equation

$$w'''(z) = w(z).$$

(b) Solve the IVP

$$w'''(z) = w(z), \quad w(0) = 1, \quad w'(0) = w''(0) = 0.$$

(c) Sum the series

$$\sum_0^\infty \frac{z^{3n}}{(3n)!} = 1 + \frac{z^3}{3!} + \frac{z^6}{6!} + \cdots .$$

2.7 Logarithm, Exponents

We want to introduce the complex logarithm function and explore its connections with multivaluedness of analytic functions, with the real logarithm, with the complex exponential function, and with the general power function. Before we begin, we say a little about notation. We will use the symbol "ln" to represent the real natural logarithm, familiar from elementary calculus, and defined on the positive reals by

$$\ln x = \int_1^x \frac{dt}{t}, \quad x > 0.$$

We will use the symbols "log" and "Log" for the complex logarithm and its principal branch, both of which we will define later.

We start by reconsidering Theorem 1.5c. Our notation is slightly different from that example since we now will use $e^{i\theta}$ rather than $E(\theta)$ which was used there. We suppose that C is a path (piecewise smooth curve) from α to β given by

$$\zeta = \phi(t) = r(t)e^{i\theta(t)}, \quad t \in I = [a, b] \tag{1}$$

with $\phi(a) = \alpha$, $\phi(b) = \beta$. Thus we assume that $r(t)$ and $\theta(t)$ are continuous and have piecewise continuous derivatives. We also assume that C does not pass through the origin, that is, $r(t)$ is never zero.

It follows that $\theta(a)$ is a value of $\arg \alpha$ and $\theta(b)$ is a value of $\arg \beta$ and using these values we have

$$\Delta_C \theta = \theta(b) - \theta(a) = \arg \beta - \arg \alpha$$

is the (net) change in $\theta(t) = \arg \phi(t)$ as t goes from a to b. That is, $\Delta_C \theta$ is the change in $\arg \zeta$ as the tracing point $\zeta = \phi(t)$ traces out C from α to β.

Then we calculate

$$\begin{aligned}
\int_C \frac{d\zeta}{\zeta} &= \int_a^b \frac{\phi'(t)dt}{\phi(t)} = \int_a^b \frac{r'(t)e^{i\theta t} + ir(t)e^{i\theta(t)}\theta'(t)}{r(t)e^{i\theta(t)}} \, dt \\
&= \int_a^b \frac{r'(t)}{r(t)} \, dt + i \int_a^b \theta'(t)dt = \left. (\ln r(t) + i\theta(t)) \right|_a^b \\
&= (\ln|\beta| + i \arg \beta) - (\ln|\alpha| + i \arg \alpha) \\
&= \ln|\beta| - \ln|\alpha| + i\Delta_C\theta.
\end{aligned}$$

Thus we have proved the following.

2.7a Lemma. *If P is a path from α to β not passing through the origin, then, taking the integral over P,*

$$\int_\alpha^\beta \frac{d\zeta}{\zeta} = \ln|\beta| - \ln|\alpha| + i\Delta_P\theta$$

This means, as in Theorem 1.5c just referred to, that *the value of the integral depends on the path of integration* in that $\Delta_P\theta$ can differ from one path to another. We are particularly interested in the case where $\alpha = 1$ and $\beta = z$. Then we define the **logarithm function** by

$$\log z \equiv \int_1^z \frac{d\zeta}{\zeta} = \ln|z| + i \arg z. \tag{2}$$

The triple barred equality in (2) means definition, and the doubled barred one is by the lemma. If we set $z = re^{i\theta}$, this yields

$$\log z = \ln r + i\theta. \tag{3}$$

Since θ is determined only up to a multiple of 2π, the values of $\log z$ arising from different paths of integration differ by multiples of $2\pi i$. Hence we can write this equation in the form

$$\log z = \ln r + i\Theta + 2k\pi i$$

where $\Theta = \operatorname{Arg} z$, and the dependence on the path is now expressed by k. The figures illustrate cases where $k = 0, -1, 2$. We leave it to you to identify which figure goes with which value of k.

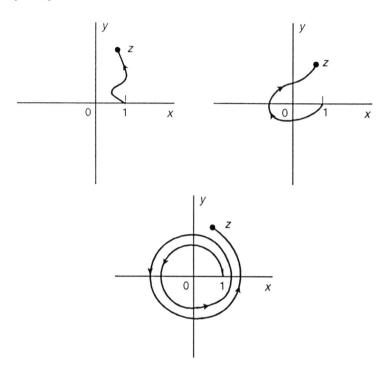

2.7b Theorem. $\log z\zeta = \log z + \log \zeta$.

This means that given a value of $\log z$ and a value of $\log \zeta$, there is a value of $\log z\zeta$ for which this equation is correct.

Proof. Let $z = re^{i\theta}$, $\zeta = \rho e^{i\phi}$. Then $z\zeta = r\rho e^{i(\theta+\phi)}$ and so

$$\log(z\zeta) = \ln(r\rho) + i(\theta + \phi)$$
$$= \ln r + \ln \rho + i\theta + i\phi$$
$$= (\ln r + i\theta) + (\ln \rho + i\phi)$$
$$= \log z + \log \zeta. \quad \square$$

2.7c Corollary. $\log \dfrac{z}{\zeta} = \log z - \log \zeta$.

The proof is left to the exercises.

2.7d Lemma. $\log z$ *is continuous except for* $z = 0$.

Proof. Even though $\log z$ is an ambiguous symbol this lemma makes sense in the following way. Let $z_0 \neq 0$ be given and $\log z_0$ be one value for $\log z$ at z_0. Then within a neighborhood of z_0 of radius less than $|z_0|$, (to avoid the origin) we determine values of $\log z$ by

$$\log z - \log z_0 = \int_{z_0}^{z} \frac{d\zeta}{\zeta}$$

where the integral is over the segment from z to z_0. Then $\log z - \log z_0 \to 0$ as $z \to z_0$ since the integral has limit zero. $\square$

2.7e Theorem. $\dfrac{d}{dz} \log z = \dfrac{1}{z}$.

Proof. Just as in the previous lemma, we examine

$$\frac{\Delta \log z}{\Delta z} - \frac{1}{z_0} = \frac{\log z - \log z_0}{z - z_0} - \frac{1}{z_0} = \frac{1}{z - z_0} \int_{z_0}^{z} \frac{d\zeta}{\zeta} - \frac{1}{z_0}$$

$$= \frac{1}{z - z_0} \int_{z_0}^{z} \left(\frac{1}{\zeta} - \frac{1}{z_0} \right) d\zeta = \frac{1}{z - z_0} \int_{z_0}^{z} \frac{z_0 - \zeta}{z_0 \zeta} \, d\zeta$$

Now on the segment from z to z_0, we have $|z_0 - \zeta| \leq |z - z_0|$, and for $|z| > \frac{1}{2}|z_0|$ we have $\left| \dfrac{1}{\zeta z_0} \right| \leq \dfrac{2}{|z_0|^2}$ so we get

$$\left| \frac{\Delta \log z}{\Delta z} - \frac{1}{z_0} \right| \leq \frac{1}{|z - z_0|} \cdot |z - z|0| \cdot \frac{2}{|z_0|^2} \cdot |z - z_0|$$

$$= \frac{2|z - z_0|}{|z_0|^2},$$

and this $\to 0$ as $z \to z_0$. $\square$

In both the meaning and proofs of 2.7d and e we used the idea of separating out and isolating a single valued piece of the multivalued function $\log z$. This is a very fruitful way to study multivalued functions, and we proceed to examine this notion more carefully.

When we isolate a single valued piece of $\log z$, and, as we will later, of other multivalued functions, it is called a **branch** of the function. The standard way to isolate single valued branches is by the use of **branch cuts** to separate different branches. For $\log z$ the ambiguity — multivaluedness — arises when z goes around the origin, for then θ changes by 2π. Such a point is called a **branch point**. So if we were prevented from going around a branch point we would not encounter this ambiguity.

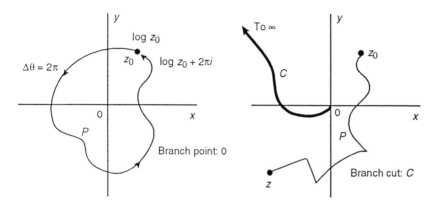

Let C be any simple curve from 0 to ∞, so that z cannot go around 0 without crossing C. For any point z_0 not on C we choose any one of the possible values of $\log z_0$, and for any other z not on C we have exactly one value of $\log z$ determined by

$$\log z - \log z_0 = \int_{z_0}^{z} \frac{d\zeta}{\zeta} \tag{4}$$

where the path of integration P *does not cross* C. Such a curve C is called a **branch cut**. For each cut C there is a branch of $\log z$ corresonding to each separate value for $\log z_0$. Since the values of $\log z_0$ differ by multiples of $2\pi i$, we have that any two branches of $\log z$ given by (4) with the same branch cut C differ by a multiple of $2\pi i$.

Since $\log z = \ln r + i\theta$, it is frequently convenient to take a ray from 0 to ∞ as our branch cut. For example if we take the positive y axis as a cut, we have that

$$\log z = \ln r + i\theta, \qquad -\frac{3\pi}{2} < \theta < \frac{\pi}{2}$$

is a single valued branch. So are

$$\log z = \ln r + i\theta, \qquad -\frac{11\pi}{2} < \theta < -\frac{7\pi}{2}$$

and

$$\log z = \ln r + i\theta, \qquad \frac{\pi}{2} < \theta < \frac{5\pi}{2}$$

We define the **principal value** or **principal branch**, denoted by Log z, by taking the negative real axis as a branch cut. Thus we define Log z by

$$\text{Log } z = \ln r + i\theta, \qquad -\pi < \theta \le \pi \tag{5}$$

or, equivalently

$$\text{Log } z = \ln r + i\Theta$$

where Θ is the principal value of $\arg z$. You should note that we have included the cut itself ($\Theta = \pi$) so that, for example

$$\text{Log}(-2) = \ln 2 + i\pi,$$

and

$$\text{Log}(-1) = i\pi,$$

while

$$\log(-2) = \ln 2 + i\pi + 2k\pi i = \ln 2 + (2k+1)\pi i$$

and

$$\log(-1) = i\pi + 2k\pi i = (2k+1)\pi i.$$

In the real case ln and exp are inverses of each other. That is

$$\ln e^x = x, \quad \text{and} \quad e^{\ln x} = x \quad \text{for} \quad x > 0.$$

With appropriate adjustments for the multivaluedness of $\log z$ and the periodicity of e^z, similar relations hold.

2.7f Theorem.
  (a) $e^{\log z} = z$, *for all* $z \ne 0$;
  (b) $\log e^z = z + (2k\pi)i$, *for all* z.

Proof. (a) Let $z = re^{i\theta}$. Then

$$\log z = \ln r + i\theta + 2k\pi i$$

and

$$e^{\log z} = e^{\log r + i\theta + 2k\pi i}$$
$$= e^{\log r} \cdot e^{i\theta} \cdot e^{2k\pi i}$$
$$= r \cdot e^{i\theta} \cdot 1 = re^{i\theta} = z. \quad \square$$

(b) Let $z = x + iy$. Then $e^z = e^x \cdot e^{iy}$. and

$$\log e^z = \log(e^x \cdot e^{iy}) = \ln e^x + iy + 2k\pi i$$
$$= x + iy + 2k\pi i = z + 2k\pi i.$$

Example 1. Discuss the multivaluedness of $f(z) = \log[z/(z-1)]$ and place cuts to get single valued branches.

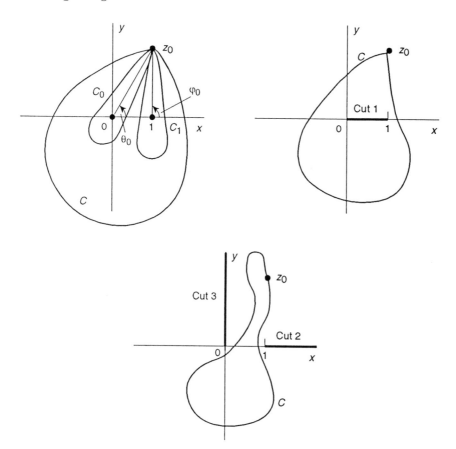

Solutions. We write both z and $z - 1$ in polar form:

$$z = re^{i\theta}, \qquad z - 1 = \rho e^{i\phi}.$$

Then we can write

$$f(z) = \log(re^{i\theta}/\rho e^{i\phi}) = \log[(r/\rho)e^{i(\theta-\phi)}]$$
$$f(z) = \ln(r/\rho) + i(\theta - \phi) \tag{6}$$

We choose a branch of $f(z)$ at a point z_0 by choosing values θ_0 of θ and ϕ_0 of ϕ. Then at z_0, $f(z)$ has the value

$$\ln(r/\rho) + i(\theta_0 - \phi_0) \equiv \alpha_0$$

which we have called α_0. As z circumscribes (encircles, goes around) the origin 0 on a curve such as C_0 in the figure so that $0 \in I(C_0)$ and $1 \in E(C_0)$, then $\theta = \arg z$ returns to the value $\theta_0 + 2\pi$, while ϕ returns to ϕ_0 so that $f(z)$ returns to the value

$$\ln(r/\rho) + i(\theta_0 + 2\pi - \phi_0) = \alpha_0 + 2\pi i$$

Similarly if z circumscribes the point 1 but not 0, as on C_1, then $f(z)$ returns to the value $\alpha_0 - 2\pi i$. And if z circumscribes both 0 and 1, it returns to the value

$$\ln(r/\rho) + i[(\theta_0 + 2\pi) - (\phi_0 + 2\pi)] = \alpha_0$$

So a branch determined at z_0 is single valued if we introduce a branch cut between the branch points 0 and 1 as in the middle figure. Or also if we make two cuts to ∞, one from each branch point as in the right hand figure.

We make use of the logarithm to define general powers, that is, to define z^α where both z and α are complex numbers. We anticipate certain ambiguity — multivaluedness — since we already have seen that $z^{p/q}$ is in general multivalued.

Let us recall that for a real exponent a acting on a positive number x we have $x^a = \exp(a \ln x)$. To see this we set

$$b = x^a$$

then, taking logarithms of both sides,

$$\ln b = \ln x^a = a \ln x$$

and so, since the exponential is the inverse of the logarithm,

$$b = e^{\ln b} = e^{a \ln x}$$

But $b = x^a$ so that

$$x^a = e^{a \ln x}.$$

We also recall that we have defined the complex exponential by

$$e^\alpha = \sum_0^\infty \frac{\alpha^n}{n!} = 1 + \alpha + \frac{\alpha^2}{2!} + \cdots + \frac{\alpha^n}{n!} + \cdots$$

and so we explicitly except e from the following.

Definition. *For $z \neq 0$, and for $z \neq e$ we define z^α by*

$$z^\alpha \equiv e^{\alpha \log z}$$

and its principal value by
$$e^{\alpha \mathrm{Log}\, z}$$

(*Unfortunately there is no standard notation for the principal value of z^α.*)

It is clear that if we had taken e as an admissible value of z in our definition of z^α, then the usual definition, i.e., the power series for e^α, would correspond to the principal value. For a number of reasons, mostly notational clarity, it is desirable that the symbol e^α be unequivocal.

There is immediately the question of compatibility: Does the definition just given agree with z^n and $z^{p/q}$ as previously defined. The general cases we leave to the exercises. Now we consider a very special case.

Example 2. We have previously defined z^2 to be $z \cdot z$ and now to be $e^{2 \log z}$. Are these in fact the same?

Solution. Let $\Theta = \mathrm{Arg}\, z$ and $r = |z|$. Then

$$e^{2 \log z} = e^{2 \ln r + 2i\Theta + 4k\pi i}$$
$$= e^{2 \ln r} \cdot e^{i(\Theta + \Theta)} \cdot 1$$
$$= r^2 e^{i\Theta} e^{i\Theta} = (re^{i\Theta}) \cdot (re^{i\Theta}) = z \cdot z.$$

In this calculation we have used $e^{2 \ln r} = r^2$ which is known for real r.

2.7g Theorem. *Each branch of z^α is analytic as a function of z for $z \neq 0$, and as a function of α, and*

$$\text{(i)} \quad \frac{d}{dz} z^\alpha = \alpha z^{\alpha - 1}; \qquad \text{(ii)} \quad \frac{d}{d\alpha} z^\alpha = z^\alpha (\log z).$$

where the same branch of $\log z$ occurs on both sides of (i), and the same branch of $\log z$ occurs on both sides of (ii).

Proof. If a branch of $\log z$ is chosen, then

$$\frac{d}{dz}(z^\alpha) = \frac{d}{dz}(e^{\alpha \log z}),$$

which, by the chain rule gives

$$\frac{d}{dz}(z^\alpha) = e^{\alpha \log z}\alpha\frac{1}{z} = \alpha\frac{z^\alpha}{z} = \alpha z^{\alpha-1}.$$

Thus the *formula* is the same as in elementary calculus. We must just be careful of the meaning of the symbols. Further

$$\frac{d}{d\alpha}(z^\alpha) = \frac{d}{d\alpha}(e^{\alpha \log z}) = e^{\alpha \log z}(\log z)$$
$$= z^\alpha(\log z),$$

where again the same caveats apply as in the previous calculations. $\square$

You should note that once a branch of $\log z$ is determined, then z^α is an entire function of α given by

$$e^{\alpha \log z} = \sum_0^\infty \frac{\alpha^n(\log z)^n}{n!}.$$

Let us examine the general formula for z^α. From the definition we have

$$z^\alpha = e^{\alpha \log z} = e^{(a+ib)(\ln r + i\theta)}$$
$$= e^{a \ln r - b\theta} \cdot e^{i(b \ln r + a\theta)}$$
$$= (r^a e^{-b\theta})e^{i(b \ln r + a\theta)}$$

where r^a means the real positive power. Thus

$$|z^\alpha| = r^a e^{-b\theta} = r^a e^{-b\Theta} e^{-2kb\pi},$$

so that the modulus changes as k changes, that is from one branch to another whenever $b \neq 0$. Also,

$$\arg(z^\alpha) = b \ln r + a\theta = b \ln r + a\Theta + 2ka\pi,$$

and so the argument also changes from one branch to another whenever $a \neq 0$.

We look specifically at the case $b = 0$, that is, when $\alpha = a$ is real. Then

$$z^\alpha = z^a = r^a e^{ia\theta}$$

and we want to examine how this changes as z moves in the plane.

If z starts at a point $z_0 = r_0 e^{i\theta_0}$, then as z circumscribes the origin in the positive direction, θ increases from θ_0 to $\theta_0 + 2\pi$, z^a starts with the value $r_0^a e^{ia\theta_0}$ and changes to

$$r_0^a e^{ia(\theta_0 + 2\pi)} = r_0^a e^{ia\theta_0} \cdot e^{2a\pi i}.$$

Thus it picks up a factor of $e^{2a\pi i}$, a unit vector. A second circuit around the origin brings the value of z^a to $(re^{ia\theta_0})e^{4a\pi i}$, etc.

Example 3. Let $f(z) = \sqrt{(z-1)(z-2)}$. Choose the branch of $f(z)$ that is positive for $z = x > 2$, and show that branch is single valued for $|z| > 2$.

Solution. Let us write each of $z - 1$ and $z - 2$ in polar form:

$$z - 1 = re^{i\theta}, \qquad z - 2 = \rho e^{i\phi}.$$

Then

$$f(z) = (re^{i\theta} \rho e^{i\phi})^{1/2} = [r\rho e^{i(\theta+\rho)}]^{1/2} \tag{7}$$

or

$$f(z) = (r\rho)^{1/2} e^{i(\theta+\phi)/2}$$

Now as z starts from the positive x-axis beyond 2, then $f(z)$ starts with the positive value $(r\rho)^{1/2}$. As z goes once around the origin outside the circle $|z| = 2$ and returns to the positive x-axis both θ and ϕ increase by 2π so that $f(z)$ comes back to the value

$$(r\rho)^{1/2} e^{i(2\pi+2\pi)/2} = (r\rho)^{1/2} e^{2\pi i} = (r\rho)^{1/2}.$$

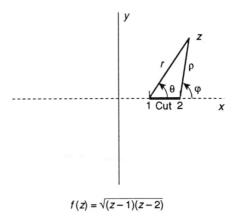

$$f(z) = \sqrt{(z-1)(z-2)}$$

From our formula (7) for $f(z)$ we easily see that it is *not* the circle $|z| = 2$ that is crucial here, but the points $z = 1$ and $z = 2$: If z goes once around either 1 or 2, but not the other, then the corresponding value of θ or ϕ changes by 2π so that by (4) $f(z)$ changes sign. But if z goes around both 1 and 2, then both θ and ϕ increase by 2π and $f(z)$ returns to its original value. Thus if we make a cut between 1 and 2, then clearly $f(z)$ is single valued outside that cut.

Example 4. Show that $\sin z$ achieves every value in $\mathbb{C}$. That is, for each α,

$$\sin z = \alpha$$

has solutions.

Solution. From the Euler formula for $\sin z$ we get

$$e^{iz} - e^{-iz} = 2i\alpha,$$

$$(e^{iz})^2 - 2i\alpha(e^{iz}) - 1 = 0,$$

$$e^{iz} = \frac{2i\alpha \pm \sqrt{(2i\alpha)^2 + 4}}{2}$$

$$= i\alpha \pm \sqrt{1 - \alpha^2}.$$

From this

$$z = -i\log(i\alpha \pm \sqrt{1 - \alpha^2}).$$

A Exercises

1. Find all values of the $\log z$ if z is
   (a) 1, (b) i, (c) $1 + i$, (d) $1 + \sqrt{3}\, i$
2. Find $\text{Log}\, z$ for each part of 1.
3. Find all values of
   (a) 1^π, (b) $1^{i\pi}$, (c) i^i, (d) $(1 - i)^{(1-i)}$, (e) $(1 + i)^i$.
4. Find the principal value of each part of 3.
5. Show that the principal value of $1^\alpha = 1$ for each α.
6. Show that $z^0 = 1$, for $z \neq 0$.

B Exercises

1. From the geometric series

$$1/(1 - z) = \sum_0^\infty z^k = 1 + z + z^2 + \cdots , \quad |z| < 1,$$

deduce

$$-\text{Log}\,(1 - z) = \sum_1^\infty z^k/k = z + \frac{z^2}{2} + \frac{z^3}{3} + \cdots , \quad |z| < 1.$$

2. Show that each branch of $\sqrt{z^2 - 1}$ is single valued in the plane cut along the real axis from -1 to $+1$.
3. We define $\sin^{-1} z = w$ to mean $\sin w = z$. We similarly define $\cos^{-1} z$, $\tan^{-1} z$, etc. Show
   (a) $\sin^{-1} z = -i \log[iz + (1 - z^2)^{1/2}]$;
   (b) $\tan^{-1} z = (i/2) \log[(i + z)/(i - z)]$.

(c) Find $\cosh^{-1} z$.

4. (a) From the formula for $\tan^{-1} z$ show that each branch is single valued in the plane cut from $-i$ to i.

  (b) Show that for each branch of $\tan^{-1} z$, $\dfrac{d}{dz} \tan^{-1} z = \dfrac{1}{1 + z^2}$, and use this formula to show that one branch is

$$\tan^{-1} z = \int_0^z \frac{dt}{1 + t^2}$$

  in the plane cut from i to ∞, and from $-i$ to ∞ along the imaginary axis.

5. Prove Corollary 2.7c.

6. Show

  (a) If any two logarithms in the formula of Theorem 2.7b are chosen, then there is a value of the other for which the formula holds.

  (b) If all the logarithms are taken as principal values, the formula is not always correct.

7. Complete the discussion of Example 1 by discussing the changes in θ and ϕ as z moves around each contour in the figures associated with that example.

8. Consider $\lim\limits_{z \to 0} z^a = L$ for a real. Show

  (a) $L = 0$ if $a > 0$.

  (b) L does not exist if $a < 0$.

9. Show that $e^{n \log z} = z^n$, $e^{(p/q) \log z} = z^{p/q}$.

10. Explain why any simple curve connecting 0 to 1 can be a branch cut in Example 1.

11. Let R be the plane cut along the negative real axis, and let $z^{1/2}$ be the principal value of $\sqrt{z}$ in R. Evaluate $\displaystyle\int_1^a z^{1/2} dz$ where the integral is over a path in R.

12. Show that the only zeros of $\sin z$ are $k\pi$, and of $\cos z$ are $(k + \frac{1}{2})\pi$, $k = 0, \pm 1, \pm 2, \ldots$.

13. If C is the circle $|z| = r$, show that

$$\frac{1}{2\pi i} \int_C \frac{\operatorname{Log} z}{z} \, dz = \ln r.$$

14. (See Example 4.) Show that

$$1/(i\alpha + \sqrt{1 - \alpha^2}) = i\alpha - \sqrt{1 - \alpha^2}.$$

15. Show that $\cos z$ achieves every value in $\mathbb{C}$.

C Exercises

1. Show that each branch of $(e^z)^{1/2}$ is entire.
2. Show that each branch of $\log(\cos z)$ is single valued in the plane cut from $-\infty$ to $-\pi/2$ and from $\pi/2$ to ∞, both cuts being on the real axis.
3. Suppose $f(z)$ is analytic and never zero in a simply connected region R. Show that each branch of $\log f(z)$ is single valued in R. Hint: Choose $\alpha \in R$ and consider $\int_\alpha^z f'(\zeta)/f(\zeta)d\zeta$.
4. Show that
   (a) $\dfrac{1}{z}\text{Log}\,(1+z) \to 1$ as $z \to 0$, and hence
   (b) $n\text{Log}\,(1+\dfrac{z}{n}) \to z$ as $n \to \infty$, and hence
   (c) $(1+\dfrac{z}{n})^n \to e^z$ as $n \to \infty$, and further that, for each $a > 0$,
   (d) $(1+\dfrac{z}{n})^n \to e^z$ uniformly for $|z| \le a$.
5. Let $P_n(z) = \sum_0^n z^k/k!$, which is a partial sum of the series defining e^z. Show that for each $a > 0$ there is an N for which all the roots of $P_n(z) = 0$ lie outside the circle $|z| = a$ if $n > N$.
6. Show that $f(z)$ in Exercise A1 of Section 2.5 is the principal value of $(1+z)^\alpha$.
7. Suppose $f(z)$ and $g(z)$ are both analytic in a simply connected region R and are never zero there. If $|f(z)| \equiv |g(z)|$, then show that there is a real constant α for which $f(z) \equiv g(z)e^{i\alpha}$. Hint: Use C3 above.
8. Suppose $L(z)$ is defined in the cut plane described by $|\arg z| < \pi$ and
   (i) $L(z\zeta) = L(z) + L(\zeta)$ whenever $|\arg \zeta| < \pi$, $|\arg \zeta| < \pi$, and $|\arg z\zeta| < \pi$.
   (ii) $\lim\limits_{z\to 0}\dfrac{1}{z} L(1+z) = 1$.
   Show that $L(z)$ is analytic in the cut plane and that $L(z) \equiv \text{Log}\,(z)$.
9. If $f(z)$ is entire and never zero and $|f(z)|$ is in the form $h(x)g(y)$, then show

$$f(z) = e^{az^2+bz+c}$$

where a is real. Hint: Use C1, Section 2.1, and C3 above.

2.8 Riemann Surfaces

In dealing with multivalued functions we have seen the utility of isolating single valued branches. We now discuss, in a heuristic way, how we can connect the branches to form what is called a **Riemann surface**. This is a device for viewing the function as a whole as being single valued on a "surface". To understand this "glue and scissors" approach we *urge* the student to use several sheets of paper and physically cut and glue, following directions as they are described.

We first discuss $\log z$ and we begin with three branches. So take three sheets of paper, draw in the coordinate axes, and physically cut them along the negative real axis corresponding to a branch cut there. We will label the sheets S_0, S_1 and S_{-1}, and superimpose them with their axes coinciding, so that when viewed from above a point $z = re^{i\theta}$ appears in the same position on all three sheets. On sheet S_0 we will assign the values

$$\text{Log } z = \ln r + i\theta, \qquad -\pi < \theta \leq \pi$$

to each position z. On S_1 we will assign

$$\log_1 z = \ln r + i\theta, \qquad \pi < \theta \leq 3\pi;$$

To each z, and on S_{-1}

$$\log_{-1} z = \ln r + i\theta, \qquad -3\pi < \theta \leq -\pi,$$

Let us choose a ppint $z = re^{i\theta}$ on the positive real axis and follow the values of $\log z$ as z movees. We start on sheet S_0, and $\log z$ then has the value $\ln r (= \text{Log } z)$ on the real axis. As θ increases, z moves counterclockwise and as it crosses the negative real axis $\log z$ changes continuously from $\text{Log } z$ to $\log_1 z$, so we glue the upper edge ($\theta < \pi$) of the cut on S_0 to the lower edge ($\theta > \pi$) of the cut on S_1. The two glued sheets form a single surface and as z moves on this surface the two branches $\text{Log } z$ and $\log_1 z$ form a one single, single-valued function on this surface.

Now start z on the real axis on S_0 again, and let it move *clockwise*. As z crosses the negative real axis, $\log z$ changes from $\text{Log } z$ to $\log_{-1} z$. Then we glue the lower edge ($\theta > -\pi$) of S_0 to the upper edge ($\theta < -\pi$) of S_{-1}. Then the three branches of $\log z$ form a single-valued function on the resulting 3-ply surface and, of course, has the value $\text{Log } z$ on S_0, $\log_1 z$ on S_1 and $\log_{-1} z$ on S_{-1}.

If we take another copy of the z-plane (another sheet of paper), cut along the negative real axis, label it S_2, and implant on it the values

$$\log_2 z = \ln r + i\theta, \qquad 3\pi < \theta \le 5\pi$$

we can attach it to S_1 so the resulting function is continuous on all four sheets. Similarly, on a fifth copy, S_{-2}, we implant

$$\log_{-2} z = \ln r + i\theta, \qquad -5\pi < \theta \le -3\pi$$

and attach S_{-2} to S_{-1}.

Continuing in this way we build up a spiral of sheets so that $\log z$ has the values

$$\log_k z = \ln r + i\theta, \qquad -\pi + 2k\pi < \theta \le \pi + 2k\pi$$

on S_k, and is continuous on this layered infinite spiral.

In retrospect, as we view the resulting construction we see that we have no longer any need for branch cuts on this surface. They were an artificial device for isolating one single valued piece, but on the surface as a whole all of $\log z$ is now a single-valued function. The figure illustrates the surface.

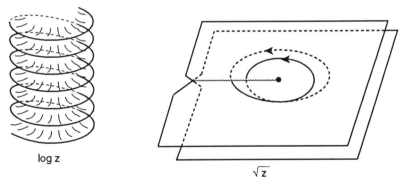

log z

$\sqrt{z}$

Riemann surfaces

The Riemann surface we have constructed for $\log z$ *can* be imbedded in Euclidean 3-space in the manner we have described. We now look at $\sqrt{z}$ for which we construct a Riemann surface which *cannot* be imbedded in Euclidean 3-space.

The function $\sqrt{z}$, with a branch cut along the negative real axis separates into two branches, the principal value given by

$$\sqrt{_1 z} = r^{1/2} e^{i\theta/2}, \qquad -\pi < \theta \le \pi$$

and the other one

$$\sqrt{_2 z} = r^{1/2} e^{i\theta/2}, \qquad \pi < \theta \le 3\pi.$$

So we take two sheets of paper with axes laid out and cut them along the branch cut, i.e., the negative real axis. On one of them which we label S_1 we implant the values of $\sqrt{_1 z}$, and on the other, S_2, the values $\sqrt{_2 z}$.

Then starting $z = re^{i\theta}$ again on the real axis of S_1 we move it with increasing θ across the cut; then $\sqrt{z}$ moves from $\sqrt{_1 z}$ to $\sqrt{_2 z}$. So we glue the upper edge ($\theta < \pi$) of S_1 to the lower edge ($\theta > \pi$) of S_2. Then $\sqrt{z}$ given by $\sqrt{_1 z}$ on S_1 and $\sqrt{_2 z}$ on S_2 is continuous on the resulting 2-ply surface. But we now have two loose edges: the lower edge ($\theta > -\pi$) of S_1 and the upper edge ($\theta < 3\pi$) of S_2.

Let us start z over again on the real axis on S_1 and this time move it *clockwise* across the cut as θ *decreases* through $-\pi$. Then $\sqrt{z}$ moves from the value $\sqrt{_1 z}$ to the values given by

$$a\sqrt{z} = r^{1/2} e^{i\theta/2}, \qquad \theta < -\pi,$$

but, for clarity let us keep $\theta > -3\pi/2$. Then for $-3\pi/2 < \theta < -\pi$, the value we have, namely,

$$\sqrt{z} = r^{1/2} e^{i\theta/2}, \qquad -3\pi/2 < \theta < -\pi$$

can be written,

$$\begin{aligned}
\sqrt{z} &= r^{1/2} e^{i(\theta - 4\pi)/2}, \quad 5\pi/2 < \theta < 3\pi \\
&= r^{1/2} e^{i\theta/2} \cdot e^{-2\pi i} \\
&= r^{1/2} e^{i\theta/2} \\
&= \sqrt{_2 z}
\end{aligned}$$

Thus as z crosses the cut this time we again go from S_1 to S_2. Furthermore, we now go from the lower edge of S_1 to the upper edge of S_2, the two loose edges. So how can we glue them together *through* the already glued part?

This is why we say that we cannot imbed this construction in Euclidean 3-space. For what we do is to consider these edges glued together as described, and they pass "through" each other in a sort of a science fiction "space warp" without touching each other. That this can be made logically rigorous is perhaps not at all obvious. However, it can be done, and done without appealing to such transcendental (in the religious sense) ideas as appear in the "space warp" description.

A Exercises

1. Construct (i.e., describe) a Riemann surface for $\sqrt[3]{z}, \sqrt[4]{z}, \ldots, \sqrt[n]{z}$.
2. Construct a Riemann surface for $\sqrt{(z-1)(z-2)}$. Hint: See Example 3, Section 2.6.
3. Construct a Riemann surface for $\log(z-1), \log(z-\alpha)$.

B Exercises

1. Construct a Riemann surface for
$$f(z) = \sqrt{z(z-1)(z-2)(z-3)}.$$
Show, by making cuts between 0 and 1 and between 2 and 3 that it is a two-ply surface.
2. Do the same for
$$f(z) = \sqrt{z(z-1)(z-2)}$$
except that the second cut is from 2 to ∞.
3. In Exercises B1 and B2, can the cuts be made elsewhere?

C Exercises

1. Construct a Riemann surface for
$$\log[z/(1-z)].$$

Chapter 3

Cauchy's Theorem

3.1 Cauchy's Integral Theorem

We have seen (Theorem 2.1e) that if $f(z)$ is continuous and is the derivative of an analytic function $F(z)$ in a region R, then for each pair of points α, β in R,

$$\int_{\alpha}^{\beta} f(z)dz = F(\beta) - F(\alpha) \tag{1}$$

whenever the curve of integration is a path in R from α to β, and so the integral is independent of the path. Conversely, if we know that $f(z)$ is continuous and that the integral in (1) is independent of the path, then we can define $F(z)$ in R by

$$F(z) = \int_{z_0}^{z} f(\zeta)d\zeta, \qquad z_0 \text{ fixed in } R,$$

from which $F(z)$ is analytic in R, $F'(z) = f(z)$, and equation (1) holds (Theorem 2.1h).

These results are not deep, but the independence of the path in (1) does have deep consequences, namely, that $f(z)$ is itself analytic, and deeper still, with proper proscription on the region involved, that this independence characterizes the analyticity of $f(z)$. The independence of the path is usually stated by asserting

$$\int_{C} f(z)dz = 0$$

150

for each closed path C in R. (Why is this equivalent to independence of the path?)

We begin by discussing the case where the curve is a triangle. Let T be a triangle with vertices at α, β, and γ, and let R be its interior. The projection of T onto the x-axis will be a segment whose ends are, say, a and b, as illustrated in the figure. For each x between a and b ($a < x < b$) the ordinate at x cuts T in exactly two points. Let the lower point have its y-coordinate $h(x)$, and the upper $g(x)$, and we extend $h(x)$ and $g(x)$ to a and b by continuity. Then the inequalities

$$a \leq x \leq b, \qquad h(x) \leq y \leq g(x)$$

describe the closed triangular set bounded by $T : \bar{R} = T \cup R$.

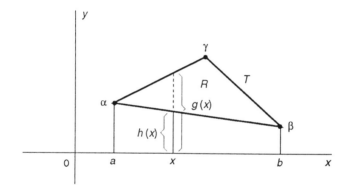

3.1a Theorem. (Cauchy's Theorem for a triangle.) *If T is a triangle and $f(z)$ is analytic in and on T, then*

$$\int_T f(z)dz = 0.$$

Remark 1. Let C be a contour, as for example the triangle T. When we say a function $f(z)$ is analytic **in and on** C we mean that it is analytic at each point of the closed set $I(C) \cup C$.

Remark 2. We give two proofs of this theorem. The first, and more elementary one, requires the assumption that $f'(z)$ is continuous. It is due, apparently, to Riemann in 1851. Cauchy's original proof in 1825 was more complicated and also required the continuity of $f'(z)$. Our second proof does not require prior knowledge that $f'(z)$ is continuous. It implies, as we shall see, that $f'(z)$ is in fact continuous and more: it is itself analytic. This proof is due to Goursat in 1900. (For these historical remarks see Copson, Functions of a Complex Variable, (Oxford, 1935), pp. 60–61.)

Proof 1. We write

$$\int_T f(z)dz = \int_T (udx - vdy) + i \int_T (vdx + udy) \equiv A + iB,$$

and show that $A = B = 0$. For the first integral in A, using the notation set up for T, we get

$$\int_T udx = \int_a^b u(x, h(x))dx + \int_b^a u(x, g(x))dx$$

$$= \int_a^b [u(x, h(x)) - u(x, g(x))]dx$$

Since $u_y = \partial u/\partial y$ is continuous in $\bar{R} = R \cup T$, we have

$$u(x, g(x)) - u(x, h(x)) = \int_{h(x)}^{g(x)} u_y(x, y)dy$$

for each x. It follows that

$$\int_T udx = -\int_a^b \left[\int_{h(x)}^{g(x)} u_y dy \right] dx$$

$$= -\iint_R u_y dx dy$$

Similarly, projecting on the y-axis, we get

$$\int_T vdy = \iint_R v_x dx dy$$

from which

$$A = \int_T (udx - vdy) = -\iint_R (u_y + v_x)dx dy$$

In the same way we compute

$$B = \int_T (vdx + udy) = \iint_R (u_x - v_y)dx dy.$$

By the Cauchy–Riemann equations both A and B are zero. $\quad\square$

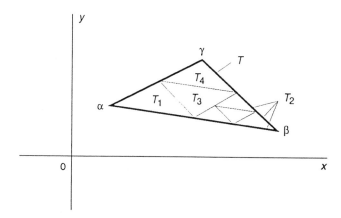

Proof 2. We join the midpoints of the sides of T by segments. This cuts the interior R of T into four similar triangular regions R_1, R_2, R_3, and R_4 with boundaries T_1, T_2, T_3 and T_4 respectively. Then since the integrals over the introduced segments cancel out, we have

$$\int_T f(z)dz = \sum_{j=1}^{4} \int_{T_j} f(z)dz$$

where all integrals are taken in the positive sense.

We now define M by

$$M \equiv \left| \int_T f(z)dz \right|.$$

Then, for at least one $j = 1, 2, 3$, or 4, we must have

$$\left| \int_{T_j} f(z)dz \right| \geq \frac{1}{4} M. \tag{2}$$

For if the left side of (2) were $< \frac{1}{4} M$ for all $j = 1, 2, 3, 4$ we would have

$$M = \left| \int_T f(z)dz \right| \leq \sum_{j=1}^{4} \left| \int_{T_j} f(z)dz \right|$$

$$< \frac{1}{4} M + \frac{1}{4} M + \frac{1}{4} M + \frac{1}{4} M = M.$$

That is, unless (2) holds for at least one j, we get $M < M$ which cannot be.

We choose one T_j for which (2) holds (we always have one, and maybe more) and rename it T^1 with interior R^1. Then we cut R^1 into four pieces as we did the original R by bisecting the sides. These we denote for now by R_j^1 with boundaries T_j^1, $j = 1, 2, 3, 4$. Arguing as before, there is at least one j for which

$$\left| \int_{T_j^1} f(z)dz \right| \geq \frac{1}{4} \left| \int_{T^1} f(z)dz \right| \geq \frac{1}{4} \cdot \frac{1}{4} M = \frac{1}{4^2} M. \tag{3}$$

Again we choose one T_j^1 for which (3) holds and rename it T^2. Continuing in this way we construct a sequence of regions R^k with boundaries T^k for which

$$\left| \int_{T^k} f(z)dz \right| \geq \frac{1}{4^k} M. \tag{4}$$

We examine this decreasing sequence of triangles. Let $D^k = R^k \cup T^k$, so that D^k is the closure of R^k. Let ℓ_k be the length of the longest side of D^k. Then $\ell_1 = \ell/2$ (where ℓ is the longest side of $D^0 = R \cup T$), $\ell_2 = \ell_1/2 = \ell/2^2$, so that finally

$$\ell_k = \ell/2^k. \tag{5}$$

Further we observe that

(i) Each D^k is a closed set.

(ii) Each D^k is contained in the preceding one:

$$D_k \subset D_{k-1},$$

(iii) by (5), $\ell_k \to 0$ as $k \to \infty$.

There is a theorem from real calculus which states that whenever the conditions (i), (ii), (iii) obtain, there is exactly one point α in all the D^k:

$$\alpha \in D^k, \qquad k = 0, 1, 2, \ldots, .$$

Since $f(z)$ is analytic in and on T, that is, in D^0, we must have $f'(\alpha)$ exists. So if we define $\omega(z)$ by

$$\begin{cases} \omega(z) = \dfrac{f(z) - f(\alpha)}{z - \alpha} - f'(\alpha) & z \neq \alpha \\[2mm] \qquad\; = 0 & z = \alpha, \end{cases} \tag{6}$$

then $\omega(z)$ is continuous in D^0 since

$$\lim_{z \to \alpha} \omega(z) = 0. \quad \text{(How?)} \tag{7}$$

Solving (6) for $f(z)$ we get

$$f(z) = f(\alpha) + f'(\alpha)(z - \alpha) + (z - \alpha)\omega(z)$$

and so

$$\int_{T^k} f(z)dz = f(\alpha) \int_{T^k} dz + f'(\alpha) \int_{T^k} (z - \alpha)dz + \int_{T^k} (z - \alpha)\omega(z)dz$$

from which

$$\int_{T^k} f(z)dz = \int_{T^k} (z - \alpha)\omega(z)dz. \quad \text{(How?)} \tag{8}$$

By (7) we have for each $\varepsilon > 0$ there is a $\delta > 0$ for which

$$|\omega(z)| < \varepsilon \quad \text{for} \quad |z - \alpha| < \delta.$$

And there is a K for which $|z - \alpha| < \delta$ for all $k > K$. (How?) Then for $k > K$ from (8) we get

$$\left| \int_{T^k} f(z)dz \right| < \ell_k \cdot \varepsilon \cdot 3\ell_k = 3\varepsilon\ell_k^2$$

$$= 3\varepsilon\ell^2/4^k.$$

Combining this with (4) we have

$$\frac{3\varepsilon\ell^2}{4^k} > \frac{M}{4^k}$$

or

$$\varepsilon > \frac{M}{3\ell^2} \geq 0.$$

Since ε can be any positive number, this implies $M = 0$. $\square$

To proceed we make use of two topological concepts. The statement that R is **simply connected** means that each simple closed curve (Jordan curve) in R has its interior also in R. Geometrically this means that a simply connected region is one without "holes" in it. For if a region had a hole in it, there would be a Jordan curve C in R circumscribing the hole, as in the figure, so that not all of $I(C)$ would be in R. This would then violate the definition of simple connectedness. Probably the simplest regions which are *not* simply connected are punctured disks; $0 < |z - \alpha| < a$, and annuli $0 < a < |z - \alpha| < b$. Regions which are not simply connected are called **multiply connected**.

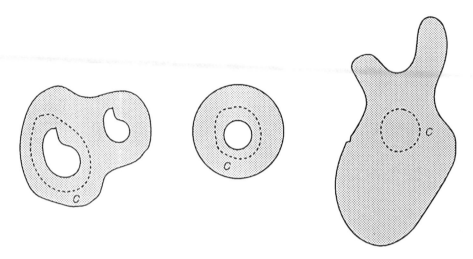

Multiply-connected regions   Simply-connected region

We consider next a special class of simply connected regions. We recall from Section 1.6 that if R is a region and α a point in R, then to say that R is **star shaped** with respect to α means that for each z in R, the segment from α to z is also in R. The point α is a **radiating center** of R. Merely saying that R is "star shaped" means that it has a radiating center with respect to which it is star shaped.

A disk $|z - z_0| < a$ is star shaped with respect to each of its points, and the same is true of an elliptical region. In fact, to say that a region is **convex** means that it is star shaped with respect to each of its points. There are regions which are star shaped with respect to only one point.

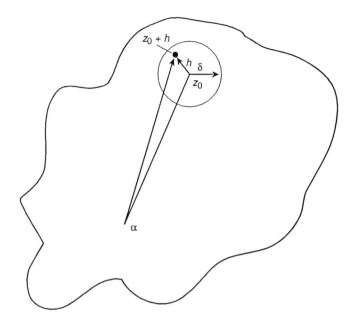

Star-shaped region

3.1b Theorem. (Cauchy's Integral Theorem) *If $f(z)$ is analytic in a simply connected region R, then*

$$\int_C f(z)dz = 0$$

for each closed path C in R.

Proof. In order to avoid some topological detail we give the proof for the restricted case where R is star shaped. The general proof is more tedious.

Thus we assume R is star shaped and α is a radiating center of R. We define $F(z)$ in R by the integral

$$F(z) = \int_\alpha^z f(\zeta)d\zeta$$

where *the path of integration is the segment from α to z.* We show that $F(z)$ is analytic in R and $F'(z) = f(z)$.

To this end let z_0 be any point in R. Then, since R is a region, there is a $\delta > 0$ for which the disk $D = \{z = z_0 + h : |h| < \delta\}$ is in R. Choose a point $z_0 + h$ in D and construct the triangle T cornering at $\alpha, z_0, z_0 + h$. Then T and its interior are in R (why?), so that Theorem 3.1a applies. Thus

$$0 = \int_T f(\zeta)d\zeta = \int_\alpha^{z_0+h} f(\zeta)d\zeta + \int_{z_0+h}^{z_0} f(\zeta)d\zeta + \int_{z_0}^{\alpha} f(\zeta)d\zeta$$

where each of the last three integrals is over the appropriate segment. Thus (how?)

$$\int_\alpha^{z_0+h} f(\zeta)d\zeta - \int_\alpha^{z_0} f(\zeta)d\zeta = \int_{z_0}^{z_0+h} f(\zeta)d\zeta$$

or, dividing by h,

$$\frac{\Delta F}{\Delta z} = \frac{F(z_0 + h) - F(z_0)}{h} = \frac{1}{h}\int_{z_0}^{z_0+h} f(\zeta)d\zeta.$$

We now use the continuity of $f(z)$: given $\varepsilon > 0$, there is an $\eta > 0$ so that $|f(z) - f(z_0)| < \varepsilon$ whenever $|z - z_0| < \eta$. Then choosing $|h| < \eta$ we compute

$$\left|\frac{\Delta F}{\Delta z} - f(z_0)\right|$$

$$= \left|\frac{1}{h}\int_{z_0}^{z_0+h} f(\zeta)d\zeta - f(z_0)\right| = \left|\frac{1}{h}\int_{z_0}^{z_0+h} f(\zeta)d\zeta - \frac{1}{h}\int_{z_0}^{z_0+h} f(z_0)d\zeta\right|$$

$$= \left|\frac{1}{h}\int_{z_0}^{z_0+h} [f(\zeta) - f(z_0)]d\zeta\right| \leq \frac{1}{|h|}\cdot\varepsilon\cdot|h| = \varepsilon$$

which proves that $F'(z) = f(z)$. Then by Theorem 2.1e with $\alpha = \beta$ we have

$$\int_C f(z)dz = 0$$

for each closed path in R. $\square$

A very useful special case is the following.

3.1c Theorem *If $f(z)$ is analytic in a region R and if C is a contour (simple closed path) in R with $I(C) \subset R$, then*

$$\int_C f(z)dz = 0.$$

Or, equivalently, if $f(z)$ is analytic in and on C, the conclusion holds.

Proof. We take it as intuitively clear that there is a subregion $R' \subset R$ which is simply connected and which contains C. The previous theorem then applies. □

We have avoided, in these results, coming to grips with topological problems of some difficulty. This was by giving the proof of Theorem 3.1b for only a special case, and by accepting an intuitive argument in Theorem 3.1c. However, our aim here is to obtain the main results and uses of complex function theory, and to avoid entanglements such as these topological difficulties, pertinent though they may be.

As an example of the use of these results, let us look again at the integral defining the logarithm. There $f(z) = 1/z$, and the region of analyticity is $R : z \neq 0$, that is, the plane punctured at 0. Thus R is *neither* star shaped (why?) *nor* simply connected (why?) so Theorem 3.1b does not apply, and indeed we know that

$$\int_C \frac{dz}{z} = 2\pi i$$

if C is any positively oriented contour with $0 \in I(C)$. On the other hand Corollary 3.1c does apply: If C is any contour with $0 \in E(C)$, then

$$\int_C \frac{dz}{z} = 0.$$

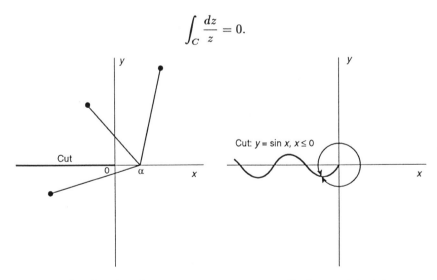

Star-shaped cut plane   Simply-connected cut plane

If we introduce a branch cut, say along the negative real axis, as we did to define the principal value of the logarithm, the resulting cut plane is star shaped with respect to any point on the positive real axis, and so 3.1b applies in the cut plane. This is, of course, equivalent to branches of the logarithm being single valued in the cut plane. Further, if we make a cut from 0 to ∞ along the negative end $(x < 0)$ of the sine curve

$$y = \sin x,$$

then the resulting cut plane, though not star shaped, is simply connected, and Theorem 3.1b again applies, and is again equivalent to the existence of single valued branches of the logarithm in this cut plane.

Let us also note, as we have before, that integrals can be independent of the path by Theorem 2.1e or, equivalently, Corollary 2.1g, rather than Cauchy's theorem. For example, if we take $f(z) = 1/z^2$, with R again being the punctured plane, then as with $1/z$, Cauchy's theorem does not apply. But Theorem 2.1e does apply with $F(z) = -1/z$ so that

$$\int_\alpha^\beta \frac{dz}{z^2} = \frac{1}{\alpha} - \frac{1}{\beta}$$

where the integration is over any path from α to β which does not go through the origin. If $\alpha = \beta$, the path is closed and the integral vanishes.

If $f(z)$ is a single valued analytic function defined in a multiply connected region R, one can ask in what ways can the Cauchy theorem fail when one considers

$$\int_C f(z)dz$$

where $C \subset R$ is a contour for which $I(C) \not\subset R$. The answer, perhaps surprisingly, is that it can only fail logarithmically. To amplify this we consider such a contour. Let $\alpha \in C$ and $z_0 \in I(C)$. Then on and near C the function $F(z)$ defined by

$$F(z) = \int_\alpha^z f(\zeta)d\zeta$$

must have the structure

$$F(z) = \phi(z) + A\log(z - z_0)$$

where $\phi(z)$ is single valued and $A = \dfrac{1}{2\pi i} \displaystyle\int_C f(\zeta)d\zeta$. We will not prove these statements.

Let C_1 and C_2 be two contours with $C_2 \subset I(C_1)$. Then $E(C_2) \cap I(C_1)$ is a region, called the **annular region between** C_1 and C_2. A form of Cauchy's Theorem holds for such regions.

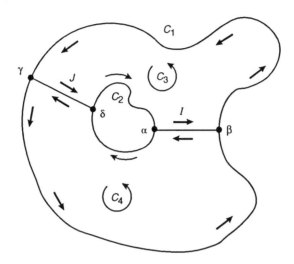

Cauchy's Theorem for two contours

3.1d Theorem. (Cauchy's Theorem for Two Contours) *If C_1 and C_2 are two contours with $C_2 \subset I(C_1)$ and if $f(z)$ is analytic in the annular region between them, and on both C_1 and C_2, then*

$$\int_{C_1} f(z)dz = \int_{C_2} f(z)dz$$

where both integrals are taken in the positive sense.

Proof. We construct two segments as illustrated connecting C_1 to C_2. (See Exercise C3). We label one of them I with ends α on C_2 and β on C_1, and the other J with ends γ on C_1 and δ on C_2. Then

  (i) α to β on I,
  (ii) β to γ counterclockwise along an arc of C_1,
  (iii) γ to δ on J,
  (iv) δ to α clockwise along an arc of C_2

describes a contour which we will call C_3, and $f(z)$ is analytic in and on C_3. Similarly

  (i) β to α along I,

(ii) α to δ clockwise around the complementary arc of C_2,

(iii) δ to γ on J,

(iv) γ to β counterclockwise around the complementary arc of C_1

describes a contour C_4, and $f(z)$ is analytic in and on C_4. Then by Cauchy's theorem we have

$$\int_{C_3} f(z)dz = 0 \quad \text{and} \quad \int_{C_4} f(z)dz = 0,$$

so

$$\int_{C_3} f(z)dz + \int_{C_4} f(z)dz = 0. \tag{9}$$

But in these integrals the parts over I and J cancel (why?) and so (9) reduces to

$$\int_{C_1} f(z)dz + \int_{(-C_2)} f(z)dz = 0$$

where the integral around C_1 is in the positive sense, and that around C_2 is in the negative sense [hence the $(-C_2)$]. This is equivalent to

$$\int_{C_1} f(z)dz - \int_{C_2} f(z)dz = 0$$

which proves the theorem. $\square$

Example 1. It is known (see Exercise C 6(b), Section 3.6) that
$\int_{-\infty}^{\infty} e^{-x^2} dx = \sqrt{\pi}$. Show that $\int_{-\infty}^{\infty} e^{-(x+ib)^2} dx$ is independent of b and deduce

$$\int_{-\infty}^{\infty} e^{-x^2} \cos 2bx\, dx = \sqrt{\pi}\, e^{-b^2}.$$

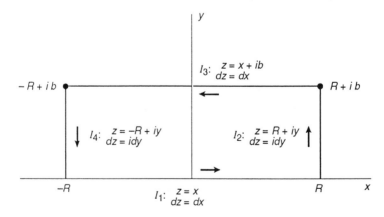

Example 1

Solution. We assume $b > 0$ since the cosine is even. Let C be the rectangle cornering at $\pm R$ and $\pm R + ib$ as shown in the figure. Then integrating in the positive sense around C we have, since e^{-z^2} is entire,

$$0 = \int_C e^{-z^2} dz = \left\{ \int_{I_1} + \int_{I_2} + \int_{I_3} + \int_{I_4} \right\} e^{z^2} dz.$$

Now let

$$J_k = \int_{I_k} e^{-z^2} dz, \qquad k = 1, 2, 3, 4.$$

Then

$$|J_2| = \left| \int_{I_2} e^{-z^2} dz \right| = \left| \int_0^b e^{-(R+iy)^2} i\, dy \right|$$

$$\leq \int_0^b |e^{-R^2 - 2iRy + y^2}|\, dy$$

$$= e^{-R^2} \int_0^b e^{y^2} dy \leq e^{-R^2} \cdot e^{b^2} \cdot b.$$

Thus for b fixed, $J_2 \to 0$ as $R \to \infty$. Similarly, $J_4 \to 0$ and so

$$\int_{-\infty}^{\infty} e^{-z^2} dz + \int_{\infty+ib}^{-\infty+ib} e^{-z^2} dz = 0,$$

but this is

$$0 = \int_{-\infty}^{\infty} e^{-x^2} dx + \int_{\infty}^{-\infty} e^{-(x+ib)^2} dx$$

or

$$\int_{-\infty}^{\infty} e^{-(x+ib)^2} dx = \int_{-\infty}^{\infty} e^{-x^2} dx = \sqrt{\pi}.$$

Hence the integral is independent of b as asserted. Going further we see that

$$e^{-(x+ib)^2} = e^{-x^2} e^{-2ibx} e^{b^2}$$

$$= e^{-x^2} e^{b^2} [\cos 2bx - i \sin 2bx]$$

so that

$$e^{b^2} \int_{-\infty}^{\infty} e^{-x^2} [\cos 2bx - i \sin 2bx]\, dy = \sqrt{\pi}.$$

Then dividing by e^{b^2} and equating real parts completes the solution.

The next theorem is an extension of the Cauchy theorem to a case where analyticity breaks down on (but not inside of) a contour C. The general result is rather difficult and the proof is left to the C Exercises for the special case of a circle.

3.1e Theorem *Suppose C is a contour and $f(z)$ is continuous in and on C, and is analytic inside, that is, in $I(C)$. Then*

$$\int_C f(z)dz = 0.$$

Example 2. By the theorem

$$\int_{|z|=1} \sqrt{1-z}\ dz = 0$$

for either branch of the root.

A Exercises

1. In the last step of Example 1 we equated real parts. If we equate imaginary parts, what do we get? How do we know this result independently of this example?

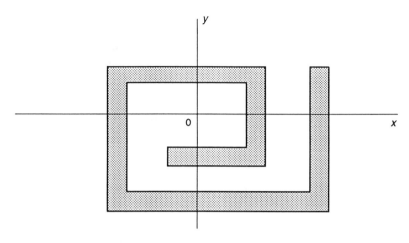

Exercise A2

2. Let R be the interior of the polygon illustrated in the figure. Is $\int dz/z$ independent of the path in R?

3. By a heuristic argument conclude that

$$-\frac{i}{2}\int_C \bar{z}\,dz$$

is the area of $I(C)$ if C is a contour.

B Exercises

1. Let C be a contour with $\alpha \in I(C)$.

   (a) Evaluate $\displaystyle\int_C \frac{e^z}{z-\alpha}\,dz$.

   Hint: By Exercise A1 of Section 2.5 and the definition of the exponential

   $$\frac{e^z}{z-\alpha} = \frac{e^\alpha e^{z-\alpha}}{z-\alpha} = \frac{e^\alpha}{z-\alpha} + e^\alpha \sum_1^\infty \frac{(z-\alpha)^{n-1}}{n!}$$

   $$= \frac{e^\alpha}{z-\alpha} + g(z)$$

   where $g(z)$ is entire.

   (b) Evaluate $\displaystyle\int_C \frac{e^z}{(z-\alpha)^2}\,dz$

(c) Evaluate $\displaystyle\int_C \frac{e^z}{(z-\alpha)^{k+1}}$ where k is a nonnegative integer.

2. Using the technique of Exercise B1 above evaluate

(a) $\displaystyle\int_C \frac{\sin z}{z-\alpha}\, dz$ (b) $\displaystyle\int_C \frac{\cos z}{z-\alpha}\, dz$

where C is a contour with $\alpha \in I(C)$.

3. In Exercises B1, 2 above evaluate the integrals if $\alpha \in E(C)$.

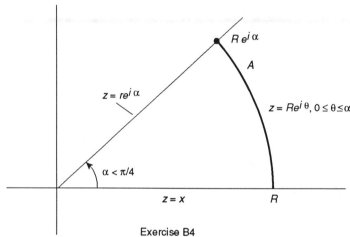

Exercise B4

4. Let C be the contour illustrated in the figure. By considering $\displaystyle\int_C e^{-z^2}dz$, and letting $R \to \infty$ deduce the values of

$$\int_0^\infty e^{-r^2 \cos 2\alpha} \cos(r^2 \sin 2\alpha)dr \ \text{ and } \ \int_0^\infty e^{-r^2 \cos 2\alpha} \sin(r^2 \sin 2\alpha)dr,$$

for $0 < \alpha < \pi/4$.

5. In the next chapter we will show

$$\int_A e^{-z^2}dz \to 0 \quad \text{as} \quad R \to \infty$$

where A is the arc $z = Re^{i\theta}$, $0 \le \theta \le \pi/4$. Use this to evaluate

$$\int_0^\infty \cos r^2 dr \quad \text{and} \quad \int_0^\infty \sin r^2 dr$$

and deduce the values of the improper integrals

$$\int_0^\infty t^{-\frac{1}{2}} \cos t\, dt \quad \text{and} \quad \int_0^\infty t^{-\frac{1}{2}} \sin t\, dt.$$

6. Show that there is a single valued branch of $\log z$ in the polygonal region of Exercise A2.

C Exercises

1. (a) Show that $\displaystyle\int_{|z|=r} e^{1/z} dz$ is independent of r for $r > 0$.

   (b) Take $r = 1$ and evaluate the integral.

   (c) Deduce the values of

   $$\int_0^{2\pi} e^{\cos\theta} \cos(\sin\theta - \theta)d\theta$$

   and

   $$\int_0^{2\pi} e^{\cos\theta} \sin(\sin\theta - \theta)d\theta.$$

2. To see how I and J can be constructed in the proof of Theorem 3.1d we can proceed as follows. Choose $z_0 \in I(C_2)$ and consider a ray L_1 given by

   $$z(t) = z_0 + te^{i\theta}, \qquad t \geq 0,$$

   where θ is a constant. Show that there is a largest t, denoted by t_0, for which $z(t_0) \in C_2$, and a smallest $t > t_0$, for which $z(t) \in C_1$. Set $\alpha = z(t_0)$, $\beta = z(t_1)$ and show that the segment I, described by

   $$I : z(t) = z_0 + te^{i\theta}, \qquad t_0 \leq t \leq t_1$$

   is a suitable segment. By taking $\phi \neq \theta + 2k\pi$ let L_2 be the ray

   $$z(t) = z + te^{i\phi}, \qquad t > 0$$

   and similarly construct a suitable J.

3. Prove Theorem 3.1e in the case where C is a circle. Hint: $f(z)$ is uniformly continuous in the closed disk consisting of C and $I(C)$.

3.2 The Cauchy Integral Formula

3.2a Theorem *If $f(z)$ is analytic in and on a contour C, then*

$$\frac{1}{2\pi i} \int_C \frac{f(\zeta)}{\zeta - z} d\zeta = \begin{cases} f(z) & z \in I(C) \\ 0 & z \in E(C) \end{cases} \qquad (1)$$

This is a very remarkable formula which depends heavily on the analyticity of $f(z)$. It is in fact equivalent to analyticity. Its remarkable feature is that from a partial knowledge of $f(z)$—the values on C which enter in the integral—we recapture all of $f(z)$ in $I(C)$. Such a formula is called a reproducing formula. We will see other reproducing formulas. It is also clear that it is a generalization of Theorems 1.5g and 1.5j.

Proof. Choose a fixed z outside C, that is, in $E(C)$. Then the function of ζ defined by $f(\zeta)/(\zeta - z)$ is analytic in and on C so Cauchy's theorem (Theorem 3.1c) applies and the integral is zero.

Now choose a fixed z in $I(C)$. There is then a positive minimal distance from z to C. Call that distance a:

$$a = \text{dist}\,(z, C).$$

For $r < a$, the closed disk $|\zeta - z| \leq r$ is in $I(C)$, and by Cauchy's theorem for two contours

$$\frac{1}{2\pi i} \int_C \frac{f(\zeta)d\zeta}{\zeta - z} = \frac{1}{2\pi i} \int_{|\zeta-z|=r} \frac{f(\zeta)d\zeta}{\zeta - z}. \tag{2}$$

Since f is continuous at z, if $\varepsilon > 0$ is given, there is a $\delta > 0$ for which

$$|f(\zeta) - f(z)| < \varepsilon \quad \text{if} \quad |\zeta - z| < \delta. \tag{3}$$

Choosing $r < \delta$ we have

$$\frac{1}{2\pi i} \int_{|\zeta-z|=r} \frac{f(\zeta)}{\zeta - z}\, d\zeta - f(z)$$

$$= \frac{1}{2\pi i} \int_{|\zeta-z|=r} \frac{f(\zeta)}{\zeta - z}\, d\zeta - \frac{1}{2\pi i} \int_{|\zeta-z|=r} \frac{f(z)d\zeta}{\zeta - z} \tag{4}$$

$$= \frac{1}{2\pi i} \int_{|\zeta-z|=r} \frac{f(\zeta) - f(z)}{\zeta - z}\, d\zeta.$$

This is because

$$\frac{1}{2\pi i} \int_{|\zeta-z|=r} \frac{d\zeta}{\zeta - z} = 2\pi i.$$

Combining (2) and (4) we have

$$\left| \frac{1}{2\pi i} \int_C \frac{f(\zeta)}{\zeta - z}\, d\zeta - f(z) \right| = \frac{1}{2\pi} \left| \int_{|\zeta-z|=r} \frac{f(\zeta) - f(z)}{\zeta - z}\, d\zeta \right|$$

$$\leq \frac{1}{2\pi} \cdot \varepsilon \cdot \frac{1}{r} \cdot 2\pi r = \varepsilon.$$

Since ε is an arbitrary positive number, we conclude

$$\frac{1}{2\pi i} \int_C \frac{f(\zeta)}{\zeta - z} \, d\zeta - f(z) = 0. \qquad \square$$

We now consider some examples which review the techniques used in Section 1.5 for the evaluation of integrals. Though the Cauchy formula is much more general here, we keep to relatively simple functions.

Example 1. Evaluate $I = \displaystyle\int_{|z|=5} \frac{z^2 + 2}{z + 3} \, dz.$

Solution. We can write

$$I = \int_{|\zeta|=5} \frac{\zeta^2 + 2}{\zeta + 3} \, d\zeta$$

so (1) applies with $z = -3$ and we get

$$I = 2\pi i(z^2 + 2)\Big|_{z=-3} = 2\pi i((-3)^2 + 2)$$
$$= 2\pi i(11) = 22\pi i.$$

Example 2. Evaluate

$$I = \frac{1}{2\pi i} \int_{|z|=2} \frac{3z - 5}{z^2 - 2z - 3} \, dz.$$

Solution 1. We write, by partial fractions

$$\frac{3z - 5}{z^2 - 2z - 3} = \frac{3z - 5}{(z - 3)(z + 1)}$$
$$= \frac{1}{z - 3} + \frac{2}{z + 1}$$

so that

$$I = \frac{1}{2\pi i} \int_{|z|=2} \frac{dz}{z - 3} + \frac{2}{2\pi i} \int_{|z|=2} \frac{dz}{z + 1}$$
$$= 0 + 2 = 2.$$

Solution 2. We write

$$\frac{3z - 5}{z^2 - 2z - 3} = \left(\frac{3z - 4}{z - 3}\right) \Big/ (z + 1)$$

where $(3z - 5)/(z - 3)$ is analytic in and on $|z| = 2$, so by Theorem 3.2a we have

$$I = \frac{3z - 5}{z - 3}\bigg|_{z=-1} = \frac{-8}{-4} = 2.$$

Example 3. Evaluate

$$I = \frac{1}{2\pi i} \int_{|z|=5} \frac{4e^z dz}{(z - 3)(z + 1)}.$$

Solution 1. Again by partial fractions,

$$\frac{4e^z}{(z - 3)(z + 1)} = \frac{e^z}{z - 3} - \frac{e^z}{z + 1}$$

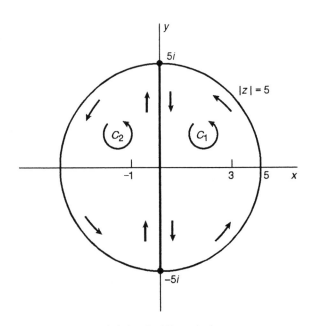

Solution 2 of Example 3:

$$\int_{|z|=5} = \int_{C_1} + \int_{C_2}$$

so that

$$I = \frac{1}{2\pi i} \int_{|z|=5} \frac{e^z}{z-3} \, dz - \frac{1}{2\pi i} \int_{|z|=5} \frac{e^z}{z+1} \, dz$$
$$= e^3 - e^{-1} = (e^4 - 1)/e$$

Solution 2. Let C_1 be the right semicircle of $|z| = 5$ and the segment from $5i$ to $-5i$, and let C_2 be the left semicircle of $|z| = 5$ and the same segment but now from $-5i$ to $5i$. Then

$$I = \frac{1}{2\pi i} \int_{C_1} \left(\frac{4e^z}{z+1} \right) dz \Big/ (z-3) + \frac{1}{2\pi i} \int_{C_2} \left(\frac{4e^z}{z-3} \right) dz \Big/ (z+1)$$
$$= \frac{4e^z}{z+1} \Big|_{z=3} + [\frac{4e^z}{z-3} \Big|_{z=-1} = e^3 - e^{-1}.$$

We now widen the scope of the discussion. We consider a path C and a function $\phi(\zeta)$ continuous on C. We *define* a function $f(z)$ for z *not* on C by the Cauchy-type integral

$$f(z) = \frac{1}{2\pi i} \int_C \frac{\phi(\zeta)}{\zeta - z} \, d\zeta.$$

If, for example, C were as illustrated, then z could be in any of the three connected components of the plane which form the complement of C which are labeled I, II, and III. In each of these components $f(z)$ is analytic *and more*.

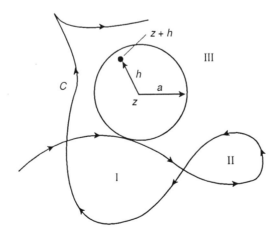

3.2b Theorem *If $\phi(z)$ is continuous on a path C, and $f(z)$ is defined by*

$$f(z) = \frac{1}{2\pi i} \int_C \frac{\phi(\zeta)}{\zeta - z} \, d\zeta$$

for $z \notin C$, then $f(z)$ is analytic and the derivative can be computed by differentiating under the integral sign:

$$f'(z) = \frac{1}{2\pi i} \int_C \frac{\phi(\zeta)}{(\zeta - z)^2} \, d\zeta.$$

Proof. Choose any $z \notin C$ and keep it fixed. Then there is a positive minimal distance from z to C. Call this distance a:

$$a = \text{dist}\,(z, C).$$

For $0 < |h| < a$, $z + h$ is in the neighborhood $N(z, a)$. We form the difference quotient

$$\frac{\Delta f}{\Delta z} = \frac{f(z+h) - f(z)}{h} = \frac{1}{2\pi i} \int \phi(\zeta) \frac{1}{h} \left[\frac{1}{\zeta - z - h} - \frac{1}{\zeta - z} \right] d\zeta$$

$$= \frac{1}{2\pi i} \int_C \phi(\zeta) \frac{1}{(\zeta - z - h)(\zeta - z)} \, d\zeta.$$

Thus, if we could let $h \to 0$ under the integral sign, we see that this would result in the integral of the conclusion. To justify this we argue as follows. Let us for now call that integral $g(z)$:

$$g(z) = \frac{1}{2\pi i} \int \frac{\phi(\zeta)}{(\zeta - z)^2} \, d\zeta$$

And we want to show that $f'(z) = g(z)$, that is we want to show $\Delta f / \Delta z \to g(z)$ as $h \to 0$. So we examine the difference

$$\frac{\Delta f}{\Delta z} - g(z) = \frac{1}{2\pi i} \int_C \phi(\zeta) \left[\frac{1}{(\zeta - z - h)(\zeta - z)} - \frac{1}{(\zeta - z)^2} \right] d\zeta$$

$$= \frac{1}{2\pi i} \int_C \frac{h\phi(\zeta) d\zeta}{(\zeta - z - h)(\zeta - z)^2}$$

We recall that $a = \text{dist}\,(z, C)$ so that $|\zeta - z| \geq a$ for $\zeta \in C$. Also, since h is going to zero anyway, we can suppose $|h| < a/2$. Then, for ζ on C,

$$|\zeta - z - h| \geq |\zeta - z| - |h| \geq a - a/2 = a/2,$$

so that for ζ on C and $|h| < a/2$ we have

$$\frac{1}{|\zeta - a|} \leq \frac{1}{a} \quad \text{and} \quad \frac{1}{|\zeta - z - h|} \leq \frac{2}{a}.$$

And, finally, since $\phi(\zeta)$ is continuous on C, there is a constant M for which

$$|\phi(\zeta)| \leq M.$$

So we get, if L is the length of C,

$$\left| \frac{\Delta f}{\Delta z} - g(z) \right| \leq \frac{|h|}{2\pi} \cdot \frac{M2}{a^2 a} L = \frac{ML}{\pi a^3} |h|$$

which clearly goes to zero as $h \to 0$. Thus

$$f'(z) = \lim_{h \to 0} \frac{\Delta f}{\Delta z} = g(z) = \frac{1}{2\pi i} \int \frac{\phi(\zeta) d\zeta}{(\zeta - z)^2}. \qquad \square$$

3.2c Theorem *Under the same conditions as the previous theorem*

$$f''(z) = \frac{2!}{2\pi i} \int_C \frac{\phi(\zeta)}{(\zeta - z)^3} \, d\zeta$$

Proof. Following the ideas of the previous proof we consider

$$D = \frac{f'(z + h) - f'((z)}{h} - \frac{1}{\pi i} \int_C \frac{\phi(\zeta) dS}{(\zeta - z)^3}.$$

After a little algebra we compute

$$D = \frac{h}{2\pi i} \int \frac{\phi(\zeta)[3(\zeta - a) - 2h] d\zeta}{(\zeta - z - h)^2 (\zeta - a)^3}$$

which is easily estimated to show that $D \to 0$ as $h \to 0$. The details we leave to the exercises. $\square$

We have now (by this last simple theorem whose proof we have shunted to the exercises) one of the big results of the subject of complex variables: If $f(z)$ is analytic, so is $f'(z)$, and so, by induction, is $f^{(n)}(z)$.

3.2d Theorem *If $f(z)$ is analytic in a region R, then $f(z)$ has derivatives of all orders, and, if C is a contour for which $f(z)$ is analytic in and on C, then*

$$\frac{n!}{2\pi i} \int_C \frac{f(\zeta)d\zeta}{(\zeta - z)^{n+1}} = \begin{cases} f^{(n)}(z), & z \in I(C) \\ 0, & z \in E(C). \end{cases} \tag{5}$$

Proof. Let α be an arbitrary point in R. Then, for $r < \mathrm{dist}\,(\alpha, \partial R)$ we have

$$f(z) = \frac{1}{2\pi i} \int_{|\zeta - z| = r} \frac{f(\zeta)}{\zeta - z}\, d\zeta, \qquad |z - \alpha| < r$$

by Theorem 3.1a. Then by Theorem 3.1b

$$f'(z) = \frac{1}{2\pi i} \int_{|\zeta - \alpha| = r} \frac{f(\zeta)d\zeta}{(\zeta - z)^2}, \qquad |z - \alpha| < r,$$

which gives a formula for $f'(z)$, which we already knew existed, for $f(z)$ is analytic. But now Theorem 3.1c asserts

$$f''(z) = \frac{2}{2\pi i} \int_{|\zeta - \alpha| = r} \frac{f(\zeta)}{(\zeta - z)^3}\, d\zeta, \qquad |z - \alpha| < r,$$

which says that $f'(z)$ has a derivative at α, which we did not previously know. But α is any point in R so $f'(z)$ has a derivative at each point in R and so $f'(z)$ is analytic in R. This says that if $f(z)$ is analytic so is $f'(z)$. Then we apply this to $f'(z)$: Since $f'(z)$ is analytic, so is $f''(z)$, and, by induction, so is $f^{(n)}(z)$ for each positive integer n.

It remains to establish the integral formula (5). If $f(z)$ is analytic in and on a contour C, then (5) holds for $n = 0$, that is, for $f(z)$ itself, by Theorem 3.1a, and for $n = 1$ and 2 by 3.1b and c respectively. We can establish the general formula by an induction argument based on the methods used to establish those theorems. However, it is simpler to use integration by parts based on the fact that we now know that $f^{(n)}(z)$ is analytic. Thus by Theorem 3.2a

$$f^{(n)}(z) = \frac{1}{2\pi i} \int_C \frac{f^{(n)}(\zeta)}{\zeta - z}\, d\zeta,$$

and integrating by parts, integrating $f^{(n)}(z)$ and differentiating $1/(\zeta - z)$, we get

$$f^{(n)}(z) = \frac{1}{2\pi i} \int_C \frac{f^{(n-1)}(\zeta)}{(\zeta - z)^2}\, d\zeta$$

and integrating by parts again

$$f^{(n)}(z) = \frac{2}{2\pi i} \int_C \frac{f^{(n-2)}(\zeta)}{(\zeta - z)^3} \, d\zeta$$

and, by induction,

$$f^{(n)}(z) = \frac{n!}{2\pi i} \int_C \frac{f(\zeta)d\zeta}{(\zeta - z)^{n+1}}.$$

We leave it to you (Exercise A11) to show that the integrated terms are zero for each integration by parts. ☐

Example 4. Evaluate

$$I = \int_{|z|=3} \frac{e^{iz}}{(z - 1)^3} \, dz$$

Solution. By Theorem 3.2d with $n = 2$ we get

$$\frac{2}{2\pi i} I = \frac{2!}{2\pi i} \int_{|z|=3} \frac{e^{iz}}{(z - 1)^3} \, dz$$

$$= \frac{d^2}{dz^2} \left(e^{iz} \right)\Big|_{z=1} = -e^{-i}$$

so that

$$I = -\pi i e^{-i}.$$

The analyticity of the derivative of an analytic function enables us to strengthen a previous theorem (2.1h). We have the following, known as **Morera's Theorem.**

3.2e Theorem *If $f(z)$ is continuous in a region R, and if $\int_C f(z)dz = 0$ for every closed path C in R, then $f(z)$ is analytic in R.*

Proof. Choose α in R, and define $F(z)$ in R by

$$F(z) = \int_\alpha^z f(\zeta)d\zeta$$

where the integral is over any path in R from α to z. Then $F(z)$ is analytic and $F'(z) = f(z)$ in R by Theorem 2.1h. By Theorem 3.2d $f(z)$ is analytic in R. ☐

There are important estimates on the size of the derivatives of an analytic function which follow from the Cauchy formulas (5) for the derivatives. They are called the **Cauchy estimates.** We describe the circumstances in which they apply. If $f(z)$ is analytic in a region R, and α is a point in R, then there is a positive minimal distance, which we will call a, from α to the boundary of R:

$$a = \text{dist}\,(\alpha, \partial R).$$

For each r, $0 < r < a$, $f(z)$ is continuous in the closed disk $D = \{z : |z - \alpha| \le r\}$, and is therefore bounded in D: there is a constant M for which

$$|f(z)| \le M \quad \text{for} \quad |z - \alpha| \le r.$$

3.2f Theorem. *In the notation described above*

$$|f^{(n)}(\alpha)| \le \frac{Mn!}{r^n}, \qquad r < a. \tag{6}$$

Proof. By (5)

$$f^n(\alpha) = \frac{n!}{2\pi i} \int_{|z-\alpha|=r} \frac{f(z)}{(z - \alpha)^{n+1}}\, dz.$$

From which

$$|f^{(n)}(\alpha)| \le \frac{n!}{2\pi} \frac{M}{r^{r+1}}\, 2\pi r = \frac{Mn!}{r^n}. \qquad \Box$$

From the Cauchy estimate for $n = 1$ there follows a famous theorem.

3.2g Theorem. (Liouville) *If $f(z)$ is an entire function and is bounded for all z (there is a constant M for which $|f(z)| \le M$), then $f(z)$ is a constant.*

Proof. Let α be any point in the plane. By the Cauchy estimate for $n = 1$ we have

$$|f'(\alpha)| \le M/r \tag{7}$$

for any $r > 0$. By letting $r \to \infty$ we get $f'(\alpha) = 0$. This holds for each α, so $f'(z) \equiv 0$, from which $f(z)$ is constant. $\Box$

Example 5. Suppose $f(z)$ is entire and there are constants $a > 0$ and $b > 0$ for which

$$|f(z)| \le a + b|z|.$$

Show that $f(z)$ is linear, that is,, there are constants α and β such that

$$f(z) = \alpha + \beta z.$$

Solution. Let z_0 be any point. Then in the disk $|z - z_0| \leq r$ we have

$$|f(z)| \leq a + b|z| = a + b|z - z_0 + z_0|$$
$$\leq a + b|z_0| + br \equiv M$$

We apply the Cauchy estimate at z_0 for $n = 2$:

$$|f''(z_0)| \leq \frac{2M}{r^2} = 2\,\frac{a + b|z_0|}{r^2} + \frac{2b}{r} \to 0$$

as $r \to \infty$. Thus $f''(z_0) = 0$. Since z_0 is arbitrary, we have $f''(z) \equiv 0$, so that

$$f(z) - f(0) = \int_0^z f'(\zeta)d\zeta = f'(0)[z - 0]$$
$$= zf'(0)$$

and we have

$$f(z) = f(0) + f'(0)z.$$

Choosing $\alpha = f(0)$, $\beta = f'(0)$ completes the solution.

We are now in a position to prove the **Fundamental Theorem of Algebra**. The main tool in the proof is Liouville's Theorem.

3.2h Theorem. *If $p(z)$ is a polynomial of degree $n \geq 1$, then $p(z) = 0$ has at least one solution.*

Proof. We will give a proof by contradiction. We suppose $p(z)$ is never zero, that is, $p(z) = 0$ has no solutions.

The form of $p(z)$ is

$$p(z) = a_n z^n + a_{n-1} z^{n-1} + \cdots + a_1 z + a_0, \quad a_n \neq 0$$

Then (how?)

$$|p(z)| \geq |a_n||z|^n - [|a_{n-1}||z|^{n-1} + \cdots + |a_1||z| + |a_0|]$$
$$= |z|^n \left\{ |a_n| - \left[\frac{|a_{n-1}|}{|z|} + \cdots + \frac{|a_0|}{|z|^n} \right] \right\}$$

Each term in the square brackets goes to zero as $|z| \to \infty$, so there is an r, for which

$$\frac{|a_{n-1}|}{|z|} + \cdots |\frac{|a_0|}{|z|^n} \leq \frac{1}{2}|a_n|$$

for $|z| \geq r$. Then

$$\begin{cases} |p(z)| > |z|^n[|a_n| - |a_n|/2] = |z|^n|a_n|/2 \\ \quad > r^n|a_n|/2, \quad \text{for} \quad |z| \geq r. \end{cases} \tag{7}$$

Also, for $|z| \leq r$, we have $p(z)$ is never zero, and so

$$\min_{|z| \leq r} |p(z)| = m \tag{9}$$

where m is some positive number. Combining (8) and (9) we have

$$|p(z)| \geq M = \min[m, r^n|a_n|/2] \tag{10}$$

Thus, since $p(z)$ is never zero, $1/p(z)$ is entire and by (10), $|1/p(z)| \leq 1/M$. Liouville's theorem applies, and implies $1/p(z)$ is constant and hence $p(z)$ is constant. But this is a contradiction. (How?) □

Theorem 3.2d shows that the real and imaginary parts of an analytic function have partial derivatives of all orders, and hence validates the discussion of harmonic functions in Section 2.1. We now want to reconsider the problem of finding conjugate harmonic functions in general. Thus given a harmonic function u, finding the harmonic conjugate v is equivalent to finding an analytic f for which $u = \operatorname{Re} f$. So knowing u, if v, and therefore f, exists we have

$$f' = u_x + iv_x = u_x - iu_y.$$

Then integrating f' we should get f.

3.2i Theorem. *Suppose $u(x, y)$ is harmonic in a simply connected region R. Set*

$$f(z) = \int_{z_0}^{z} \left[u_x(\xi, \eta) - iu_y(\xi, \eta) \right] d\zeta + u(x_0, y_0)$$

where $\zeta = \xi + i\eta$ is the point of integration and the integral is over any path from z_0 to z in R. Then $f(z)$ is analytic in R and $\operatorname{Re} f(z) = u(x, y)$.

Proof. We examine $g(z)$ defined by

$$g(z) = u_x(x, y) - iu_y(x, y).$$

Set $U = u_x$, $V = -u_y$. Then

$$U_x = u_{xx}, \quad V_y = -u_{yy},$$

and so $U_x = V_y$ since $u_{xx} + u_{yy} = 0$. And

$$U_y = u_{xy} \quad V_x = -u_{xy}$$

and so $U_y = -V_x$. Thus U and V satisfy the Cauchy–Riemann equations, which shows that $g(z)$ is analytic in R. Thus the integral defining $f(z)$ is independent of the path, and hence $f(z)$ depends only on z, once z_0 is chosen, and so $f(z)$ is analytic in R by Morera's theorem (3.2e).

Now suppose $\varphi(t) = h(t) + ik(t)$, $0 \le t \le 1$, is a path in R from z_0 to z. Then

$$f(z) = \int_0^1 \Big[u_x(h(t), k(t)) - iu_y(h(t), k(t)) \Big] \Big[h'(t) + ik'(t) \Big] dt + u(x_0, y_0)$$

so that

$$\text{Re } f(z) = \int_0^1 \Big[u_x(h(t), k(t))h'(t) + u_y(h(t), k(t))k'(t) \Big] dt + u(x_0, y_0)$$

$$= \int_0^1 \frac{d}{dt} u(h(t), k(t))dt + u(x_0, y_0)$$

$$= u(h(t), k(t)) \Big|_0^1 + u(x_0, y_0)$$

$$= u(x, y). \quad \square$$

We observe that the substitutions we used in Examples 5 and 6 of Section 1.4 can now be written:

$$\begin{cases} e^{i\theta} = z, \quad d\theta = \dfrac{dz}{iz}, \quad |z| = 1, \\[2mm] \cos\theta = \dfrac{e^{i\theta} + e^{-i\theta}}{2} = \dfrac{1}{2}\left(z + \dfrac{1}{z}\right) \\[2mm] \sin\theta = \dfrac{e^{i\theta} - e^{-i\theta}}{2i} = \dfrac{1}{2i}\left(z - \dfrac{1}{z}\right). \end{cases} \tag{11}$$

Example 6. Evaluate $I = \displaystyle\int_0^\pi \frac{d\theta}{1 - a\cos\theta}$, $-1 < a < 1$.

Solution. Since the integrand is an even function of θ, we have

$$I = \frac{1}{2} \int_{-\pi}^{\pi} \frac{d\theta}{1 - a \cos \theta}.$$

By the substitution (11) we get, after simplifying

$$
\begin{aligned}
I &= -\frac{1}{ai} \int_{|z|=1} \frac{dz}{z^2 - \frac{2}{a} z + 1} \\
&= -\frac{1}{ai} \int_{|z|=1} \frac{dz}{[z - (1 - \sqrt{1 - a^2})/a][z - (1 + \sqrt{1 - a^2})/a]} \\
&= -\frac{1}{ai} \, 2\pi i \, \frac{1}{z - (1 + \sqrt{1 - a^2})/a} \Big|_{z=(1-\sqrt{1-a^2})/a} \\
&= -\frac{1}{ai} \, 2\pi i \, \frac{1}{-(2\sqrt{1 - a^2})/a} = \frac{\pi}{\sqrt{1 - a^2}}.
\end{aligned}
$$

Significant Topics

A Exercises

Evaluate the following integrals

1. $\displaystyle \int_{|z|=2} \frac{z^2 + 1}{z^2 - 1} \, dz$

2. $\displaystyle \int_{|z|=2} \frac{z^2 - 1}{z^2 + 1} \, dz$

3. $\displaystyle \int_C \frac{e^z}{z(z - 1)} \, dz$

(a) C:

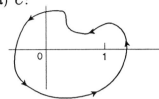

(b) C:

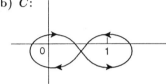

(c) C:

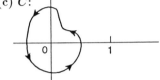

4. $\displaystyle\int_C \frac{e^z \cos z\, dz}{z^2(z-1)^2}$ (a), (b), (c) as in 3 above.

5. $\displaystyle\int_C \frac{z\, dz}{(1-z)(z+2)^4}$ C:

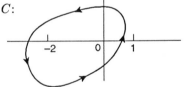

6. $\displaystyle\int_{|z|=2a} \frac{e^z dz}{(z-a)^5}$, $a > 0$.

7. (a) $\displaystyle\int_{|z|=5} \frac{dz}{(z^2-\pi^2)^5}$ (b) $\displaystyle\int_{|z|=5} \frac{\cos z\, dz}{(z+\pi)^5}$

8. $\displaystyle\int_0^{2\pi} \frac{\cos\theta\, d\theta}{10+8\sin\theta}$

9. Show (a) $\displaystyle\frac{1}{2\pi i}\int_{|z|=2} \frac{e^{az} dz}{z^2+1} = \sin a$

        (b) $\displaystyle\frac{1}{2\pi i}\int_{|z|=2} \frac{z e^{az} dz}{z^2+1} = \cos a$

10. If $f(z)$ is analytic for $|z| \leq 1$, and $|f(z)| < 1$, then show $|f^{(n)}(0)| \leq n!$.
11. In the proof of Theorem 3.2d we integrated by parts. Why are the integrated terms zero?

B Exercises

1. If one tries to apply the proof of Liouville's theorem to an arbitrary entire function, it fails. Say exactly where and how it fails.
2. In the last step in the proof of the fundamental theorem of algebra we asserted that the conclusion $p(z)$ is a constant is a contradiction. How?
3. If C is a contour and $f(z)$ is analytic in and on C, and $\alpha \in I(C)$, $\beta \in I(C)$, with $\alpha \neq \beta$, then show

$$I = \int_C \frac{f(z)dz}{(z-\alpha)(z-\beta)} = \int_{|z-\alpha|<\rho} \frac{f(z)dz}{(z-\alpha)(z-\beta)}$$
$$+ \int_{|z-\beta|<\rho} \frac{f(z)dz}{(z-\alpha)(z-\beta)}$$

where ρ is sufficiently small. How big can ρ be? Conclude

$$I = 2\pi i \frac{f(\alpha) - f(\beta)}{\alpha - \beta}.$$

4. (Compare A10 above) If $f(z)$ is analytic for $|z| < 1$ and $|f(z)| \leq 1$, then show $|f^n(0)| \leq n!$.
5. Suppose $f(z) = u(z) + iv(z)$ is an entire function, and there is a constant M for which $u(z) \leq M$. Apply Liouville's theorem to $e^{f(z)}$ and conclude that $f(z)$ is a constant.
6. Show $\int_{|z|=1} \frac{e^{az}}{z} dz = 2\pi i$, and deduce that if a is real, then

$$\int_0^\pi e^{a \cos \theta} \cos(a \sin \theta) = \pi$$

7. Show that

$$\int_0^{2\pi} \frac{e^{3e^{i\theta}} d\theta}{5 - 3\cos \theta} = \frac{\pi e}{2}$$

and deduce the value of

$$\int_0^\pi \frac{e^{3 \cos \theta} \cos(3 \sin \theta) d\theta}{5 - 3\cos \theta}.$$

8. Fill in the details in the proof of Theorem 3.2c.

9. Suppose $f(z)$ is an entire function and that it is periodic with period $p > 0$ and with period $q = a + ib$ with $b \neq 0$. Show that $f(z)$ is a constant. Hint. Consider the parallelogram cornering at $0, p, q, p + q$, and its translates.

C Exercises

1. Suppose $f(z)$ is analytic for $|z| \leq 1$. Show that for $|z| < 1$

$$f(z) = \frac{1}{2\pi i} \int_{|\zeta|=1} \frac{1 - z\bar{z}}{(\zeta - z)(1 - \zeta\bar{z})} \, f(\zeta)d\zeta.$$

2. (Continuation) Deduce that for $z = re^{i\theta}$ with $r < 1$

$$f(z) = \frac{1}{2\pi} \int_0^{2\pi} \frac{1 - r^2}{1 - 2r\cos(\theta - t) + r^2} \, f(e^{it})dt$$

3. Suppose C is a contour and that $f(z)$ is continuous in and on C and is analytic in $I(C)$. Show that for $z \in I(C)$

$$f^{(n)}(z) = \frac{1}{2\pi i} \int_C \frac{f(\zeta)}{(\zeta - z)^{n+1}} \, d\zeta.$$

Hint: Assume Theorem 3.1e (see C4, Section 3.1) which was established only for the case of a circle.

3.3 Taylor and Laurent Series

We commented without proof in Section 2.2 that representability by power series characterizes analytic functions. There we showed that any power series

$$\sum_{k=0}^{\infty} a_k(z - \alpha)^k, \tag{1}$$

whose sum we denote by $f(z)$, is analytic within its disk of convergence, that its derivatives can be computed termwise in that disk, and that the coefficients are related to the sum by

$$a_n = f^{(n)}(\alpha)/n! \tag{2}$$

We are now in a position to prove the rest of the statement, namely that any function $f(z)$ which is analytic at a point α has a power series expansion about α with a non-zero radius of convergence. This series is in the form (1) with coefficients given by (2).

One should understand clearly the difference between the two previous paragraphs. In the first we are *given a power series and we discuss properties of its sum.* In the second *we are given an analytic function* (by whatever means) and we assert that it can be represented by a power series; that is, it can be expressed as the sum of a power series.

This expansion of an analytic function in a power series in the form (1) is called the **Taylor series** of $f(z)$ about α. If $\alpha = 0$ it is sometimes called the **Maclaurin series** of $f(z)$, but we will call it the *Taylor series about 0.*

3.3a Theorem. *Let $f(z)$ be analytic in a region R with $\alpha \in R$ and with $\delta = dist(\alpha, \partial R)$, where ∂R is the boundary of R. Then*

$$f(z) = \sum_{0}^{\infty} c_n (z - \alpha)^n, \quad |z - \alpha| < \delta \qquad (3)$$

where

$$c_n = \frac{f^n(\alpha)}{n!} = \frac{1}{2\pi i} \int_{|\zeta - \alpha| = a} \frac{f(\zeta)}{(\zeta - \alpha)^{n+1}} \, d\zeta, \quad 0 < a < \delta. \qquad (4)$$

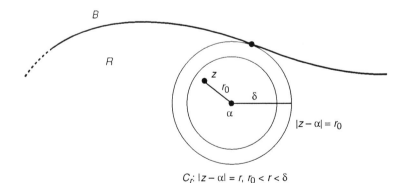

$C_r: |z - \alpha| = r, \ r_0 < r < \delta$

Taylor's Theorem

Proof. Choose z so that $|z - \alpha| < \delta$ and denote $|z - \alpha|$ by r_0. Then choose r between r_0 and δ, for example, $r = (r_0 + \delta)/2$. By the Cauchy integral formula we have

$$f(z) = \frac{1}{2\pi i} \int_{|\zeta - \alpha| = r} \frac{f(\zeta)}{\zeta - z} \, d\zeta.$$

The idea is to expand $1/(\zeta - z)$ in powers of $(z - \alpha)$ and integrate termwise. To do this we write

$$\frac{1}{\zeta - z} = \frac{1}{(\zeta - \alpha) - (z - \alpha)} = \frac{1}{\zeta - \alpha} \frac{1}{1 - \left(\frac{z - \alpha}{\zeta - \alpha}\right)}$$

$$= \frac{1}{\zeta - \alpha} G\left(\frac{z - \alpha}{\zeta - \alpha}\right) = \sum \frac{(z - \alpha)^n}{(\zeta - \alpha)^{n+1}}$$

where G is the geometric series, formula (5) of Section 2.2. The series converges since its argument is $(z - \alpha)/(\zeta - \alpha)$ and

$$\left|\frac{z - \alpha}{\zeta - \alpha}\right| = \frac{r_0}{r} < 1.$$

In fact the series converges uniformly with respect to ζ and so can be integrated termwise by Theorem 3.5b which will be proved in Section 3.5. This gives

$$f(z) = \frac{1}{2\pi i} \int_{|\zeta - \alpha| = r} \frac{f(\zeta)}{\zeta - z} \, d\zeta = \frac{1}{2\pi i} \int_{|\zeta - \alpha| = r} \frac{f(\zeta)}{\zeta - \alpha} G\left(\frac{z - \alpha}{\zeta - \alpha}\right) d\zeta$$

$$= \frac{1}{2\pi i} \int_{|\zeta - \alpha| = r} \frac{f(\zeta)}{\zeta - \alpha} \sum_{n=0}^{\infty} \frac{(z - \alpha)^n}{(\zeta - \alpha)^n} \, d\zeta$$

$$= \sum c_n (z - \alpha)^n$$

where

$$c_n = \frac{1}{2\pi i} \int_{|\zeta - \alpha| = r} \frac{f(\zeta)}{(\zeta - \alpha)^{n+1}} \, d\zeta = \frac{f^{(n)}(\alpha)}{n!}.$$

Further, by Cauchy's theorem for two contours, the integral for c_n can as well be taken over $|\zeta - \alpha| = a$ for any a, $0 < a < \delta$.

Let us examine more carefully the convergence argument, and instead of citing the uniform convergence theorem which could have been proved, but has not been, we prove directly that the integrated series converges. To do this we recall that for $|\rho| < 1$,

$$G(\rho) = \frac{1}{1 - \rho} = \sum_0^n \rho^k + \frac{\rho^{n+1}}{1 - \rho}.$$

This follows from (10) section 2.2 or by multiplying both sides here by $1-\rho$.
Setting $\rho = (z - \alpha)/(\zeta - \alpha)$ we have

$$\frac{1}{\zeta - z} = \frac{1}{\zeta - \alpha}\frac{1}{1 - \left(\frac{z-\zeta}{\zeta-\alpha}\right)}$$

$$= \frac{1}{\zeta - \alpha}\sum_0^n \frac{(z - \alpha)^k}{(\zeta - \alpha)^k} + \frac{1}{\zeta - \alpha}\frac{\frac{(z-\alpha)^{n+1}}{(\zeta-\alpha)^{n+1}}}{1 - \frac{z-\alpha}{\zeta-\alpha}}$$

$$= \sum_0^n (z - \alpha)^k\frac{1}{(\zeta - \alpha)^{k+1}} + \frac{(z - \alpha)^{n+1}}{(\zeta - \alpha)^{n+1}(\zeta - z)}.$$

Then multiplying by $f(\zeta)/2\pi i$ and integrating around $|\zeta - \alpha| = r$ we get

$$f(z) = \frac{1}{2\pi i}\int_{|\zeta-\alpha|=r}\frac{f(\zeta)}{\zeta - z}\,d\zeta$$

$$= \sum_0^n c_k(z - \alpha)^k + \frac{(z - \alpha)^{k+1}}{2\pi i}\int_{|\zeta-\alpha|=r}\frac{f(\zeta)d\zeta}{(\zeta - \alpha)^{n+1}(\zeta - z)}$$

or

$$f(z) = \sum_0^n c_k(z - \alpha)^k + R_n \tag{7}$$

with

$$R_n = \frac{(z - \alpha)^{n+1}}{2\pi i}\int_{|\zeta-\alpha|=r}\frac{f(\zeta)d\zeta}{(\zeta - \alpha)^{n+1}(\zeta - z)}. \tag{8}$$

The formula (7) is **Taylor's formula with remainder**, and (8) is
the **remainder** after $n + 1$ terms, that is, after the nth power of $(z - \alpha)$.

To complete the proof we show that $R_n \to 0$ as $n \to \infty$. To see this
we observe that when ζ is on the circle $|\zeta - \alpha| = r$, we have

$$|\zeta - z| = |(\zeta - \alpha) - (z - \alpha)| \geq |\zeta - \alpha| - |z - \alpha| = r - r_0.$$

Furthermore, by the continuity of $f(\zeta)$, there is a constant M for which

$$|f(\zeta)| \leq M \quad \text{for} \quad |\zeta - \alpha| = r.$$

Then, from (8),

$$|R_n| = \frac{|z - \alpha|^{n+1}}{2\pi}\left|\int_{|\zeta-\alpha|=r}\frac{f(\zeta)d\zeta}{(\zeta - \alpha)^{n+1}(\zeta - z)}\right|$$

$$< \frac{r_0^{n+1}}{2\pi}\cdot\frac{M}{r^{n+1}(r - r_0)}\cdot 2\pi r = \frac{rM}{r - r_0}\left(\frac{r_0}{r}\right)^{n+1} \to 0$$

as $n \to \infty$ since $r_0/r < 1$. $\square$

This theorem shows that the radius of convergence of the Taylor series centered at α is at least as great at δ, the distance from α to the boundary of R, the known region of analyticity. There are examples that show that *it can be greater.*

Note that the formula for the coefficients $c_n = f^{(n)}(\alpha)/n!$ is the same as for any power series. This has the consequence that any power series must be the Taylor series of its sum. Thus

$$G(z) = 1 + z + z^2 + \cdots + z^n + \cdots$$

is the Taylor series at 0 of $1/(1-z)$, though we computed the series by other means.

This observation can be useful in both directions: If we have a power series expansion, as in the case of $G(z)$, then we can read off the coefficients, and from (4), we have the nth derivative of $f(z)$ evaluated at α, and, of course, the integrals evaluated. Conversely, given an analytic function we can use (4) to compute the coefficients and hence obtain the Taylor series.

Example 1. Find the Taylor series of $\sin z$ about $\pi/2$.

Solution 1. We compute derivatives and evaluate at $\pi/2$:

n	at z	at $\pi/2$	coefficients
0	$\sin z$	1	1
1	$\cos z$	0	0
2	$-\sin z$	-1	$-1/2!$
3	$-\cos z$	0	0
4	$\sin z$	1	$1/4!$
$\vdots$	$\vdots$	$\vdots$	$\vdots$
$2k$	$(-1)^k \sin z$	$(-1)^k$	$(-1)^k/(2k)!$
$2k+1$	$(-1)^k \cos z$	0	0
$\vdots$	$\vdots$	$\vdots$	$\vdots$

The series we want is then

$$\sin z = 1 - \frac{1}{2!}\left(z - \frac{\pi}{2}\right)^4 + \cdots + \frac{(-1)^k}{(2k)!}\left(z - \frac{\pi}{2}\right)^{2k} + \cdots$$

$$= \sum_0^\infty \frac{(-1)^k}{(2k)!}\left(z - \frac{\pi}{2}\right)^{2k}.$$

Solution 2. We can write, by the addition formula for the sine,

$$\sin z = \sin\left[\frac{\pi}{2} + \left(z - \frac{\pi}{2}\right)\right] = \cos\left(z - \frac{\pi}{2}\right).$$

By the definition of the cosine we get

$$\sin z = \sum_{0}^{\infty} \frac{(-1)^k}{(2k)!}\left(z - \frac{\pi}{2}\right)^{2k}$$

If we proceed this way to get the power series for $\sin z$, how do we know a priori that the result is the Taylor series? See the two paragraphs just preceding this example.

Example 2. Suppose $1/(z-2)$ is expanded in a Taylor series centered at $1 + 2i$. Find its radius of convergence without actually finding the series itself.

Solution. The radius of convergence ρ is the distance from the center $(1 + 2i)$ to the nearest point where the function fails to be analytic. Since $1/(z-2)$ fails to be analytic only at 2, we have

$$\rho = |(1 + 2i) - 2| = |2i - 1| = \sqrt{5}.$$

The Taylor series is a power series and so converges inside a circle. We then observe that a series of negative powers such as

$$f(z) = \sum_{1}^{\infty} a_n(z - \alpha)^{-n} = \sum_{1}^{\infty} \frac{a_n}{(z - \alpha)^n}$$

must converge, if it converges at all, for those values of $z - \alpha$ for which

$$1/|z - \alpha| < R$$

for some $R > 0$. This is then, for

$$|z - \alpha| > 1/R$$

and so is the *exterior* of a circle.

If such a series is added to a power series (with the same center α) we get a series of the form

$$\sum_{-\infty}^{\infty} c_n(z - \alpha)^n = \sum_{-\infty}^{-1} c_n(z - \alpha)^n + \sum_{0}^{\infty} c_n(z - \alpha)^n$$

$$= \sum_{1}^{\infty} \frac{c_{-n}}{(z - \alpha)^n} + \sum_{0}^{\infty} c_n(z - \alpha)^n$$

$$(9)$$

The first series on the right of (9) converges *outside* some circle of radius a, say, and the second *inside* a circle of radius b. If $a < b$, the full series on the left of (9) then converges in a ring region or *annulus* A described by

$$A = \{z : a < |z - \alpha| < b\}.$$

When we say that a doubly infinite series such as (9) converges, we mean that both series on the right converge. The series (9) then represents an analytic function in the annulus of convergence.

The converse of this, namely, that a function analytic and single valued in such an annulus A has such an expansion valid in A is the content of the next theorem. This is **Laurent's Theorem**. The adjectival phrase "single valued" is, strictly speaking, redundant, but it is included for emphasis and as a mnemonic device to help one avoid such multivalued functions as $\sqrt{z - \alpha}$ or $\log(z - \alpha)$ in $A = \{z : a < |z - \alpha| < b\}$. Any branch of these is locally analytic in A, but there is no single valued branch analytic in all of A.

3.3c Theorem (Laurent). *Suppose* $0 \le a < r < b$, *and that* $f(z)$ *is analytic in* $A = \{z : a < |z - \alpha| < b\}$. *Then*

$$f(z) = \sum_{-\infty}^{\infty} c_n (z - \alpha)^n, \quad z \in A \tag{10}$$

where

$$c_n = \frac{1}{2\pi i} \int_{|z-\alpha|=r} \frac{f(\zeta)}{(\zeta - \alpha)^{n+1}} \, d\zeta, \quad -\infty < n < \infty \tag{11}$$

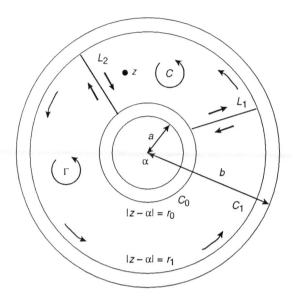

Laurent's Theorem

Proof. Choose $z \in A$, and then choose r_0 and r_1 so that $a < r_0 < |z - \alpha| < r_1 < b$. Let C_0 be the circle $|z - \alpha| = r_0$, and C_1 the circle $|z - \alpha| = r_1$, both taken in the positive sense. Connect C_0 with C_1 by two radial setments L_1 and L_2 which do not pass through z. These two segments, together with arcs of C_0 and C_1 form two contours C and Γ with z interior to one, say C as illustrated. Then

$$f(z) = \frac{1}{2\pi i} \int_C \frac{f(\zeta)}{\zeta - z} \, d\zeta$$

by the Cauchy integral formula, and

$$0 = \frac{1}{2\pi i} \int_\Gamma \frac{f(\zeta)}{\zeta - z} \, d\zeta$$

by Cauchy's theorem. If we add these two equations the contribution to the integrals arising from the segments L_1 and L_2 cancel and we are left with

$$f(z) = -\frac{1}{2\pi i} \int_{C_0} \frac{f(\zeta)}{\zeta - z} \, d\zeta + \frac{1}{2\pi i} \int_{C_1} \frac{f(\zeta)}{\zeta - z} \, d\zeta$$
$$\equiv f_0(z) + f_1(z)$$

respectively. We expand $f_1(z)$ into a power series, *exactly repeating the argument* in the proof of Taylor's theorem. This gives

$$f_1(z) = \sum_0^\infty c_n(z - \alpha)^n, \quad c_n = \frac{1}{2\pi i} \int_C \frac{f(\zeta)}{(\zeta - \alpha)^{n+1}} \, d\zeta$$

We make the following observations: This power series converges for $|z - \alpha| < b$ (how?) and so represents a function analytic *inside* the circle of center α and radius b. Secondly by Cauchy's theorem for two contours the integrals for the coefficients, which are over C_1, can just as well be taken over C_r:

$$\frac{1}{2\pi i} \int_{C_1} \frac{f(\zeta)d\zeta}{(\zeta - \alpha)^{n+1}} = \frac{1}{2\pi i} \int_{C_r} \frac{f(\zeta)d\zeta}{(\zeta - \alpha)^{n+1}}, \quad n = 0, 1, 2, \dots \,.$$

The integral which we denoted by $f_0(z)$ we handle in an analogous, but significantly different way. In that integral ζ is nearer to α than z is, and so we expand as follows:

$$-\frac{1}{\zeta - z} = \frac{1}{z - \zeta} = \frac{1}{(z - \alpha) - (\zeta - \alpha)}$$

$$= \frac{1}{(z - \alpha)} \cdot \frac{1}{1 - \left(\frac{\zeta - \alpha}{z - \alpha}\right)}.$$

We can use the geometric series for the second factor since $|(\zeta - \alpha)/(z - \alpha)| < 1$ for $\zeta \in C_0$. Thus we can integrate termwise and get

$$f_0(z) = \sum_{n=0}^\infty \frac{c_{-(n+1)}}{(z - \alpha)^{-(n+1)}} = \sum_1^\infty \frac{c_{-n}}{(z - \alpha)^{-n}} = \sum_{-\infty}^{-1} c_n(z - \alpha)^n$$

where

$$c_n = \frac{1}{2\pi i} \int_{C_0} \frac{f(\zeta)d\zeta}{(\zeta - \alpha)^{n+1}}, \quad n = -1, -2, -3, \dots \,.$$

The termwise integration can be justified by the uniform convergence theorem, to come in Section 3.5, or we can give estimates as we did in the proof of Taylor's theorem. $\square$

Example 3. Find the Laurent series for $(z - 2)/[(z + 1)(z - 3)]$ in the annulus $1 < |z| < 3$.

Solution 1. The center of the annulus is the origin, and so by definition of the coefficients we have

$$c_n = \frac{1}{2\pi i} \int_{|z|=2} \frac{2z - 2}{(z + 1)(z - 3)} \frac{1}{z^{n+1}} \, dz$$

where any circle $|z| = r$, if $1 < r < 3$, would work as well as $|z| = 2$. (Why?) Thus

$$c_n = \frac{1}{2\pi i} \int_{|z|=2} \left[\frac{1}{z+1} + \frac{1}{z-3} \right] \frac{1}{z^{n+1}} \, dz$$

$$c_n = \frac{1}{2\pi i} \int_{|z|=2} \frac{dz}{(z+1)z^{n+1}} + \frac{1}{2\pi i} \int_{|z|=2} \frac{dz}{(z-3)z^{n+1}} \qquad (12)$$

For $n \geq 0$ the first integral in (12) is (how?)

$$\frac{1}{z^{n+1}}\bigg|_{z=-1} + \frac{1}{n!} \left(\frac{d}{dz} \right)^n \frac{1}{z+1}\bigg|_{z=0} = (-1)^{n+1} + (-1)^n = 0.$$

And the second integral in (12) is

$$\frac{1}{n!} \left(\frac{d}{dz} \right)^n \frac{1}{z-3}\bigg|_{z=0} = -\frac{1}{3^{n+1}}.$$

So $c_n = -1/3^{n+1}$ for $n \geq 0$.

Then for $n < 0$ the second integral in (12) is 0 (how?) and the first (again, how?) is $1/z^{n+1}\big|_{z=-1} = (-1)^{n+1}$. Thus

$$c_n = \begin{cases} (-1)^{n+1} & n < 0 \\ -1/3^{n+1} & n \geq 0, \end{cases}$$

so that the desired series is

$$\sum_{-\infty}^{-1} (-1)^{n+1} z^n + \sum_{0}^{\infty} (-1/3^{n+1}) z^n = \sum_{1}^{\infty} \frac{(-1)^{n+1}}{z^n} - \sum_{0}^{\infty} \frac{z^n}{3^{n+1}}.$$

Solution 2. Since

$$\frac{2z - 2}{(z+1)(z-3)} = \frac{1}{z+1} + \frac{1}{z-3}$$

we expand $1/(z+1)$ for $|z| > 1$ by using the geometric series as follows:

$$\frac{1}{z+1} = \frac{1}{z} \frac{1}{1 + (1/z)} = \frac{1}{z} \sum_{0}^{\infty} (-1)^n \left(\frac{1}{z} \right)^n$$

$$= \sum_{0}^{\infty} \frac{(-1)^n}{z^{n+1}} = \sum_{1}^{\infty} \frac{(-1)^{n+1}}{z^n}.$$

And we expand $1/(z-3)$ for $|z| < 3$, also using the geometric series:

$$\frac{1}{z-3} = -\frac{1}{3}\frac{1}{1-(z/3)} = -\frac{1}{3}\sum_0^\infty \left(\frac{z}{3}\right)^n$$

$$= -\sum_0^\infty \frac{z^n}{3^{n+1}}.$$

Adding these again yields

$$\sum_1^\infty \frac{(-1)^{n+1}}{z^n} - \sum_0^\infty \frac{z^n}{3^{n+1}}$$

as the required series.

If $f(z)$ is analytic at α and $f(\alpha) \neq 0$, then $1/f(z)$ is analytic there and so has a Taylor series at α. This remark, together with Exercise C3 of Section 2.3, provides the proof of Theorem 2.3h.

Significant Topics   Page

A Exercises

1. Find the Taylor Series of the following functions about the indicated centers:

    (a) $1/z, 1$;   (b) $1/z, 2$:   (c) $\cos z, \pi/2$;   (d) $\sin z, \pi$;

    (e) $(\sin z - z)/z^3, 0$;   (f) $\int_0^z \frac{\sin \zeta}{\zeta}\, d\zeta, 0$;   (g) $\text{Log}\,(1+z), 0$;

    (h) $\text{Log}\,[(1+z)/(1-z)], 0$;   (i) principal value of $\sqrt{1+z}, 0$;

    (j) principal value of $\sqrt{z}, 2i$;

    (k) $\sin^2 z, 0$ (Hint: $\sin^2 z = (1 - \cos 2z)/2$);

    (ℓ) $\cos^2 z, 0$.

2. State the radius of convergence of each series in A1. Use Theorem 3.3a.

3. Find the Laurent series of the following functions in the indicated annuli.

(a) $1/(1 - z), |z| > 1$;
(b) $\cos(1/z), |z| > 0$;
(c) $1/[z(1 - z)], 0 < |z| < 1$;
(d) $1/[z(1 - z)], |z| > 1$;
(e) $1/[z(1 - z)], 0 < |z - 1| < 1$;
(f) $1/[z(1 - z)], |z - 1| > 1$;
(g) $(z + 1)/z, |z| > 0$;
(h) $1/z, |z| > 0$;
(i) $(\sin z)/z^3, |z| > 0$;
(j) $(\cos z)/z^3, |z| > 0$.

4. Find the first few terms of the Laurent expansion of

(a) $\tan z$; $0 < \left| z - \frac{\pi}{2} \right| < \pi$,

(b) $\csc z$; $0 < |z| < \pi$.

5. Find the Laurent series of the following and discuss the character of the functions at the centers:

(a) $(\sin z)/z, |z| > 0$;
(b) $(1 - \cos z)/z^2, |z| > 0$

(c) $z \cot z, |z| > 0$ (to 3 non-zero terms);

(d) $(z^3 - 3z^2 + 2z)/z, |z| > 0$.

6. Find the first three nonzero terms of the Taylor series about 0 of

(z) $e^{\sin z}$
(b) $\arcsin z$ (i.e., $\sin^{-1} z$)

B Exercises

1. Suppose $f(z)$ is defined by

$$f(z) = \sum_{-\infty}^{\infty} c_n (z - \alpha)^n$$

and the series converges in $a < |z - \alpha| < b$, where $a < b$, of course. Show that $f(z)$ is analytic in its annulus of convergence and the given series is its Laurent expansion there.

2. The function $\dfrac{1}{z + 1} + \dfrac{1}{z + 3} + \dfrac{1}{z - 4}$ has a Laurent expansion *centered* at 4 which *converges* at $z = -2$. Find it and describe fully the set on which it converges.

3. Suppose $f(z)$ is an entire function which satisfies

$$|f(z)| \leq A + B|z|^k$$

where A, B and k are positive constants. Show that $f(z)$ is an polynomial of degree no larger than k. Hint: Estimate the integral for the coefficient c_n in the Taylor series for $f(z)$ about 0.

4. Suppose $f(z)$ is analytic for $|z| \leq 1$.

   (a) Show that

   $$\int_{|z|=1} \overline{f(z)} \, z^k \, dz = 0, \quad k = 2, 3, \ldots \, .$$

   (b) If $f(z) = u(z) + iv(z)$ show that

   $$f'(0) = \frac{1}{\pi} \int_0^{2\pi} u(e^{i\theta}) e^{-i\theta} \, d\theta.$$

   Hint: Use (a) with $k = 2$, and the Cauchy formula for $f'(0)$.

5. (a) Show that each branch of

   $$f(z) = \sqrt{z(z-1)(z-2)(z-3)}$$

   has a Laurent expansion in $1 < |z| < 2$ and another for $|z| > 3$.

   (b) Where does $f(z)$ have a Laurent expansion centered at 1? At -1?

6. Show $|\text{Log}\,(1-z)| < |z|/(1-|z|)$ for $0 < |z| < 1$.

7. Let R be a region, $\alpha \in R$, $\rho = \text{dist}\,(\alpha, \partial R)$ and $u(z)$ be harmonic in R. Set $z - \alpha = re^{i\theta}$ and show that

   $$u(\alpha + re^{i\theta}) = \sum_0^\infty r^n \left[a_n \cos n\theta - b_n \sin n\theta \right]$$

   for $0 \leq r < \rho$. What are a_n and b_n?

C Exercises

1. Show $\displaystyle\int_0^{2\pi} e^{2\cos\theta} \, d\theta = 2\pi \sum_0^\infty \frac{1}{(n!)^2}$.

2. Evaluate $\displaystyle\int_0^{2\pi} e^{2\sin\theta} \, d\theta$.

3. Suppose $f(z)$ is analytic for $|z| < 1$ and $|f(z)| \leq M$. If $f(z) = \sum_0^\infty a_n z^n$,

   show that $\displaystyle\sum_0^\infty |a_n|^2 \leq M^2$. Hint: For $r < 1$

   $$|f(re^{i\theta})|^2 = f(re^{i\theta}) \overline{f(re^{i\theta})}$$

   $$= \left(\sum_0^\infty a_n r^n e^{in\theta} \right) \left(\sum_0^\infty \bar{a}_k r^k e^{-ik\theta} \right)$$

and integrate from 0 to 2π with respect to θ.

4. (See Exercise B3 above.) Show that if $f(z)$ is entire and

$$|f(z)| \leq M|z|^k$$

where M is a positive constant and k is a positive integer, then $f(z) = \alpha z^k$ where α is a constant.

5. In C4 above, if k is a positive constant, and *not an integer*, show that $f(z) \equiv 0$.

6. Suppose $f(z)$ is analytic in the strip $|y| < a$, and that $f(z)$ is periodic with period 2π, i.e., $f(z + 2\pi) = f(z)$ in the strip.

   (a) By using the substitution $w = e^{iz}$ and Laurent's theorem show that

   $$f(z) = \sum_{-\infty}^{\infty} c_n e^{inz}, \quad c_n = \frac{1}{2\pi} \int_0^{2\pi} f(x)e^{-inx} dx$$

   where the series converges in the strip.

   (b) Show further that

   $$c_n = \frac{1}{2\pi} \int_{ib}^{2\pi+ib} f(z)e^{-inz} dz$$

   for any real b with $|b| < a$.

7. If we apply Laurent's theorem to $e^{\frac{w}{2}(z-\frac{1}{z})}$ as a function of z, the coefficients then are a function of the parameter w:

   $$\exp\left\{\frac{w}{2}(z - \frac{1}{z})\right\} = \sum_{-\infty}^{\infty} J_n(w)z^n, \quad |z| > 0.$$

   The coefficients J_n are called Bessel functions. Show

   (a) for each w the series converges for all $z \neq 0$;

   (b) $J_n(w) = \frac{1}{\pi} \int_0^{\pi} \cos(w \sin\theta - n\theta)d\theta$;

   (c) $J_n(w) = (-1)^n J_{-n}(w)$;

   (d) $J_0(w) = \sum_0^{\infty} \left(-\frac{w^2}{4}\right)^k \frac{1}{(k!)^2}$.

   (e) $J_0(2i) = \frac{1}{2\pi} \int_0^{2\pi} e^{2\cos\theta} d\theta$

8. If we apply Taylor's theorem at the origin to the principal value of $[1 - 2tz + z^2]^{-1/2}$, as a function of z, the coefficients are functions of t:

$$[1 - 2tz + z^2]^{-1/2} = \sum_0^\infty P_n(t)z^n.$$

The coefficients are called Legendre polynomials. Show
   (a) The series converges for $|z|$ less than the smaller of $|t + \sqrt{t^2 - 1}|$ and $|t - \sqrt{t^2 - 1}|$.
   (b) Verify $P_0(t) = 1$, $P_1(t) = t$, $P_2(t) = \frac{1}{2}(3t^2 - 1)$.

3.4 Mean Values, Maximum Principle, Zeros

In this section we derive some important properties of analytic and harmonic functions most of which are consequences of the three preceding sections of this chapter. We begin with the **Gauss Mean Value Theorem**.

3.4a Theorem. *Let $f(z)$ be analytic in a region R with $\alpha \in R$. Denote by a the distance from α to the boundary of $R : a = dist(\alpha, \partial R)$. Then*

$$f(\alpha) = \frac{1}{2\pi} \int_0^{2\pi} f(\alpha + re^{i\theta})d\theta \tag{1}$$

for each r, $0 < r < a$.

Proof. Let C_r be the circle $|\zeta - \alpha| = r$. Then

$$\zeta = \alpha + re^{i\theta}, \quad d\zeta = ire^{i\theta}d\theta.$$

By the Cauchy integral formula

$$f(\alpha) = \frac{1}{2\pi i} \int_{C_r} \frac{f(\zeta)}{\zeta - \alpha} \, d\zeta = \frac{1}{2\pi i} \int_0^{2\pi} \frac{f(\alpha + re^{i\theta})ire^{i\theta}d\theta}{re^{i\theta}}$$

$$= \frac{1}{2\pi} \int_0^{2\pi} f(\alpha + re^{i\theta})d\theta. \quad \square$$

Since the element of arc length on C_r is $ds = rd\theta$, the formula (1) is sometimes written

$$f(\alpha) = \frac{1}{2\pi r} \int_{C_r} f(\zeta)ds.$$

3.4b Theorem. *Suppose $u(z) = u(x, y)$ is harmonic in a region R with $\alpha = a + ib \in R$. Let $\rho = dist(\alpha, \partial R)$. Then*

$$u(z) = \frac{1}{2\pi} \int_0^{2\pi} u(\alpha + re^{i\theta}) d\theta$$

$$= \frac{1}{2\pi} \int_0^{2\pi} u(a + r\cos\theta, b + r\sin\theta) d\theta \tag{2}$$

for each r, $0 < r < \rho$.

Proof. There is an analytic function defined in $|z - \alpha| < a$ for which $u(z) = \operatorname{Re} f(z)$. Then (2) follows from (1) by taking the real parts of both sides. $\square$

We next get the **maximum modulus theorem** which is also called the **maximum principle** for analytic functions.

3.4c Theorem. *Let $f(z)$ be analytic in a region R. If there is a point z_0 in R for which*

$$|f(z_0)| \geq |f(z)| \qquad z \in R, \tag{3}$$

then $f(z) \equiv constant$ in R.

Proof. First we show that if such a point z_0 exists, and if $|f(z)|$ is not a constant, we are led to a contradiction. So let $|f(z_0)| = M$. Then $|f(z)| \leq M$ in R, by (3). Now suppose there is a point z_1 in R for which $|f(z_1)| < M$. We connect z_0 to z_1 by a polygonal line L in R. Then there is a last point α on L, going from z_0 to z_1 at which $|f(z_0)| = M$. (See Exercise C1). This point α might be z_0 itself, or it might be further along L as illustrated, but it does occur on L before z_1.

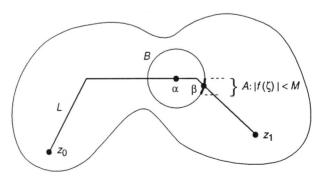

Proof of Theorem 3.4c

We choose a so that $|z - \alpha| \leq a$ is in R, and apply the Gauss mean value theorem

$$f(\alpha) = \frac{1}{2\pi a} \int_{C_a} f(\zeta) ds$$

where C_a is the circle $|z - \alpha| = a$. Then

$$M = |f(\alpha)| \leq \frac{1}{2\pi r} \int_{C_a} |f(\zeta)| ds$$

The circle C_a cuts L at a point β between α and z_1, and $|f(\beta)| < M$ (why?). Then, since $|f(\zeta)|$ is continuous, there is an arc A of the circle on which $|f(\zeta)| < M$. Call the complementary arc B. Then

$$M \leq \frac{1}{2\pi a} \int_A |f(\zeta)| ds + \frac{1}{2\pi a} \int_B |f(\zeta)| ds \tag{4}$$

But, as we noted $|f(\zeta)| < M$ on A and also $|f(\zeta)| \leq M$ on B. So

$$\frac{1}{2\pi a} \int_A |f(\zeta)| ds < \frac{1}{2\pi a} \cdot M \cdot a \, (\text{Angle } A)$$

and

$$\frac{1}{2\pi a} \int_B |f(\zeta)| ds \leq \frac{1}{2\pi a} \cdot M \cdot a \, (\text{Angle } B)$$

so that, from (4),

$$M < \frac{1}{2\pi} M \, (\text{Angle } A) + \frac{1}{2\pi} M \, (\text{angle } B)$$
$$= \frac{1}{2\pi} M \, (\text{Angle } A + \text{Angle } B) = \frac{1}{2\pi} M \cdot 2\pi$$
$$= M.$$

This says $M < M$, which is a contradiction, and so $|f(z)|$ must be constant. Then it follows that $f(z) = \text{constant}$. $\square$

3.4d Corollary. *Suppose $f(z)$ is analytic in a region R and never zero in R. If there is a point $z_0 \in R$ for which*

$$|f(z_0)| \leq |f(z)| \qquad z \in R,$$

then $f(z)$ is a constant in R.

For the proof see Exercise A3. See also Exercise A4.

The maximum principle for harmonic functions takes a slightly different form.

3.4e Theorem. *Suppose $u(z)$ is harmonic in a region R, and suppose there is a point z_0 in R for which either (i) $u(z_0) \geq u(z)$ or (ii) $u(z_0) \leq u(z)$ for all $z \in R$. Then $u(z)$ is constant in R.*

The proof is almost identical with that of Theorem 3.4c and so is left as an exercise. You should note that unless a harmonic function u is a constant *it can achieve neither a maximum nor a minimum* in its region of harmonicity.

The following theorem is called the **identity theorem**. It is very useful, as its name implies, in establishing certain identities. It also is the basis for the very important concept of analytic continuation (Section 3.6).

3.4f Theorem. *If $f(z)$ is analytic in a region R, $\alpha \in R$, and $\{z_k\}^\infty$, is a sequence of distinct points in R for which $z_k \to \alpha$, and $f(z_k) = 0$ for $k = 1, 2, \ldots$, then $f(z) \equiv 0$ in R.*

Proof. Since $\alpha \in R$, there is a neighborhood of α in R for which

$$f(z) = \sum_0^\infty a_k(z - \alpha)^k = a_0 + a_1(z - \alpha) + a_2(z - \alpha)^2 + \cdots . \tag{5}$$

then

$$0 = a_0 + a_1(z_k - \alpha) + a_2(a_k - \alpha)^2 + \cdots \tag{6}$$

for each k. We let $k \to \infty$ so that $z_k \to \alpha$, and we conclude that $a_0 = 0$. Thus (6) reduces to

$$0 = a_1(z_k - \alpha) + a_2(z_k - \alpha)^2 + a_3(z_k - \alpha)^3 + \cdots .$$

We divide by $(z_k - \alpha)$ to get

$$0 = a_1 + a_2(z_k - \alpha) + a_3(z_k - \alpha)^2 + \cdots ,$$

let $k \to \infty$ and conclude $a_1 = 0$. Continuing in this way (we could give an induction proof) we conclude that $a_k = 0$ for $k = 1, 2, 3, \ldots$.

Thus the Taylor series (5) has all its coefficients equal to zero. This then says that $f(z) \equiv 0$ in a neighborhood of α. By an argument similar to that in Theorem 3.4c (see Exercise C2) we conclude $f(z) \equiv 0$ in R. $\square$

3.4g Corollary. *Suppose $f(z)$ and $g(z)$ are both analytic in a region R, and either*

(i) $$f(z) = g(z) \quad \text{in a disk} \quad |z - \alpha| < a \quad \text{in} \quad R$$

or

(ii) $$f(z) = g(z) \quad \text{on a path} \quad C \quad \text{in} \quad R$$

then $f(z) \equiv g(z)$ in R.

Proof. In either case there is a sequence of points to which 3.4f applies. (How?) □

Example 1. Does Theorem 3.4f apply to $\sin 1/z$ at $z = 0$?

Solution. We observe that $\sin 1/z = 0$ when $z = 1/k\pi$. Then choosing $a_k = 1/k\pi$ we see that $1/k\pi \to 0$. So we try to apply the identity theorem with $\alpha = 0$. However, 0 is not in the region of analyticity of $\sin 1/z$, so the theorem does not apply.

Example 2. Prove that $\sin 2z = 2 \sin z \cos z$.

Solution. For $z = x$, i.e., on the real axis, we have

$$f(z) = \sin 2z - 2 \sin z \cos z = 0.$$

Thus taking any convergent real sequence, for example $z_k = 1/k$, we have $f(z_k) = 0$. Further $z_k \to 0$, and this time 0 is in the region of analyticity. Thus $f(z) = 0$, that is, $\sin 2z = 2 \sin z \cos z$ for all z.

There are two properties of an analytic function $f(z)$ concerning points where $f(z) = 0$ which we now describe.

3.4h Theorem. *If $f(z)$ is analytic in a region R, and there is a point $\alpha \in R$ for which $f(\alpha) = 0$, then either $f(z) \equiv 0$ in R, or there is a deleted neighborhood of α in which $f(z) \neq 0$.*

Proof. Let us consider the set Z of all points besides α in R where $f(z) = 0$. Then either (i) $\text{dist}(\alpha, Z) = a > 0$, and so $f(z) \neq 0$ for $|z - \alpha| < a$ or (ii) $\text{dist}(\alpha, Z) = 0$ so that there is a sequence $\{z_k\}$ in Z with $z_k \to \alpha$. Then $f(z) \equiv 0$ by the identity theorem. □

This says that for nontrivial analytic functions, the zeros are isolated. The next result shows that they have integral order. The number n in this theorem is called the **order** of the zero.

3.4i Theorem. *If $f(z)$ is analytic in a region R, and α is an isolated zero of $f(z)$, then there is a positive integer n for which*

$$f(z) = (z - \alpha)^n g(z)$$

where $g(z)$ is analytic in R, and $g(\alpha) \neq 0$.

Proof. By Taylor's theorem there is a neighborhood of α in which

$$f(z) = a_0 + a_1(z - \alpha) + \cdots + a_k(z - \alpha)^k + \cdots , \tag{7}$$

where $a_0 = f(\alpha) = 0$. Then either $f(z) \equiv 0$, which is not possible since α is an isolated zero, or there is a first integer n for which $a_n \neq 0$. Then the Taylor expansion (7) reduces to

$$f(z) = a_n(z - \alpha)^n + a_{n+1}(z - \alpha)^{n+1} + \cdots$$

$$= (z - \alpha)^n \sum_0^\infty a_{n+k}(z - \alpha)^k.$$

Now this says that in the neighorhood of α

$$f(z)/(z - \alpha)^n = \sum_0^\infty a_{n+k}(z - \alpha)^k,$$

except at α itself, since we cannot divide by zero. But the series is analytic at α, so we observe that

$$g(z) = \begin{cases} f(z)/(z - \alpha)^n, & z \neq \alpha \\ a_n, & z = \alpha \end{cases}$$

is analytic in R and $g(\alpha) = a_n \neq 0$. With $g(z)$ defined this way we have

$$f(z) = (z - \alpha)^n g(z)$$

where $g(z)$ has the required properties. $\square$

This argument should be compared with the discussion of removable singularities in Chapter 4.

A Exercises

1. Determine the order of the zeros of the following at the indicated points:

   (a) $e^z - 1, 0$;   (b) $\cos z - 1, 0$;

   (c) $\sin z + (z - \pi), \pi$;   (d) $\tan z, \pi$;

   (e) $\tan z - (z - \pi), \pi$;   (f) $\sinh z - z, 0$;

   (f) $\sin z^2, 0$;   (h) $\cos z^2 - 1, 0$.

2. Use the identity theorem to prove

   (a) $\cos^2 z + \sin^2 z = 1$,

   (b) $\sin(z + \alpha) = \sin z \cos \alpha + \cos z \sin \alpha$.

3. Prove Corollary 3.4d.

4. Show that $f(z) = z^2$ in the disk $|z| < 1$ illustrates the necessity of the condition $f(z) \neq 0$ in Corollary 3.4d.

5. (See Exercise B1 below). If $p(z) = (z - 2)^3(z + 3)^2$, then show

   (a) $\dfrac{p'(z)}{p(z)} = \dfrac{3}{z - 2} + \dfrac{2}{z + 3}$;

   (b) $\dfrac{1}{2\pi i} \displaystyle\int_C \dfrac{p'(z)}{p(z)}\, dz = 3$;

   if C is a contour with $2 \in I(C)$ and $-3 \in E(C)$;

   (c) if both $2 \in I(C)$ and $-3 \in I(C)$, then the integral in (b) is $= 5 =$ degree of p.

6. Prove that all real trigonometric identities extend to complex identities. (See Example 2.)

B Exercises

1. If $p(z)$ is a polynomial of degree n, then by the Fundamental Theorem of Algebra and Exercise C3 of Section 1.1, $p(z)$ can be factored as

$$p(z) = c(z - \alpha_1)^{k_1}(z - \alpha_2)^{k_2} \cdots (z - \alpha_j)^{k_j}$$

where the α's are the distinct zeros of $p(z)$, the k's are the respective orders of the zeros, and $\Sigma k_m = n$, the degree of p. Show that

(a) $\dfrac{p'(z)}{p(z)} = \dfrac{k_1}{z - \alpha_1} + \dfrac{k_2}{z - \alpha_2} + \cdots + \dfrac{k_j}{a - \alpha_j}$

(b) if C is a contour, and $\alpha_1 \in I(C)$ with all the other α's in $E(C)$, then

$$\frac{1}{2\pi i} \int_C \frac{p'(z)}{p(z)} \, dz = k_1$$

(c) if all the α's are in $I(C)$, then the integral is $= n$.

2. There is a theorem of Weierstrass which states that if $g(z)$ is *real valued and continuous in a closed and bounded set S*, then there are points z_1 and z_2 in S for which

$$g(z_1) \le g(z) \le g(z_2)$$

for all $z \in S$. That is, $g(z)$ attains a maximum and a minimum in S.

Let R be a bounded region and $f(z)$ be analytic in R and continuous in $R \cup B$ where B is the boundary of R. Use the theorem quoted above to show that in $\bar{R} = R \cup B$, $|f(z)|$ attains a maximum on B.

3. If $u(z)$ is the real part of $f(z)$ in B2 above, show that $u(z)$ attains both its maximum and its minimum on B.

4. Suppose $f(z)$ is analytic for $|z| < a$. Let $M(r)$ be the maximum of $|f(z)|$ for $|z| \le r$. Show that, for $r < a$,
   (a) $M(r)$ is attained on $|z| = r$;
   (b) $M(r)$ is nondecreasing;
   (c) improve (b) to: unless $f(z)$ is a constant, then $M(r)$ is strictly increasing.

5. Suppose $f(z)$ and $g(z)$ are both analytic in a region R and that $f(z)g(z) \equiv 0$ in R. Show that at least one of them is $\equiv 0$ in R.

C Exercises

1. (Compare C4, Section 3.3) Suppose $f(z)$ is analytic for $|z| < a$ and (see B4 above) $M(r) = r^n$. Show that $f(z) = e^{i\alpha}z^n$ where α is a real constant.

2. (a) Prove Theorem 3.4e.

   (b) Complete the proof of Theorem 3.4f.

3. Suppose $f(z)$ is analytic in a region R with all of its zeros being $\{\alpha_j\}_1^k$ with multiplicity $\{p_j\}_1^k$, respectively. Let

$$P(z) = (z - \alpha_1)^{p_1}(z - \alpha_2)^{p_2} \cdots (z - \alpha_k)^{p_k},$$

   and show that $f(z) = P(z)g(z)$ where $g(z)$ is analytic and never zero in R.

4. Extend Exercise C7, Section 2.6 as follows. Suppose $f(z)$ and $g(z)$ are analytic in a simply connected region R. Each has only a finite number of zeros in R whose locations coincide, that is, α is a zero of $f(z)$ if and only if it is a zero of $g(z)$. Furthermore $|f(z)| \equiv |g(z)|$ in R. Show that there is a real constant a for which $f(z) = g(z)e^{ia}$. Hint: Show that the zeros of $f(z)$ and $g(z)$ have the same order.

5. Suppose $f(z)$ is analytic and bounded for $|z| < 1$, and that $f(re^{i\theta}) \to 0$ uniformly for $0 \le \theta \le \alpha$ for some fixed α, $0 < \alpha < 2\pi$. Show that $f(z) \equiv 0$. Hint: Take n so large that $2\pi/n < \alpha$, set $\omega = e^{2\pi i/n}$ and form

$$F(z) = f(z)f(\omega z)f(\omega^2 z) \cdots f(\omega^{n-1}(z)).$$

3.5 Uniform Convergence

Let us recall the meaning of uniform convergence. If $\{\phi_n(z)\}$ is a sequence of complex valued functions on a set D, then $\{\phi_n(z)\}$ converges uniformly to a limit function $\phi(z)$ on D means that for each $\varepsilon > 0$ there is an $N = N(D, \varepsilon)$ for which

$$|\phi_n(z) - \phi(z)| < \varepsilon \quad \text{for all} \quad z \in D \quad \text{when} \quad n > N.$$

We sometimes write this as

$$\phi_n(z) \to \phi(z) \quad \text{uniformly on} \quad D.$$

And a series $S(z) = \sum_0^\infty w_k(z)$ converges uniformly on D means that the sequence of partial sums converges uniformly on D:

$$S_n(z) \to S(z) \quad \text{uniformly on} \quad D$$

where

$$S_n(z) = \sum_0^n w_k(z).$$

The concept of uniform convergence is a very useful tool in analysis, especially in complex variables. We now prove two important consequences of uniform convergence.

3.5a Theorem. *If $\{u_n(z)\}$ converges uniformly to $u(z)$ on a set D, and if $u_n(z)$ are all continuous at $z_0 \in D$, then $u(z)$ is continuous at z_0.*

Proof. Let $\varepsilon > 0$ be given. There is an $N = N(D, \varepsilon)$ for which

$$|u_n(z) - u(z)| < \varepsilon/3 \qquad z \in D \text{ if } n > N. \tag{1}$$

Choose such an n and keep it fixed. Then there is a $\delta > 0$ for which

$$|u_n(z_0) - u_n(z)| < \varepsilon/3 \quad \text{if} \quad |z - z_0| < \delta \tag{2}$$

since the chosen u_n is continuous at z_0. So

$$|u(z_0) - u(z)| \le |u(z_0) - u_n(z_0)| + |u_n(z_0) - u_n(z)| + |u_n(z) - u(z)|$$
$$< \varepsilon/2 + \varepsilon/3 + \varepsilon/3 = \varepsilon \text{ for } |z - z_0| < \delta,$$

the first and last $\varepsilon/3$ by (1) and the middle one by (2). $\quad\square$

3.5b Theorem. *Suppose $\{u_n(z)\}$ is a sequence of continuous functions on a path C, and $u_n(z) \to u(z)$ uniformly on C. Then*

$$\lim \int_C u_n(z)dz = \int_C u(z)dz$$

Proof. Since $u(z)$ is continuous by Theorem 3.5a, all the integrals exist. Furthermore, if $\varepsilon > 0$ is given, and L is the length of C, then there is an N such that

$$|u_n(z) - u(z)| < \varepsilon/L \quad \text{if} \quad z \in C \quad \text{and} \quad n > N.$$

For $n > N$ we have

$$\left| \int_C u_n(z)dz - \int_C u(z)dz \right| = \left| \int_C \left[u_n(z) - u(z) \right] dz \right| \le \frac{\varepsilon}{L} \cdot L = \varepsilon$$

which proves the theorem. □

These two theorems of course apply to a series in the form that
(a) *the sum of a uniformly convergent series of continuous functions is continuous, and*
(b) *a uniformly convergent series of continuous functions can be integrated termwise:*

$$\int_C \left(\sum_0^\infty u_n(z) \right) dz = \sum_0^\infty \left(\int_C u_n(z)dz \right).$$

We now establish a very useful result.

3.5c Theorem. *Suppose $\{\phi_n(z)\}$ is a sequence of analytic functions defined in a region R. If*

$$\phi_n(z) \to \phi(z) \quad \text{uniformly in} \quad R$$

then $\phi(z)$ is analytic in R and

$$\phi_n^{(k(}(z) \to \phi^{(k)}(z), \quad \text{for each } k, \quad \text{and each } z \in R.$$

When applied to series of course this means that $\sum w_n(z)$ is analytic in R and $\left[\sum w_n(z) \right]' = \sum \left[w'_n(z) \right]$.

Proof. Let α be an arbitrary point in R, and the disk $D : |z - \alpha| < a$ be a neighborhood of α in R. Then $\{\phi_n(z)\}$ are all analytic in D and converge uniformly there. Let C be a contour in D. We have

$$\int_C \phi_n(z)dz = 0$$

by Cauchy's theorem. Hence

$$0 = \lim \int_C \phi_n(z)dz = \int_C \phi(z)dz$$

by Theorem 3.5b. By Morera's theorem $\phi(z)$ is analytic in D, in particular $\phi'(\alpha)$ exists. Since α is arbitrary, $\phi(z)$ is analytic in R.

Further, for $0 < r < a$,

$$\phi_n^{(k)}(z) = \frac{1}{2\pi i} \int_C \frac{\phi_n(\zeta)}{(\zeta - z)^{k+1}} \, d\zeta$$

if $|z - \alpha| < r$, and C_r is the circle $|z - \alpha| = r$. Again, by uniform convergence on C_r,

$$\lim \phi_n^{(k)}(z) = \frac{1}{2\pi i} \int_{C_r} \frac{\phi(z)}{(\zeta - z)^{k+1}} \, d\zeta = \phi^{(k)}(z), \quad |a - \alpha| < r, \qquad (3)$$

which holds at α since $\alpha \in D$. But again α is arbitrary, so $\phi_n^{(k)}(z) \to \phi_n^{(k)}(z)$ for each $z \in R$. $\square$

Example 1. Show that $f(z) = \sum_{0}^{\infty} \dfrac{z^n}{1 + z^{2n}}$ is analytic for $|z| < 1$ and also for $|z| > 1$.

Solution. Let R_a be the region $|z| < a$ where $a < 1$. In R_a we have

$$\left| \frac{z^n}{1 + z^{2n}} \right| < \frac{a^n}{1 - a^{2n}}$$

and there is an N for which $a^{2n} < 1/2$ if $n > N$. For $n > N$ we have

$$\left| \frac{z^n}{1 + z^{2n}} \right| \leq \frac{a^n}{1 - 1/2} = 2a^n$$

and $\sum 2a^n$ converges. By the M-test the original series converges uniformly in R_a and is therefore analytic in R_a. Now let z_0 be any point in $|z| < 1$. Choose a between $|z_0|$ and 1, say $a = (|z_0| + 1)/2$. Then $f(z)$ is analytic at z_0. Since z_0 is arbitrary, $f(z)$ is analytic for $|z| < 1$.

Next consider $|z| > 1$. If we set $\zeta = 1/z$, then $|\zeta| < 1$, and

$$f(z) = \sum_{0}^{\infty} \frac{z^n}{1 + z^{2n}} = \sum_{0}^{\infty} \frac{z^n / z^{2n}}{(1 + z^{2n})/z^{2n}}$$

$$= \sum \frac{1/z^n}{1/z^{2n} + 1} = \sum \frac{\zeta^n}{1 + \zeta^{2n}} = f(\zeta).$$

and so $f(z)$ is also analytic for $|z| > 1$.

Example 2. Show that

$$f(z) = \sum_{0}^{\infty} \frac{(-1)^n}{n!} \frac{1}{z + n}$$

is analytic in the plane punctured at 0 and the negative integers.

Solution. We choose α to be any complex number $\alpha \neq 0, -1, -2, \ldots$, and we show that $f(z)$ is analytic at α. Now let $2a = \min\{\text{dist}\,(\alpha, n),\ n = 0, -1, -2, \ldots\}$ and examine $f(z)$ in $|z - \alpha| < a$. For such z we have

$$\left| \frac{(-1)^n}{n!} \frac{1}{z + n} \right| \leq \frac{1}{a \cdot n!} = M_n$$

and so the series converges uniformly for $|z - \alpha| < a$ by the M-test since $\sum 1/n!$ converges, and hence is analytic at α. Since α is arbitrary, we are finished.

We will consider integrals as well as series. As a first example we consider an extension of Theorem 3.2b to certain improper integrals.

3.5d Theorem. *If $\phi(t)$ is continuous for t real and if $\displaystyle\int_1^\infty |\phi(t)/t|dt$ and $\displaystyle\int_{-\infty}^{-1} |\phi(t)/t|dt$ both converge, then $f(z)$ defined by*

$$f(z) = \frac{1}{2\pi i} \int_{-\infty}^\infty \frac{\phi(t)dt}{t - z}$$

is analytic for $\text{Im}\, z \neq 0$.

Proof. Let

$$f_n(z) = \frac{1}{2\pi i} \int_{-n}^n \frac{\phi(t)}{t - z}\, dt, \quad n = 1, 2, 3, \ldots\ .$$

These functions are all analytic for $\text{Im}\, z \neq 0$ by Theorem 3.2b. And

$$f(z) - f_n(z) = \frac{1}{2\pi i} \int_{-\infty}^{-n} \frac{\phi(t)dt}{t - z} + \frac{1}{2\pi i} \int_n^\infty \frac{\phi(t)dt}{t - z}$$

$$\equiv I(z) + J(z) \quad \text{respectively.}$$

Now let $\alpha = a + ib$ with $b \neq 0$, and let D be the disk $D = \{z : |z - \alpha| < |b/2|\}$ so that D is a bounded region. Then for $z \in D$

$$|b|/2 > |z - \alpha| \geq |z| - |\alpha|$$

and hence $|z| \leq |\alpha| + |b|/2 \equiv M$. We now take $n > 2M$ so that in both I and J we have

$$|t - z| \geq |t| - |z| > |t| - M > |t| - \frac{n}{2} > |t| - \frac{|t|}{2} = \frac{|t|}{2}.$$

Putting this last estimate into I we get

$$|I| \leq \frac{1}{2\pi} \int_{-\infty}^{-n} \frac{|\phi(t)|}{|t|/2} \, dt = \frac{1}{\pi} \int_{-\infty}^{-n} \left| \frac{\phi(t)}{t} \right| \, dt$$

and similarly

$$|J| \leq \frac{1}{\pi} \int_{n}^{\infty} \left| \frac{\phi(t)}{t} \right| \, dt.$$

thus, for $z \in D$,

$$|f(z) - f_n(z)| \leq \frac{1}{\pi} \int_{-\infty}^{-n} \left| \frac{\phi(t)}{t} \right| \, dt + \frac{1}{\pi} \int_{n}^{\infty} \left| \frac{\phi(t)}{t} \right| \, dt$$

from which we get $f_n(z) \to f(z)$ uniformly in D (how?). Thus $f(z)$ is analytic at α, and since α is arbitrary, $f(z)$ is analytic for $|y| = |\operatorname{Im} z| > 0$. $\square$

One of the objects of this section is to discuss analytic functions given in the form

$$F(z) = \int_{a}^{b} f(z, t) dt.$$

There are general theorems which describe circumstances under which such an integral represents an analytic function. We will not investigate these more general situations but will stay with rather specific formulas which will still yield some useful results.

3.5e Theorem. *If $f(t)$ is continuous for $a \leq t \leq b$, then*

$$F(z) = \int_{a}^{b} e^{-zt} f(t) dt \tag{4}$$

is an entire function and

$$F'(z) = - \int_{a}^{b} e^{-zt} t f(t) dt.$$

Proof. First suppose $a = 0$, $b > 0$. We then choose any z and keep it fixed, and form the difference quotient at z with $h = \Delta z$:

$$\frac{\Delta F}{\Delta z} = \frac{F(z + h) - F(z)}{h}$$

$$= \frac{1}{h} \int_{0}^{b} [e^{-(z+h)t} - e^{-zt}] f(t) dt$$

$$= \int_{0}^{b} e^{-zt} t f(t) [(e^{-ht} - 1)/ht] dt.$$

Next we subtract the formula asserted to be $F'(z)$ to get the difference D to be

$$D \equiv \frac{\Delta F}{\Delta z} + \int_0^b e^{-zt} t f(t) dt$$

$$= \int_0^b e^{-zt} t f(t) [(e^{-ht} - 1)/ht + 1] dt.$$

We now let

$$B = \max\{|(e^{-ht} - 1)/ht + 1| : 0 \le t \le b\}$$

and

$$M = \max\{|f(t)| : 0 \le t \le b\},$$

and estimate the last integral for D to get

$$|D| \le e^{b|z|} b M B b = (b^2 M e^b |z|) B. \tag{5}$$

By (5) we only need show that $B \to 0$ as $h \to 0$. To do this we calculate as follows:

$$e^{-ht} = 1 - ht + \sum_2^\infty (-1)^n (ht)^n / n!,$$

$$e^{-ht} - 1 = -ht + \sum_2^\infty (-1)^n (ht)^n / n!,$$

$$\frac{e^{-ht} - 1}{ht} = -1 + \sum_2^\infty (-1)^n (ht)^{n-1} / n!,$$

$$\frac{e^{-ht} - 1}{ht} + 1 = \sum_2^\infty (-1)^n (ht)^{n-1} / n!,$$

so, finally we can estimate to get

$$B \le \sum_2^\infty |h|^{n-1} b^{n-1} / n!$$

and the sum does indeed $\to 0$ as $h \to 0$ (how?). Since z is arbitrary, we have both conclusions for the case $a = 0$, $b > 0$. We leave it as an exercise to complete the proof. $\square$

The integral (4) is close to the form of an important type of integral. If $f(t)$ is defined for $t \ge 0$, then the **Laplace transform** of $f(t)$ is defined by

$$F(z) = \int_0^\infty e^{-zt} f(t) dt. \tag{6}$$

whenever this integral exists.

3.5f Theorem. *If $f(t)$ is continuous for $t \geq 0$, $M > 0$, a is real, and*

$$|f(t)| \leq Me^{at}, \qquad t \geq 0$$

then the Laplace transform exists, is analytic for $\operatorname{Re} z > a$, and

$$F'(z) = -\int_0^\infty e^{-zt} t f(t)\,dt$$

Proof. We define

$$F_n(z) = \int_0^n e^{-zt} f(t)\,dt, \qquad n = 1, 2, 3, \ldots$$

each of which is an entire function by Theorem 3.5d. Let us look at the region R defined by $x = \operatorname{Re} z > a + s$ for fixed $s > 0$. Then for $z \in R$ we have

$$
\begin{aligned}
|F_n(z) - F(z) &= \left| \int_n^\infty e^{-zt} f(t)\,dt \right| \\
&< \int_n^\infty e^{-xt} Me^{at}\,dt \\
&< \int_n^\infty e^{-(a+s)t} Me^{at}\,dt \\
&= M \int_n^\infty e^{-st}\,dt = \frac{M}{s}\, e^{-sn} \\
&\to 0 \quad \text{as} \quad n \to \infty.
\end{aligned}
$$

So given $\varepsilon > 0$ there is an N for which

$$|F_n(z) - F(z)| < \varepsilon \quad \text{if} \quad z \in R \quad \text{and} \quad n > N.$$

Thus $F(z)$ is analytic for $x > a + s$ and so for $x > a$, since s is arbitrary. And, also by Theorem 3.5c,

$$F_n'(z) \to -\int_0^\infty e^{-zt} t f(t)\,dt = F'(z). \quad \square$$

As an example of the power of these results and of the identity theorem let us examine the simple case of the Laplace transform of 1; let us call it $F(z)$:

$$F(z) = \int_0^\infty e^{-zt} \cdot 1 \cdot dt.$$

This function is analytic for $\operatorname{Re} z > 0$ by Theorem 3.5e. We evaluate it easily for $\operatorname{Re} z > 0$

$$F(z) = \lim_{a \to \infty} \int_0^a e^{-zt} dt = \lim_{a \to \infty} \left(-\frac{1}{z} e^{-zt} \right) \Big|_0^a$$

$$= \lim_{a \to \infty} \left[\frac{1}{z} - \frac{1}{z} e^{-za} \right] = \frac{1}{z}.$$

If we set $z = x + iy$, $(x > 0)$ we get

$$\int_0^\infty e^{-xt} e^{-iyt} dt = \frac{1}{z} = \frac{x - iy}{x^2 + y^2}$$

Then equating real and imaginary parts we get

$$\left\{ \begin{array}{l} \displaystyle\int_0^\infty e^{-xt} \cos yt\, dt = \dfrac{x}{x^2 + y^2} \\[4mm] \displaystyle\int_0^\infty e^{-xt} \sin yt\, dt = \dfrac{y}{x^2 + y^2} \end{array} \right\} \quad x > 0. \tag{7}$$

Example 3. Compute the Laplace transforms of $\sin at$ and $\cos at$ for $a > 0$.

Solution. The transform of $\cos at$ is analytic for $\operatorname{Re} z > 0$ by Theorem 3.5e. It is

$$F(z) = \int_0^\infty e^{-zt} \cos at\, dt, \qquad \operatorname{Re} z > 0.$$

Furthermore, if we define $G(z)$ by

$$G(z) = \frac{z}{z^2 + a^2},$$

it too is analytic for $\operatorname{Re} z > 0$. And by the first integral of (7) $F(z) = G(z)$ on the positive real axis. Therefore by the identity theorem they are the same:

$$\int_0^\infty e^{-zt} \cos at\, dt \equiv \frac{z}{z^2 + a^2}, \qquad \operatorname{Re} z > 0.$$

Similarly $\displaystyle\int_0^b e^{-zt} \sin at\, dt = \frac{a}{z^2 + a^2}.$

Also, by the formula for the derivative in Theorem 3.5e we get

$$\int_0^\infty e^{-zt} t \cos at\, dt = \frac{z^2 - a^2}{(z^2 + a^2)^2}$$

and

$$\int_0^\infty e^{-zt} t \sin at\, dt = \frac{2az}{(z^2 + a^2)^2}$$

3.5g Lemma. *If $0 < a < b < \infty$, $g(t)$ is continuous on $a \le t \le b$, and if t^z is the principal value, then*

$$f(z) = \int_a^b t^z g(t)\, dt$$

is an entire function and

$$f'(z) = \int_a^b t^z \ln t\, g(t)\, dt.$$

Proof. Set $t = e^{-\tau}$. Then

$$f(z) = -\int_{\ln 1/a}^{\ln 1/b} e^{-z\tau} g(e^{-\tau}) e^{-\tau}\, d\tau$$

$$= \int_{\ln 1/b}^{\ln 1/a} e^{-z\tau} g(e^{-\tau}) e^{-\tau}\, d\tau$$

so that Theorem 3.5e applies. $\square$

One of the most interesting and useful special functions is the Gamma function $\Gamma(z)$ defined by

$$\Gamma(z) = \int_0^\infty e^{-t} t^{z-1}\, dt.$$

3.5h Theorem. $\Gamma(z)$ *is analytic for $\operatorname{Re} z > 0$ and* $\Gamma'(z) = \displaystyle\int_0^\infty e^{-t} t^{z-1} \ln t\, dt.$

Proof. We write

$$\Gamma(z) = \int_0^1 e^{-t} t^{z-1}\, dt + \int_1^\infty e^{-t} t^{z-1}\, dt$$

$$= \Gamma_1(z) + \Gamma_2(z), \quad \text{respectively.}$$

We leave it as an exercise that $\Gamma_2(z)$ is an entire function. We examine $\Gamma_1(z)$. We take an arbitrary z_0 with $\operatorname{Re} z_0 > 0$, and then choose a so that

$0 < a < \operatorname{Re} z_0$ and define $R = \{z : \operatorname{Re} z > a\}$. It suffices to show $\Gamma_1(z)$ analytic in R. We set

$$f_n(z) = \int_{1/n}^{1} t^z [e^{-t}(1/t)]dt$$

which is entire by Lemma 3.5f. So it is enough to show $f_n(z) \to \Gamma_1(z)$ uniformly in R. But

$$|\Gamma_1(z) - f_n(z)| = \left| \int_0^{1/n} t^{z-1} e^{-t} dt \right|$$

$$\leq \int_0^{1/n} |t^{z-1}| e^{-t} dt$$

$$\leq \int_0^{1/n} t^{z-1} dt < \int_0^{1/n} t^{a-1} dt \quad \text{(how?)}$$

$$= \frac{1}{a} t^a \Big|_0^{1/n} = \frac{1}{a} \left(\frac{1}{n} \right)^a \to 0.$$

Thus $f_n(z) \to \Gamma_1(z)$ uniformly in R. $\square$

3.5i Theorem.

$$\Gamma(z+1) = z\Gamma(z) \quad \text{for} \quad \operatorname{Re} z > 0.$$

Proof. By definition of $\Gamma(z+1)$ we have

$$\Gamma(z+1) = \int_0^{\infty} e^{-t} t^z dt = \lim_{a \to \infty} \int_0^{a} e^{-t} t^z dt$$

which we integrate by parts to get

$$\Gamma(z+1) = \lim_{a \to \infty} \left[(-e^{-t}) \cdot t^z \Big|_0^a - \int_0^a (-e^{-t}) z t^{z-1} dt \right]$$

$$= z \int_0^{\infty} e^{-t} t^{z-1} dt - \lim_{a \to \infty} e^{-a} a^z$$

$$= z\Gamma(z). \quad \square$$

3.5j Theorem. $\Gamma(1) = 1$.

Proof.

$$\Gamma(1) = \int_0^\infty e^{-t} dt = \lim_{a \to \infty} \int_0^a e^{-t} dt$$

$$= \lim_{a \to \infty} [-e^{-t}\big|_0^a] = \lim_{a \to \infty} [1 - e^{-a}] = 1. \quad \square$$

3.5k Theorem. $\Gamma(n + 1) = n!, \, n > 1.$

Proof. From Theorem 3.5j we have

$$\Gamma(2) = 1 \cdot \Gamma(1) = 1, \quad \Gamma(3) = 2 \cdot \Gamma(2) = 2 \cdot 1,$$
$$\Gamma(4) = 3 \cdot \Gamma(3) = 3 \cdot 2 \cdot 1,$$

and by induction

$$\Gamma(n + 1) = n \cdot (n - 1)(n - 2) \cdots 2 \cdot 1 = n! \quad \square$$

Notice also that the formula of this theorem agrees with the common usage that $0! = 1$, for using this formula and Theorem 3.5i we get $0! = \Gamma(1) = 1$. This is not the primary reason that $0!$ is used as 1, but it is probably more than just a happy coincidence.

We briefly outline the discussion of one more integral, namely

$$g(z) = \int_0^\infty \frac{t^{z-1}}{1+t} \, dt.$$

3.5ℓ Theorem. $g(z)$, *as defined above, is analytic for* $0 < Re\, z < 1$.

The proof goes as follows: we set

$$\int_0^\infty \frac{t^{z-1}}{1+t} \, dt = \int_0^1 t^z \cdot \frac{1}{t(1+t)} \, dt + \int_1^\infty t^z \frac{1}{t(1+t)} \, dt$$
$$= g_1(z) + g_2(z) \quad \text{respectively.}$$

Then we show that $g_1(z)$ is analytic for $Re\, z > 0$, and that $g_2(z)$ is analytic for $Re\, z < 1$, so their sum is analytic in the strip $0 < Re\, z < 1$. The details are left to the Exercises.

A Exercises

1. Show that the sequences given below converge uniformly on the indicated sets.

   (a) $\dfrac{1}{n+z}$, $|z| \le 1$;   (b) $\dfrac{1}{n}\, e^{-nz}$, $\operatorname{Re} z \ge 0$;

   (c) $\dfrac{1}{nz+1}$, $|z| \ge 2$;   (d) $nz^n(1-z)^n$, $|z| \le \frac{1}{2}$;

   (e) $\dfrac{z^{2n}}{1+z^{2n}}$, $|z| \le a < 1$;   (f) $\dfrac{z^{2n}}{1+z^{2n}}$, $|z| \ge b > 1$.

2. Show that the following series converge uniformly on the indicated sets.

   (a) $\displaystyle\sum_{1}^{\infty} \dfrac{1}{n^2}\, e^{-nz}$, $\operatorname{Re} z \ge 0$;

   (b) $\displaystyle\sum_{1}^{\infty} \dfrac{1}{n}\, e^{-nz}$, $\operatorname{Re} z \ge a > 0$;

   (c) $\displaystyle\sum_{1}^{\infty} \dfrac{1}{n(n-z)}$, $|\operatorname{Im} z| \ge a > 0$;

   (d) $\displaystyle\sum_{1}^{\infty} \left(\dfrac{z}{z+1}\right)^n$, $\left|\dfrac{z}{z+1}\right| \le a < 1$;

   (e) $\displaystyle\sum_{1}^{\infty} \dfrac{z^{2n}}{1+z^{2n}}$, $|z| \le a < 1$;

3. In A2 above show that the series is analytic in the region indicated here.

      (a) $\operatorname{Re} z > 0$;   (b) $\operatorname{Re} z > 0$;   (c) $\operatorname{Im} z \ne 0$;
      (d) $\operatorname{Re} z > -\frac{1}{2}$;   (e) $|z| < 1$;

4. Compute the derivative of each series above in A2. Where is the result valid?

5. Note that

$$f(z) \equiv \frac{1}{2\pi i} \int_{-\infty}^{\infty} \frac{1}{1+t^2} \frac{1}{t-z} \, dt$$

is analytic for $y = \text{Im } z \neq 0$. (How?)

(a) Show that

$$f'(z) = \frac{1}{2\pi i} \int_{-\infty}^{\infty} \frac{1}{1+t^2} \frac{1}{(t-z)^2} \, dt$$

(b) Show that

$$f(z) = \frac{-1}{2\pi i} \int_{-\infty}^{\infty} \frac{\arctan t}{(t-z)^2} \, dt$$

6. (a) Show that

$$\Gamma\left(\frac{1}{2}\right) = 2 \int_0^{\infty} e^{\tau^2} \, d\tau,$$

which is known to be $\sqrt{\pi}$. (See Exercise C6 (b), Section 3.6.)

(b) Deduce $\Gamma\left(\dfrac{3}{2}\right) = \frac{1}{2}\sqrt{\pi}$, $\Gamma\left(\dfrac{5}{2}\right) = \frac{3}{2} \cdot \frac{1}{2}\sqrt{\pi}$.

(c) By induction show

$$\Gamma\left(n + \frac{1}{2}\right) = \frac{1 \cdot 3 \cdot 5 \cdots (2n-1)}{2^n} \sqrt{\pi}.$$

7. Show by induction

$$\Gamma(z+n+1) = (z+n)(z+n-1)\cdots(z+1)z\Gamma(z)$$

for $\text{Re } z > 0$.

B Exercises

1. Improve the result of A3 (c) above by showing that $\displaystyle\sum_{1}^{\infty} \frac{1}{n(n-z)}$ is analytic for $z \neq 1, 2, 3, \ldots$.

2. Show that $\dfrac{1}{z} + \displaystyle\sum_{1}^{\infty} \frac{2z}{z^2 - k^2}$ is analytic for $z \neq 0, \pm 1, \pm 2, \ldots$.

3. Show that $\displaystyle\sum_{1}^{\infty} \sin\left(\frac{z}{n^2}\right)$ is entire.

4. If n^z means the principal value, show that $\displaystyle\sum_{1}^{\infty} 1/n^z$ is analytic for $x = \text{Re } z > 1$.

5. Suppose that $f(z,t)$ is continuous for all $t \geq 0$ and all z in a set S, and that there is a function $M(t)$ for which $|f(z,t)| \leq M(t)$ for all $z \in S$. If

$$\int_0^\infty M(t)dt$$

converges, show that

$$\int_0^n f(z,t)dt \rightarrow \int_0^\infty f(z,t)dt$$

uniformly for $z \in S$.

6. Complete the proof of Theorem 3.5d by discussing the remaining cases. Hint: Use the part proved. No more calculations are necessary.

7. Find the Laplace transform of $\sinh at$, $t \sinh at$, $\cosh at$, $t \cosh at$ if $a > 0$. Say where these transforms are analytic.

8. In the proof of 3.5h, show that $\Gamma_2(z)$ is entire.

9. Show $\Gamma(z) = \int_0^1 (\ln 1/t)^{z-1} dt$.

10. Use A6 (c) above to show that

$$\Gamma\left(n + \frac{1}{2}\right) = \frac{(2n)!}{n!2^{2n}} \sqrt{\pi}.$$

C Exercises

1. Prove Theorem 3.5ℓ by following the outline given with its statement.

2. Show that

   (a) $\displaystyle\int_0^\infty t^{z-1} \cos t\, dt$ is analytic for $0 < \operatorname{Re} z < 1$.

   (b) $\displaystyle\int_0^\infty t^{z-1} \sin t\, dt$ is analytic for $-1 < \operatorname{Re} z < 1$.

   Hint: In each case write $\displaystyle\int_0^\infty = \int_0^1 + \int_1^\infty$.

3. Show that

$$f(z) \equiv \int_{-\pi}^\pi \frac{dt}{z - \cos t}$$

is analytic in the plane cut along the real axis from -1 to 1, and that

$$f'(z) = -\int_{-\pi}^\pi \frac{dt}{(z - \cos t)^2}.$$

Hint: Follow the same procedure as in the proof of Theorem 3.5d.

4. Choose z with $0 < \mathrm{Re}\, z < 1$ and integrate $\zeta^{z-1} e^{-\zeta}$ around a quarter circle consisting of the segment from 0 to a, the arc $\zeta = a e^{i\theta}$, $0 \le \theta \le \pi/2$, and the segment from ia to 0. Deduce the values of the integrals in C2 above.

3.6 Analytic Continuation

In the case of Taylor series we saw that if a function $f(z)$ were analytic in a region R, and α a point of R, then we could represent $f(z)$ by a power series in a disk D of radius $a = \mathrm{dist}(\alpha, \partial R)$. This usually results in a shrinkage of the region R on which we consider $f(z)$ to the smaller region D on which we consider the series. In this section we want to consider what is in some sense an inverse problem. If we know $f(z)$ in a region R in which it is analytic, can we enlarge the region of analyticity?

We recall some point set notation from Section 1.6. If R and S are point sets, then

(i) $$R \cap S = \{z : z \in R \text{ and } z \in S\}$$

is the **common part** or **intersection** or **meet** of R and S.

(ii) $$R \cup S = \{z : z \in R \text{ or } z \in S\}$$

is the **union** or **sum** of R and S.

Definition. *Suppose R and S are regions and $R \cap S$ is not empty — the regions actually overlap. If $f(z)$ is analytic in R, $g(z)$ is analytic in S, and $f(z) \equiv g(z)$ in $R \cap S$, then $g(z)$ is called the* **analytic continuation** *of $f(z)$ into S.*

Notice that the definition is symmetric so that $f(z)$ is the analytic continuation of $g(z)$ into R.

3.6a Theorem. *Analytic continuation, as defined above, is unique.*

Proof. Suppose both $g_1(z)$ and $g_2(z)$ are analytic continuations of $f(z)$ from R into S. Then $g_1(z) \equiv g_2(z)$ in $R \cap S$, and so, by the identity theorem, they are the same throughout S. $\square$

So under these circumstances it makes sense to consider the result as defining a single function, which we might call $F(z)$, which is analytic in

$R \cup S$ and which is given by the formula $F(z) = f(z)$ in R and the formula $F(z) = g(z)$ in S.

Example 1. Suppose we have $f(z)$ defined by

$$f(z) = \int_0^\infty e^{-zt} dt.$$

We observe that $f(z)$ is analytic for $\operatorname{Re} z > 0$. When we evaluate the integral, as we did in the previous section, we find that

$$f(z) = \frac{1}{z} \quad \text{for} \quad \operatorname{Re} z > 0.$$

Let us take $g(z) = 1/z$. Now $g(z)$ is analytic in the whole plane punctured at the origin, and so provides the analytic continuation of $f(z)$ from the right half-plane into the punctured plane.

Example 2. The Laplace transform of $\cos at$ (Example 3, Section 3.5) was shown to be analytic for $\operatorname{Re} z > 0$ if $a > 0$. Its evaluation, namely $z/(z^2 + a^2)$, provides the analytic continuation into the whole plane punctured at $\pm ia$.

Example 3. Consider the function $f(z)$ given by

$$f(z) = \int_0^\infty \frac{t^{z-1}}{1+t} dt.$$

In the previous section this was shown to be analytic for $0 < \operatorname{Re} z < 1$. In the next chapter we will evaluate this for $z = a$, $0 < a < 1$, to be $\pi \csc \pi a = \pi / \sin \pi a$. If we take $g(z) \equiv \pi \csc \pi z$, then we have $g(z)$ analytic in $0 < \operatorname{Re} z < 1$, and $f(z) = g(z)$ on the segment of the real axis between 0 and 1. So by the identity theorem they are the same for $0 < \operatorname{Re} z < 1$. That is

$$\int_0^\infty \frac{t^{z-1}}{1+t} = \pi \csc \pi z, \qquad 0 < \operatorname{Re} z < 1.$$

However, $\pi \csc \pi z$ is analytic in the whole plane except at the integers. Thus $\pi \csc \pi z$ is the analytic continuation of $f(z)$ into the whole plane punctured at the integers.

Example 4. Let us consider the Gamma function from the previous section:

$$\Gamma(z) = \int_0^\infty e^{-t} t^{z-1} dt, \qquad \operatorname{Re} z > 0.$$

We write this as

$$\Gamma(z) = \int_0^1 e^{-t}t^{z-1}\,dt + \int_1^\infty e^{-t}t^{z-1}\,dt$$
$$= \Gamma_1(z) + \Gamma_2(z).$$

We have $\Gamma_2(z)$ is an entire function (from the proof of Theorem 3.5g) and $\Gamma_1(z)$ is analytic for $\operatorname{Re} z > 0$. For such z we have

$$\Gamma_1(z) = \int_0^1 e^{-t}t^{z-1}\,dt = \int_0^1 \left(\sum_0^\infty \frac{(-1)^n t^n}{n!} \right) t^{z-1}\,dt$$
$$= \sum_0^\infty \frac{(-1)^n}{n!} \int_0^1 t^{n+z-1}\,dt \quad \text{(how?)}$$
$$= \sum_0^\infty \frac{(-1)^n}{n!} \frac{1}{n+z}, \quad \operatorname{Re} z > 0.$$

But, by Example 2 of Section 3.5, this series is analytic in the whole plane punctured at 0 and the negative integers. Thus the formula

$$\Gamma(z) = \sum_0^\infty \frac{(-1)^n}{n!} \frac{1}{n+z} + \int_1^\infty e^{-1}t^{z-1}\,dt,$$

provides the analytic continuation of $\Gamma(z)$ into the plane punctured at $0, -1, -2, \dots$.

Suppose $f(z)$ is analytic in a region containing a segment of the real axis, and is real valued on the segment, then all the examples we can examine have the property that f takes conjugate values at conjugate points, that is, $\overline{f(z)} = f(\bar{z})$. This observation was used by H. A. Schwarz as a basis for analytic continuation across the real axis when the function is originally known on only one side.

Let S be an open ended segment of the real axis $a < x < b$, and let C be a simple path connecting a to b but otherwise lying in the upper half-plane. Then $\Gamma = S \cup C$ is a contour whose interior is a region R lying in the upper half-plane. We reflect this configuration in the real axis to create the region R' lying in the lower half-plane and bounded by C' and S. Then the set $\mathcal{R} = R \cup S \cup R'$ is a region which is symmetric about the x-axis. If $f(z)$ is continuous in $R \cup S$, is analytic in R, and is real valued on S, then $F(z)$ defined by

$$F(z) = \begin{cases} f(z) & z \in R \cup S \\ \overline{f(\bar{z})} & z \in R' \end{cases}$$

is analytic in $\mathcal{R} = R \cup S \cup R^\iota$. This is called the **Schwarz Reflection Principle** and its proof is outlined in the C Exercises.

Finally, if $f(z)$ is analytic on one side of a curve C but cannot be continued analytically across it, then C is called a **natural boundary** for $f(z)$.

Example 5. Show that the circle $|z| = 1$ is a natural boundary for

$$f(z) = \sum_0^\infty z^{2^k}.$$

Solution. Clearly the radius of convergence of the series is 1, so that $f(z)$ is analytic for $|z| < 1$. To see that $f(z)$ cannot be continued analytically across the circle $|z| = 1$ we argue as follows. If it were possible to continue $f(z)$ analytically across any arc of the circle, then $f(z)$ would be continuous at each point of the arc. Thus the limit of $f(z)$, as z approaches any point on the arc would exist as a finite complex number. Further, any arc of the circle must contain points of the form $z_0 = e^{2m\pi i/2^n}$ (how?). We select one such point, that is, we choose m and n. Then we examine $f(z)$ along the segment from 0 to z_0. For $0 \le r < 1$, we write

$$\begin{aligned}
f(rz^0) &= \sum_0^\infty r^{2^k}(e^{2m\pi i/2^n})^{2^k} \\
&= \sum_0^n r^{2^k} e^{2m\pi i/2^{n-k}} + \sum_{n+1}^\infty r^{2^k} e^{2m\pi i 2^{k-n}} \\
&= \sum_0^n r^{2^k} e^{2m\pi i/2^{n-k}} + \sum_{n+1}^\infty r^{2^k},
\end{aligned}$$

the last step being correct since $m2^{k-n}$ is an integer in the second sum. The first sum is a polynomial in r and so is continuous for all $r \ge 0$. Denoting the second sum by $\sigma(r)$ we have, for $0 < r < 1$

$$\sigma(r) = \sum_{n+1}^\infty r^{2^k} > \sum_{n+1}^{n+p} r^{2^k}$$

for any positive integer p. So, if $\lim_{r \to 1-0} \sigma(r)$ exists, we have

$$\lim_{r \to 1-0} \sigma(r) \ge \sum_{n+1}^{n+p} 1 = p,$$

that is if $\lim_{r \to 1-0} \sigma(r)$ exists, it must be $\geq p$ for *every* positive integer p. This is a contradiction, so the limit cannot exist.

Suppose $f(z)$ is analytic in a neighborhood of a point α. It then has a power series centered at α of radius, say r, which we suppose to be finite. If β is a point in the disk $|z - \alpha| < r$, then $f(z)$ will have a power series development centered at β whose radius of convergence is ρ, say, and clearly $\rho \geq r - |\alpha - \beta|$. If in fact $\rho > r - |\alpha - \beta|$, this second power series will provide an analytic continuation of $f(z)$ outside its original disk of convergence.

Now suppose the point α is connected to a point γ outside the original disk by a path C. If there is a connecting chain of disks along C by which $f(z)$ can be analytically continued from one to the other to give a continuation which includes γ, then $f(z)$ is said to be continued analytically along C. This process of continuation along paths is an extremely powerful theoretical tool for investigating the process of continuation and the properties of the resulting functions. However, it is in general not of much immediate value for providing a continuation of specific functions such as those in the examples we have seen.

Significant Topics **Page**

B Exercises

1. If $f(z)$ is analytic at α, show that $\overline{f(\bar{z})}$ is analytic at $\bar{\alpha}$.
2. The function

$$f(z) \equiv \sum_{1}^{\infty} \left(\frac{z}{1 + z} \right)^k$$

was shown, in Exercise A3 (d) of Section 3.5, to be analytic for $\operatorname{Re} z > -\frac{1}{2}$. Sum the series and so continue $f(z)$ analytically into the whole plane.

3. Show that

$$g(z) \equiv \sum_0^\infty z^{2^k+1}/(2^k+1)$$

is continuous for $|z| \le 1$ and analytic for $|z| < 1$. Use Example 5 to show that $|z| = 1$ is a natural boundary for $g(z)$.

4. Suppose $f(z), g(z)$, and $h(z)$ are all analytic in a region R and all can be continued analytically into an intersecting region S. Show that if any one of the following formulas holds in R, then it also holds in S.

(a) $f(z)g(z) = h(z)$;   (b) $f(z) + g(z) = h(z)$;
(c) $f(z)/g(z) = h(z)$;   (d) $f(z) = g'(z)$;

(e) $\displaystyle\int_\alpha^z f(\zeta)d\zeta = h(z)$, $\alpha \in R$ and the integral is over an arbitrary path in R from α to z.

This property of analytic continuation, which holds for other formulas as well, was once called *permanence of form*. The term is no longer in common usage.

5. Use Exercise A7 of Section 3.5 to continue $\Gamma(z)$ analytically (by a method different from that of the text) into the plane punctured at 0, $-1, -2, \ldots$.

C Exercises

1. Suppose $f(z)$ is analytic for $|z| < a$ and satisfies

$$f(2z) = 2f(z)f'(z), \qquad |z| < a/2.$$

(For example $f(z) = z$, $f(z) = \sin z$.) Show that $f(z)$ can be continued analytically into the whole plane.

2. Let $f(z) \equiv \sum_0^\infty z^{n!}$. Show that $f(z)$ has $|z| = 1$ as a natural boundary.

3. Show that $g(z) \equiv \sum_0^\infty z^{n!+1}/(n!+1)$ is continuous for $|z| \le 1$, analytic for $|z| < 1$, and has $|z| = 1$ as a natural boundary.

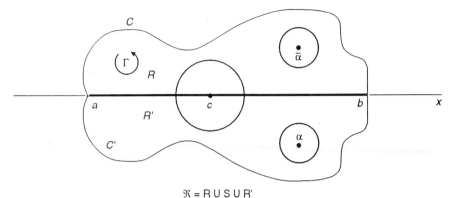

$$\mathfrak{R} = R \cup S \cup R'$$

Schwarz reflection principle

4. Take $S, R, R', \mathcal{R}, f(z)$, and $F(z)$ as described in the discussion of the Schwarz reflection principle.

   (a) Show that $F(z)$ is continuous in $\mathcal{R}$, real on the segment S, and analytic in R and R'. (See Exercise B1 above.)

   (b) Let c be a point of S. There is an $r > 0$ for which $|z - c| \le r$ is in $\mathcal{R}$. Define $g(z)$ for $|z - c| < r$ by

$$g(z) = \frac{1}{2\pi i} \int_{|z-c|=r} \frac{F(\zeta)d\zeta}{\zeta - z}.$$

   Show that $g(z)$ is analytic for $|z - c| < r$.

   (c) Show that $g(z) \equiv F(z)$ for $|z - c| < r$ and conclude that $F(z)$ is analytic in $\mathcal{R}$.

   Hint: Use Exercise C3 of Section 3.2 and the figure.

Properties of Γ and B.

5. The Beta function $B(z, \zeta)$ is defined by

$$B(z, \zeta) = \int_0^1 t^{z-1}(1 - t)^{\zeta-1}dt.$$

   Show that, for $\operatorname{Re} z > 0$, $\operatorname{Re} \zeta > 0$,

   (a) $B(z, \zeta) = B(\zeta, z)$;

   (b) as a function of z, $B(z, \zeta)$ is analytic for each fixed ζ, and conversely;

(c) if $t = \tau/(1 + \tau)$ so that $dt/d\tau = 1/(1 + \tau)^2$,

$$B(z, \zeta) = \int_0^\infty \frac{\tau^{z-1} d\tau}{(1 + \tau)^{z+\zeta}};$$

(d) $B(z, 1 - z) = \pi \csc \pi z$ (see Example 3);
(e) if $t = \sin^2 \theta$,

$$B(z, \zeta) = 2 \int_0^{\pi/2} (\sin \theta)^{2z-1} (\cos \theta)^{2\zeta-1} d\theta.$$

6. (a) Give a formal proof (calculation without justification) that for $\operatorname{Re} \alpha > 0$, $\operatorname{Re} \beta > 0$,

$$\Gamma(\alpha)\Gamma(\beta) = \Gamma(\alpha + \beta)B(\alpha, \beta).$$

Start with

$$\Gamma(\alpha)\Gamma(\beta) = \int_0^\infty \int_0^\infty e^{-(t+\tau)} t^{\alpha-1} \tau^{\beta-1} dt d\tau.$$

Set $t = x^2$, $\tau = y^2$, change to polar coordinates, separate as a product of two single integrals.
(b) Deduce $\Gamma(\alpha)\Gamma(1 - \alpha) = \pi \csc \pi \alpha$ and hence $\Gamma\left(\frac{1}{2}\right) = \sqrt{\pi}$.
(c) Extend (b) to all $\alpha \neq 0, \pm 1, \pm 2, \ldots, \pm n, \ldots$.
(d) Show that $\Gamma(\alpha) = 0$ has no solutions.
(e) Show that $B(\alpha, \beta) = \Gamma(\alpha)\Gamma(\beta)/\Gamma(\alpha + \beta)$ for $\operatorname{Re} \alpha > 0$, $\operatorname{Re} \beta > 0$ and then that it continues to hold for all $\alpha \neq -n$, $\beta \neq -n$, $\alpha + \beta \neq -n$, $n = 0, 1, 2, \ldots$. (See also Exercise C3 (f), Section 4.5).

Chapter 4

Mappings

4.1 Mappings by Analytic Functions

In Section 1.3 when we defined complex functions, we pointed out that the standard geometrical interpretation of an equation $w = f(z)$ is a mapping of points z is a z-plane to points w in a w-plane. We further noted that it is occasionally convenient to consider these planes as coincident, that is, we plot the z points and the w points in the same plane. In this chapter we examine these ideas in some detail by considering specific functions and by looking at general properties.

Let $f(z)$ be an analytic function (we will consider only analytic functions) with its domain being a region R and range $\mathcal{R}$. (Though we will not prove it, $\mathcal{R}$ will also be a region if R is.) If $z_0 \in R$, then $w_0 = f(z_0)$ will be in $\mathcal{R}$. We will say that w_0 is **the image** of z_0 under the mapping $w = f(z)$, and z_0 is **a pre-image** of w_0: *not the pre-image*, for there may be more than one. For example, if $f(z) = z^2$, then -1 is the image of i, and both i and $-i$ are pre-images of -1.

Let us suppose that $f(z)$ is analytic in a region R with range $\mathcal{R}$, that $f'(z) \neq 0$ in R, and that α is a point in R. We consider a smooth curve C in R given by

$$C : z = \zeta(t), \quad t \in I = [a, b],$$

which passes through $\alpha : \zeta(c) = \alpha$ where $a < c < b$. If we set $\beta = f(c)$, then the image Γ of C under the mapping $w = f(z)$ is a smooth curve (how?) in $\mathcal{R}$ given by

$$\Gamma : w = f(\zeta(t)), \quad t \in I.$$

Furthermore Γ passes through β since

$$f(\zeta(c)) = f(\alpha) = \beta.$$

We want to examine the connections between C and Γ near $t = c$, that is near the corresponding points $\alpha \in C$, $\beta \in \Gamma$.

The tangent vector to C at α is given by

$$\frac{dz}{dt}\bigg|_c = \zeta'(t)\bigg|_c = \zeta'(c) \neq 0 \tag{1}$$

($\neq 0$ since C is smooth). The tangent vector to Γ at β is given by

$$\frac{dw}{dt}\bigg|_c = \frac{d}{dt} f(\zeta(t))\bigg|_c = f'(\zeta(c))\zeta'(c) = f'(\alpha)\zeta'(c). \tag{2}$$

Thus the tangent vector to Γ at β is $f'(\alpha)$ times the tangent vector to C at α. *Clearly $f'(\alpha)$ is independent of C, that is, it is the same for all curves through α.*

To see the geometrical meaning of (2) we set $f'(\alpha) = Ae^{i\phi}$, and $\zeta'(c) = Be^{i\theta}$, and (2) becomes

$$\frac{dw}{dt}\bigg|_c = Ae^{i\phi}Be^{i\theta} = ABe^{i(\theta+\phi)}.$$

Thus, for any smooth curve C through α, the direction of the image curve Γ is rotated by the mapping through the angle $\phi = \arg f'(\alpha)$. Also an approximate form of (2) is

$$\Delta w \approx f'(\alpha)\Delta z.$$

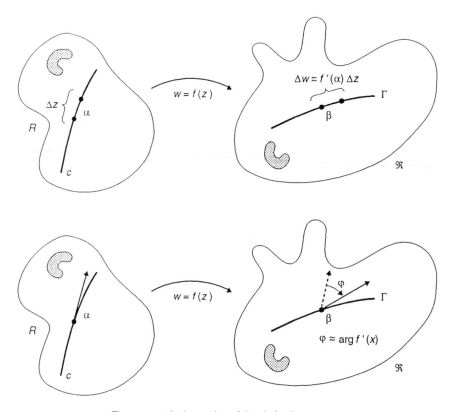

The geometrical meaning of the derivative

Thus $|\Delta w| \approx A|\Delta z|$, so that "infinitesimal" lengths are distorted (lengthened or shortened) by a factor of $A = |f'(\alpha)|$.

Again we remark that the distortion factor $|f'(\alpha)|$ and the rotation angle $\phi = \arg f'(\alpha)$ are independent of the smooth curve C. In particular, let C_1 and C_2 be two curves through α with tangent angles θ_1 and θ_2 respectively so that they have an angle of separation from C_2 to C_1 of $\psi = \theta_1 - \theta_2$. Then their images Γ_1 and Γ_2 (respectively) both go through β with **the same angle of separation**:

$$(\theta_1 + \phi) - (\theta_2 + \phi) = \theta_1 - \theta_2 = \psi \quad \text{from} \quad \Gamma_2 \text{ to } \Gamma_1.$$

Mappings which have this angle preserving property are called **conformal** maps. (A familiar example, which we will not prove, is Mercator's projection. Most maps of the earth are such projections, and they distort

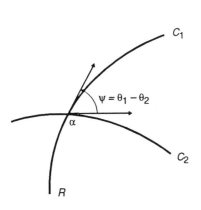

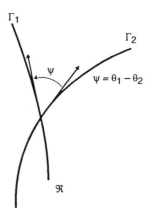

Conformality

distances but preserve angles. Thus Greenland is frequently shown much magnified.) We have therefore proved the following.

4.1a Theorem. *If $f(z)$ is analytic and $f'(z) \neq 0$ in a region R, the map produced by $w = f(z)$ is conformal.*

The basic theorem concerning conformal maps is the following, which is one of the most fundamental results in the advanced theory of analytic functions.

4.1b Theorem. (Riemann Mapping Theorem). *If R is a simply connected region which is not all of $\mathbb{C}$, then there is a function $f(z)$ analytic in R with $f'(z) \neq 0$ in R for which the mapping*

$$w = f(z)$$

takes R in a one-to-one and conformal way onto $|w| < 1$. Furthermore, if $z_0 \in R$, then there is exactly one such function for which

$$f(z_0) = 0 \quad and \quad f'(z_0) > 0.$$

We will not prove this result since it is too difficult and time consuming for a course at our level. We will however prove the uniqueness part: there can be no more than one mapping function meeting the conditions $f(z_0) = 0$, $f'(z_0) > 0$.

Before we begin our technical proof we remark that if $f(z)$ is an analytic function which is zero at a point α, then the function $\varphi(z) \equiv$

$f(z)/(z - \alpha)$ can be defined at α so as to be analytic there. For $f(z)$ has a Taylor expansion

$$f(z) = a_1(z - \alpha) + a_2(z - \alpha)^2 + a_3(z - \alpha)^3 + \cdots$$

and so

$$\varphi(z) = a_1 + a_2(z - \alpha) + a_3(z - \alpha)^2 + \cdots ,$$

which is analytic at α if we take $\varphi(\alpha) = a_1$. In Section 5.1 we will discuss similar questions in some detail.

The following famous theorem goes by the name of the **Schwarz Lemma.**

4.1c Lemma (Schwarz). *Suppose D is the disk $|z| < 1$ and (i) $f(z)$ is analytic in D, (ii) $|f(z)| < 1$ in D, and (iii) $f(0) = 0$. Then either (a) $|f(z)| < |z|$ for $0 < |z| < 1$, or (b) there is a real constant a such that $f(z) = ze^{ia}$.*

Proof. By the preceding remark (with $\alpha = 0$) $\varphi(z) = f(z)/z$ is analytic in D. Now for $z \in D$ we take r, $|z| < r < 1$. Then

$$|\varphi(z)| \le \max_{|z|=r} |\varphi(z)| = \left(\max_{|z|=r} |f(z)| \right) \Big/ r < 1/r, \tag{3}$$

The first inequality in (3) being by the maximum modulus theorem. (See also Exercise B4, Section 3.4.) Then as $r \to 1$ we get $|\varphi(z)| \le 1$ in D. Thus either $|\varphi(z)| < 1$ for all z in D (which is our conclusion (a)) or there is a $z_0 \in D$ at which $|\varphi(z_0)| = 1$. Then, again by the maximum modulus theorem, $\varphi(z)$ is constant in D with $|\varphi(z)| = 1$ (how?). This implies $\varphi(z) = e^{ia}$, that is, $f(z) = ze^{ia}$. $\square$

4.1d Theorem. *There can be no more than one mapping function as described in the Riemann Mapping Theorem.*

Proof. We can, by translation, assume that $z_0 = 0$, so that our uniqueness conditions become $f(0) = 0$, $f'(0) > 0$. Now any such mapping function will have an analytic inverse $f^{-1}(w)$ which maps $D : |w| < 1$ onto R in a one-to-one and conformal way with $f^{-1}(0) = 0$. Furthermore $df^{-1}(w)/dw$ will never be zero and will be a positive real number at $w = 0$. (See Exercise C1.)

Suppose there were two such mapping functions $f(z)$ and $g(z)$. We form the two functions

$$\zeta = \varphi(w) = f(g^{-1}(w)), \quad \chi = \psi(w) = g(f^{-1}(w)).$$

In each of these we map D onto R by one of the inverses, and then back into D by the other mapping function. We examine φ and ψ:

(i) $\varphi(w)$ is analytic in D; $\psi(w)$ is analytic in D;

(ii) $|\varphi(w)| < 1$ in D; $|\psi(w)| < 1$ in D;

(iii) $\varphi(0) = 0$; $\psi(0) = 0$.

Thus each of φ and ψ satisfies the hypotheses of the Schwarz Lemma, and so

$$|\zeta| = |\varphi(w)| \leq |w|; \quad |\chi| = |\psi(w)| \leq |w|. \tag{4}$$

Now we operate on $\zeta = \varphi(w) = f(g^{-1}(w))$ by taking f^{-1} of both sides:

$$f^{-1}(\zeta) = g^{-1}(w);$$

and now by taking g of both sides to get

$$w = g(f^{-1}(\zeta)) = \psi(\zeta).$$

Then by the second equation of (4)

$$|w| = |\psi(\zeta)| \leq |\zeta|,$$

so applying also the first equation of (4) we have

$$|\zeta| \leq |w| \leq |\zeta|$$

Similarly

$$|\chi| \leq |w| \leq |\chi|,$$

so that, in other symbols,

$$|w| = |\varphi(w)| = |\psi(w)|.$$

Then the equality conclusion of the Schwarz Lemma applies and we have

$$\varphi(w) = we^{ia}; \quad \psi(w) = we^{ib}.$$

But both $\varphi'(0) > 0$ and $\psi'(0) > 0$, that is $e^{ia} = e^{ib} = 1$ and we get

$$w = f(g^{-1}(w)) = g(f^{-1}(w)). \tag{5}$$

Taking f^{-1} of the first two terms of (5) yields

$$f^{-1}(w) = g^{-1}(w). \tag{6}$$

this really completes the argument, for the range of each of f^{-1} and g^{-1} is R, and for any w they are, by (6), the same z in R. Then the last two terms of (5) become

$$f(z) = g(z). \quad \square$$

B Exercises

1. Let $w = f(z)$ map a region R in a one-to-one and conformal way onto $D : |w| < 1$. If $C : z = z(t)$, $t \in I$, is a smooth curve in R, then $\Gamma : w = f(z(t))$, $t \in I$, is smooth in D and its length is

$$\int_I |f'(z(t))||z'(t)|dt.$$

2. In the proof of the Schwarz Lemma we had $\varphi(z) = f(z)/z$. Show that $\varphi(0) = f'(0)$, and so $|f'(0)| < 1$ or $f'(0) = e^{ia}$.

3. The Riemann Mapping Theorem specifically excludes $R = \mathbb{C}$. Show that no such map can exist for $R = \mathbb{C}$. Hint: If it did, then $f(z)$ would be entire.

C Exercises

1. In the first paragraph of the proof of Theorem 5.1 certain statements about the inverse of a mapping function were made. Prove them.
   (i) $f^{-1}(w)$ is analytic in D.
   (ii) $df^{-1}(w)/dw \neq 0$ in D.

(iii) $df^{-1}(w)/dw\big|_{w=0} > 0.$

2. Suppose R and $\mathcal{R}$ are regions, neither of which is $\mathbb{C}$. Then show that there is a function $f(z)$ analytic in R, with $f'(z) \neq 0$, such that $w = f(z)$ maps R in a one-to-one and conformal way onto $\mathcal{R}$, and that if $a \in R$ and $\alpha \in \mathcal{R}$, then there is exactly one such function for which

$$f(a) = \alpha, \qquad f'(a) > 0.$$

What is the situation if $f'(a) > 0$ is replaced by $\arg f'(a) = \pi/2$, or in general by $\arg f'(a) = \theta$?

4.2 Linear and Bilinear Transformations

The general linear transformation

$$w = \alpha z + \beta, \qquad \alpha \neq 0, \tag{1}$$

is a composition of a dilation, a rotation, and a translation as discussed in Section 1.3. This mapping takes the extended z-plane onto the extended w-plane in a one-to-one manner with $z = \infty$ going onto $w = \infty$. Furthermore it maps lines onto lines and circles onto circles (Exercises A1 and 2). The map is conformal since $dw/dz = \alpha \neq 0$.

The map provided by

$$w = \frac{\alpha z + \beta}{\gamma z + \delta}, \qquad \det \begin{bmatrix} \alpha & \beta \\ \gamma & \delta \end{bmatrix} = \alpha\delta - \beta\gamma \neq 0, \tag{2}$$

(Exercise A3) is called a **bilinear transformation** because when (2) is cleared of fractions it becomes

$$\gamma w z + \delta w - \alpha z - \beta = 0$$

which is linear in each of z and w. Other names (which are used in other books) are **linear fractional**, **Möbius**, and in older books, **homographic transformations**.

If $\gamma = 0$ equation (2) reduces to

$$w = (\alpha/\delta)z + (\beta/\delta)$$

which is linear. Thus we will be primarily interested in the case $\gamma \neq 0$. For $\gamma \neq 0$ we can write (1) in the form

$$w = \frac{\beta\gamma - \alpha\delta}{\gamma} \frac{1}{\gamma z + \delta} + \frac{\alpha}{\gamma} \tag{3}$$

(Exercise A4). If we set

$$\begin{cases} z_1 = \gamma z + \delta \\ z_2 = 1/z_1 \\ w = [(\beta\gamma - \alpha\delta)/\gamma]z_2 + (\alpha/\gamma) \end{cases} \tag{4}$$

we see that (3) is a linear transformation followed by an **inversion** ($z_2 = 1/z_1$) followed in turn by another linear transformation. Since linear mappings are relatively simple, we look next at the inversion

$$w = 1/z \tag{5}$$

which is a special case of (2) with $\alpha = \delta = 0$ and $\beta = \gamma = 1$.

Clearly (Exercise A6) $w = 1/z$ maps the punctured plane $z \neq 0$ onto the punctured plane $w \neq 0$, and if we adopt the conventions $1/0 = \infty$ and $1/\infty = 0$, it maps the extended z-plane *in a one-to-one manner* onto the extended w-plane. By composition with linear maps (4) we get the following. The details of the proof are left to the exercises.

4.2a Theorem. *A bilinear transformation maps the extended z-plane onto the extended w-plane in a one-to-one manner. The map is conformal except at the pole $z = -\delta/\gamma$.*

Writing $z = r\, e^{i\theta}$ we have

$$w = 1/(r\, e^{i\theta}) = (1/r)e^{-i\theta}, \tag{6}$$

so that the modulus of w is the reciprocal of that of z and its argument is the negative. If we keep θ fixed and let z go from 0 to ∞ along the ray $\arg z = \theta$, then by (6) w goes from ∞ to 0 along the ray $\arg w = -\theta$. If we keep r fixed and let θ go from 0 to 2π around the circle of radius r, then w goes *clockwise* around the circle of radius $1/r$.

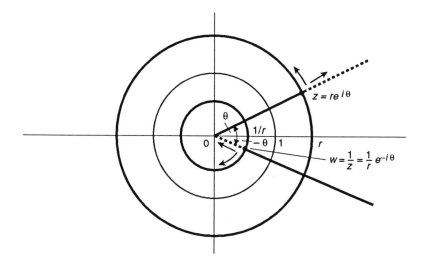

The mapping $w = 1/z$

We next examine the images of the coordinate lines in rectangular coordinates. Since $z = 1/w$, we have

$$x = \frac{u}{u^2 + v^2}, \qquad y = \frac{-v}{u^2 + v^2}. \tag{7}$$

So the lines $x = a$, $y = b$ give rise, after a little algebra, to

$$\begin{cases} \left(u - \dfrac{1}{2a}\right)^2 + v^2 = \left(\dfrac{1}{2a}\right)^2 \\[3mm] u^2 + \left(v + \dfrac{1}{2b}\right)^2 = \left(\dfrac{1}{2b}\right)^2 \end{cases} \tag{8}$$

These are circles through the origin with centers at $(1/2a, 0)$ and $(0, -1/2b)$ and with radii $1/2|a|$ and $1/2|b|$, respectively.

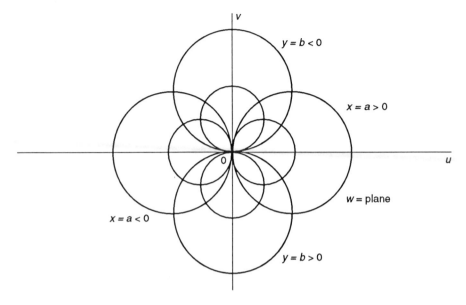

The mapping w = 1/z: the images of x = a, y = b

Fig. 4.4

A glance at the figure illustrating these circles suggests that any two which intersect (as opposed to being tangent) do so at right angles, just as do their pre-images $x = a$ and $y = b$. This is certainly true at the origin, and it is easily shown to be true at the other point of intersection of an intersecting pair (Exercise A8).

Also easy to prove (Exercise A2) are the following: $w = 1/z$ maps

  (i) a line through the origin onto a line;

  (ii) a circle through the origin onto a line;

  (iii) a line not through the origin onto a circle not through the origin;

  (iv) a circle not through the origin onto a circle not through the origin.

This suggests that circles and lines should be considered as a single collection. Also supporting this notion is the observation that a circle or a line is mapped by stereographic projection onto a circle on the sphere. Thus we state the following.

Definition. *A line is a circle through* ∞.

With this definition the previous paragraph can be restated as the next theorem.

4.2b Theorem. $w = 1/z$ *maps*

    (i) *a circle through ∞ onto a circle through the origin;*

    (ii) *a circle through the origin onto a circle through ∞;*

    (iii) *a circle through neither 0 nor ∞ onto a circle through neither.*

Further we note that $w = 1/(\gamma z + \delta)$ amounts to a linear transformation in the z-plane followed by inversion. Thus $z_0 = -\delta/\gamma$ plays the role of the origin for this function. And so, finally, we have

4.2c Theorem. *If $\gamma \neq 0$,, then Theorem 4.2b remains true for*

$$w = (\alpha z + \beta)/(\gamma z + \delta)$$

with "the origin" replaced by $z_0 = -\delta/\gamma$.

We close this section by observing that a bilinear transformation

$$w = (\alpha z + \beta)/(\gamma z + \delta)$$

has but three independent coefficients. If $\gamma = 0$, then $w = (\alpha/\delta)z + (\beta/\delta)$ which is in the *form* $w = \alpha z + \beta$, and so we can take the coefficients as $\alpha, \beta, 0, 1$. If $\gamma \neq 0$, we can divide by γ to get

$$w = [(\alpha/\gamma)z + (\beta/\gamma)]/[z + (\delta/\gamma)]$$

which is in the *form*

$$w = (\alpha z + \beta)/(z + \delta).$$

This suggests that a bilinear map should be determined if we know what happens to three points.

4.2d Theorem. *Suppose three distinct z values, z_1, z_2, z_3, and three distinct w values, w_1, w_2, w_3 (all in the extended sense) are given. Then there is exactly one bilinear transformation $w = T(z)$ which takes z_k onto w_k, $k = 1, 2, 3$.*

If $z_3 = \infty$ and $w_3 = \infty$, then $T(z)$ is linear (how?) and the rest of the argument is easy. If there is no one k for which both $z_k = \infty$ and $w_k = \infty$ a direct assault is difficult. But if we write down

$$\frac{(w - w_1)(w_2 - w_3)}{(w - w_3)(w_2 - w_1)} = \frac{(z - z_1)(z_2 - z_3)}{(z - z_3)(z_2 - z_1)} \tag{9}$$

and examine it properly we can extract the proof. See the C Exercises.

A Exercises

1. Prove that a linear transformation $w = \alpha z + \beta$, $\alpha \neq 0$, maps the extended z-plane onto the extended w-plane in a one-to-one manner with $z = \infty$ mapping onto $w = \infty$.

2. Show that under a linear transformation
   (i) a line L in the z-plane maps onto a line Λ in the w-plane;
   (ii) a circle C in the z-plane maps onto a circle Γ in the w-plane;
   (iii) $I(C)$ maps onto $I(\Gamma)$.

3. Show that (2) reduces to $w = $ constant if $\alpha\delta - \beta\gamma = 0$.

4. Establish formula (3).

5. Show that $w = 1/z$, with the conventions $1/0 = \infty$ and $1/\infty = 0$, maps the extended z-plane in a one-to-one manner onto the extended w-plane.

6. Prove Theorem 4.2a.

7. Derive (8) from (7).

8. Show that any two circles from (8) which intersect, do so at right angles.

B Exercises

1. Prove Theorem 4.2b. Hint: Observe that any circle (in our extended sense) in the z-plane has an equation in the form

$$a(x^2 + y^2) + bx + cy + d = 0$$

and show that this maps onto

$$d(u^2 + v^2) + bu - cv + a = 0.$$

2. Prove Theorem 4.2c.

3. If $T(z) = \dfrac{\alpha_1 z + \beta_1}{\gamma_1 z + \delta_1}$ and $S(z) = \dfrac{\alpha_2 z + \beta_2}{\gamma_2 z + \delta_2}$, then show that $T \circ S(z) = T(S(z))$ is in the form

$$T \circ S(z) = \frac{\alpha z + \beta}{\gamma z + \delta},$$

Also determine $\alpha, \beta, \gamma, \delta$ in terms of $\alpha_1, \beta_1, \dots, \gamma_2, \delta_2$, and show

$$\begin{bmatrix} \alpha & \beta \\ \gamma & \delta \end{bmatrix} = \begin{bmatrix} \alpha_1 & \beta_1 \\ \gamma_1 & \delta_1 \end{bmatrix} \begin{bmatrix} \alpha_2 & \beta_2 \\ \gamma_2 & \delta_2 \end{bmatrix}.$$

4. If a linear transformation $w = \alpha z + \beta$ leaves two distinct points fixed ($z = \alpha z + \beta$), then show that it is the identity mapping, that is, $\alpha = 1$, $\beta = 0$.

5. If a bilinear transformation $w = (\alpha z + \beta)/(\gamma z + \delta)$ leaves three distinct points fixed, then it is the identity mapping.

6. Since $w = T(z)$ where

$$T(z) = (\alpha z + \beta)/(\gamma z + \delta), \quad \alpha\delta - \beta\gamma \neq 0,$$

is one-to-one, there is an inverse T^{-1}, such that $T^{-1} \circ T(z) = T \circ T^{-1}(z) \equiv z$. Find $T^{-1}(z)$.

C Exercises

1. Show that (9) leads to a bilinear transformation $T(z)$ when solved for w.

2. Show that $T(z)$ maps z_k onto w_k, $k = 1, 2, 3$.

3. Show that $T(z)$ is unique: no other bilinear transformation maps z_k onto w_k, $k = 1, 2, 3$. Hint: If $S(z)$ is a second such linear transformation consider $S^{-1} \circ T(z)$.

4.3 Examples of Bilinear Transformations

Example 1. If $\alpha = a + ib$ with $b > 0$, then show that

$$w = \frac{z - \alpha}{z - \bar{\alpha}} \tag{1}$$

maps $y \geq 0$ in a one-to-one and conformal manner onto $|w| \leq 1$ with $z = \alpha$ going onto $w = 0$.

Solution. It is clear that the map is conformal (how?) and that $w = 0$ when $z = \alpha$. Further, the line $L : y = 0$ (i.e., the real axis) maps onto *some* circle (how?). When $z \in L$, we have $z = x$ and then $w = (x - \alpha)/(x - \bar{\alpha})$ and so $|w| = 1$ since the numerator and denominator are conjugates of each other so that they have the same modulus. Hence $L : |w| = 1$, the unit circle in the w-plane.

Since (1) maps the z-plane onto the w-plane in a one-to-one manner, one of the half-planes $y > 0$, $y < 0$ must go onto $I(U)$ and the other onto $E(U)$. But α is in $y > 0$ and so all of $y > 0$ must go into $I(U)$, that is, onto $|w| < 1$.

Remarks:

A. We observe that if a is a real constant, then

$$w = e^{ia} \frac{z - \alpha}{z - \bar{\alpha}} \tag{2}$$

merely rotates the w-plane so (2) also maps $y \geq 0$ onto $w \leq 1$.

B. $w = e^{ia} R \dfrac{z - \alpha}{z - \bar{\alpha}}$ maps $y \geq 0$ onto $|w| \leq R$.

C. $w = \beta + e^{ia} R \dfrac{z - \alpha}{z - \bar{\alpha}}$ maps $y \geq 0$ onto $|w - \beta| \leq R$, that is, onto a disk of radius R centered at β.

Example 2. Map the half plane $y \geq \sqrt{3}\, x + 4$ in the z-plane onto $|w| \leq 1$ by a bilinear transformation.

Solution. We outline our steps:

(i) We map the z-plane onto a z_1-plane by a rotation of $\pi/3$. This brings $y \geq \sqrt{3}\, x + 4$ onto $y_1 \geq c$, for some constant c.

(ii) We map the z_1-plane onto a z_2-plane by a translation of $-ic$. This gives $y_2 \geq 0$.

(iii) We map the z_2-plane onto the w-plane by Example 1.

(i) Thus $z_1 = e^{-i\pi/3} z$, or $z = e^{i\pi/3} z_1$, from which

$$x = (x_1 - \sqrt{3}\, y_1)/2, \qquad y = (\sqrt{3}\, x_1 + y_1)/2$$

and $y \geq \sqrt{3}\, x + 4$ becomes, after a little algebra, $y_1 \geq 2$. Thus $c = 2$.

(ii) Then $z_2 = z_1 - 2i$ or $z_1 = z_2 + 2i$ from which $y_1 = y_2 + 2$, so $y_1 \geq 2$ becomes $y_2 \geq 0$.

(iii) Finally

$$w = e^{ia} \frac{z_2 - \alpha}{z_2 - \bar{\alpha}}.$$

Putting this all together we have

$$w = e^{ia} \frac{z_1 - 2i - \alpha}{z_1 - 2i - \bar{\alpha}} = e^{ia} \frac{3^{-i\pi/3} z - 2i - \alpha}{e^{-i\pi/3} z - 2i - \bar{\alpha}}$$

$$= e^{ia} \frac{z - (2i + \alpha)e^{i\pi/3}}{z - (2i + \bar{\alpha})}$$

How should we choose α so that $z = 5i$ maps onto $w = 0$?

Example 3. Find the pre-image of $u = \operatorname{Re} w < 0$ under the transformation $w = z/(z - 2)$.

Solution. Setting $z = x + iy$, $w = u + iv$ we get

$$u + iv = \frac{x + iy}{(x - 2) + iy}$$

and taking real parts we get

$$u = \frac{x(x - 2) + y^2}{(x - 2)^2 + y^2}$$

Then $u < 0$ leads to

$$x(x - 2) + y^2 < 0$$

or

$$(x - 1)^2 + y^2 < 1.$$

Thus the pre-image of $u < 0$ is the open disk with center at 1 and radius 1.

Points w and w^* are called **inverse images** of each other with respect to the circle $|w| = a$ if $|w \, w^*| = a^2$ and $\arg w = \arg w^*$. This means w and w^* are on the same ray from the origin, and their product has modulus a^2. This is equivalent to $w^* = a^2/\bar{w}$. (Exercise A2).

Example 4. Let

$$w = a\frac{z - \alpha}{z - \bar{\alpha}}$$

where a is a positive constant. Show that $z = \beta$ and $z = \bar{\beta}$ map into $w = \gamma$ and $w = \gamma^*$ which are inverse images with respect to $|w| = a$.

Solution.
$$\gamma = a(\beta - \alpha)/(\beta - \bar{\alpha})$$
$$\gamma^* = a(\bar{\beta} - \alpha)/(\bar{\beta} - \bar{\alpha}).$$

Then
$$\gamma^* = a \Big/ \frac{\bar{\beta} - \bar{\alpha}}{\bar{\beta} - \alpha}$$
$$= a^2 \Big/ a\frac{\bar{\beta} - \bar{\alpha}}{\bar{\beta} - \alpha}$$
$$= a^2/\bar{\gamma}.$$

Example 5. Show that if a is real, and $|\alpha| < 1$, then

$$w = e^{ia}\frac{z - \alpha}{\bar{\alpha}z - 1}$$

maps $|z| < 1$ onto $|w| < 1$ with $z = \alpha$ mapping onto $w = 0$.

Solution. Clearly $w = 0$ when $z = \alpha$. And when $z = e^{i\theta}$, that is, when $|z| = 1$

$$|w| = \left|\frac{e^{i\theta} - \alpha}{e^{i\theta}\bar{\alpha} - 1}\right| = \left|\frac{e^{i\theta} - \alpha}{e^{-i\theta} - \bar{\alpha}}\right| = 1,$$

since $e^{i\theta} - \alpha$ and $e^{-i\theta} - \bar{\alpha}$ are conjugates. Thus the unit circle in the z-plane maps onto the unit circle in the w-plane. Then, since $|\alpha| < 1$, and it maps onto $w = 0$, the interior must go to the interior.

Example 6. Let a, b, c be three distinct points on the x-axis in the z-plane. Find a bilinear transformation which maps $y > 0$ onto $v > 0$ and takes a, b, c onto $0, 1, \infty$, respectively.

Solution. Clearly
$$w = \frac{z - a}{z - c}$$

takes a to 0 and c to ∞, but $z = b$ goes to

$$w = \frac{b - a}{b - c} \neq 1$$

since $a \neq c$ and so b does not go to 1. But then

$$w = \frac{z - a}{z - c}\frac{b - c}{b - a}$$

will take a, b, c to $0, 1, \infty$, respectively. And for $y > 0$ to map onto $v > 0$ we must have one point in $y > 0$ mapping onto a point in $v > 0$. Set $z = i$ and we get

$$w = u + iv = \frac{i - a}{i - c} \cdot \frac{b - c}{b - a}$$

from which we must have (how?)

$$(a - c)(b - c)(b - a) > 0.$$

This is true if $a < b < c$, $c < a < b$ or $b < c < a$, and not true for any other order. For these other orders of $a, b, c, y > 0$ maps onto $v < 0$.

Significant Topics   **Page**

A Exercises

1. How should we choose α so that $z = 5i$ maps into $w = 0$ in Example 2.

2. Show that if w and w^* are inverse images of each other with respect to the circle $|w| = a$, then $w^* = a^2/\bar{w}$.

3. Find the image of $\operatorname{Re} z < 1$ in Example 3.

4. Find a linear transformation which maps $|z| \le 1$ onto $|w - 3i| \le 5$ and takes i onto $-2i$.

5. Find a bilinear transformation which keeps 1 and 2 fixed and takes i to ∞.

6. Find the image of $x > 0$ if

   (a) $\quad w = \dfrac{z - 1}{z + 1}$;   (b) $\quad w = e^{ia} \dfrac{z - 1}{z + 1}$, a real.

7. Find a bilinear transformation which maps $|z| \le 1$ onto $v \ge 0$ and takes the origin to a point $\alpha = a + ib$, $b > 0$.

8. Map the region between $x^2 + (y - 1)^2 = 1$ and $x^2 + (y - 2)^2 = 4$ onto a strip by a bilinear transformation.

9. Find a bilinear transformation which takes $y \le 1 - x$ onto $|w| \le 1$.

B Exercises

1. (a) If $\operatorname{Re}\alpha > 0$, show that

$$w = \frac{z - \alpha}{z + \bar{\alpha}}$$

    maps $x \geq 0$ onto $|w| \leq 1$ and takes α to 0.

  (b) Discuss

$$w = e^{ia}\frac{z - \alpha}{z + \bar{\alpha}} \quad \text{and} \quad w = e^{ia}R\frac{z - \alpha}{z + \bar{\alpha}}$$

    if a is real, $R > 0$ and $\operatorname{Re}\alpha > 0$.

2. Show that if $w = (\alpha z + \beta)/(\gamma z + \delta)$ takes ∞ to ∞, then $\gamma = 0$ and so the transformation is linear.

3. Reconsider Example 6 if $c = \infty$.

4. (a) Find the inverse mapping for Example 5.

  (b) Re-examine Example 5 if $|\alpha| > 1$.

C Exercises

1. If C_1 is an arc of $(x - 2)^2 + (y + 1)^2 = 2$ and C_2 is an arc of $(x - 2)^2 + (y + 2)^2 = 5$, find a bilinear transformation which maps the shaded area between them onto an angle $0 < \arg w < \phi$, and find ϕ.

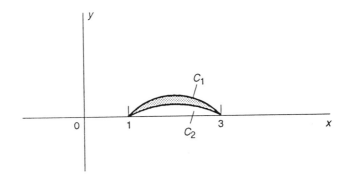

2. Suppose $w = f(z)$ is a bilinear transformation which maps $|z| < 1$ onto $|w| < 1$ with $f(0) = 0$. Show that there is a real constant a for which $f(z) = ze^{ia}$.

3. Find the image of $0 < \arg z < \pi/4$ under the mapping $w = z/(z - 2)$.

4.4 Other Elementary Mappings

The Power Function z^p. Each branch of this function is analytic and $w = z^p$ produces a conformal map except at $z = 0$. For $p > 0$ it maps the angle $0 \leq \arg z \leq \eta$ onto the angle $0 \leq \arg w \leq p\eta$ so long as both η and $p\eta$ are $\leq 2\pi$. this is easily seen in polar coordinates with $w = \rho e^{i\phi}$ and $z = r e^{i\theta}$:

$$\rho e^{i\phi} = r^p e^{ip\theta},$$

so as θ increases from 0 to η, ϕ increases from 0 to $p\eta$.

In particular, $w = z^2$ maps the first quadrant of the z-plane onto the upper half-plane in the w-plane, and of course $w = z^{1/2}$ does the reverse: it maps the upper half-plane of the z-plane onto the first quadrant of the w-plane. More generally if an angle $A : 0 \leq \arg z \leq \eta$ is given in the z-plane with $\eta < 2\pi$, then $w = z^{\pi/\eta}$ maps A onto the upper half-plane in the w-plane since

$$w = \rho e^{i\phi} = (r\, e^{i\theta})^{\pi/\eta} = r^{\pi/\eta} e^{i\theta\pi/\eta},$$

and as θ goes from 0 to η, ϕ goes from 0 to $\eta\pi/\eta = \pi$.

The Exponential Function e^z. We have seen that e^z is periodic with period $2\pi i$. We examine the mapping $w = e^z$ in a strip $S : 0 \leq y \leq h$ where $h < 2\pi$. For $z \in S$ we have, with $w = \rho e^{i\phi}$, $z = x + iy$

$$w = \rho e^{i\phi} = e^x e^{iy}$$

from which

$$\rho = e^x, \qquad \arg w = y.$$

Thus S maps onto the angle $A : 0 \leq \arg w \leq h$.

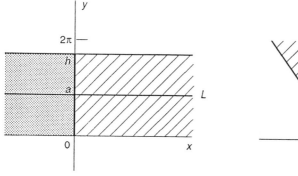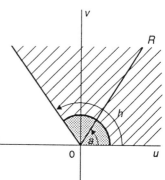

Clearly each line $L : y = a$ maps onto a ray $R : \arg w = a$ with the negative end of L going onto the segment $0 < \rho < 1$ of R. Similarly the strip $-h \leq y \leq 0$ maps onto $-h \leq \arg w \leq 0$, and more generally $a \leq y \leq b$ maps onto $a \leq \arg w \leq b$ if $b - a < 2\pi$.

We remark, without further exploration of the idea, that one can consider conformal mappings from one Riemann surface to another and by this device remove the restriction above that angles be $< 2\pi$. Since the inverse of e^z is $\log z$, if we let h move above 2π we are mapping a strip of width $> 2\pi$ onto the Riemann surface of $\ln w$.

The Function $\sin z$. We consider the mapping $w = \sin z$ on the strip $-\pi/2 < x < \pi/2$ in the z-plane. Setting $z = x + iy$, $w = u + iv$ we write

$$u + iv = w = \sin z = \sin x \cosh y + i \cos x \sinh y,$$

so that
$$u = \sin x \cosh y, \qquad v = \cos x \sinh y.$$

The image Γ of the line $L : x = a$, $0 < a < \pi/2$ is then given parametrically by
$$u = \sin a \cosh y, \qquad v = \cos a \sinh y. \tag{1}$$

We eliminate y by the identity $\cosh^2 y - \sinh^2 y = 1$ to get

$$\frac{u^2}{\sin^2 a} - \frac{v^2}{\cos^2 a} = 1, \tag{2}$$

which is a hyperbola in the w-plane with foci at $(\pm 1, 0)$. From the positivity of u in (1) (since $a > 0$) we see that Γ is the right branch of (2), with the negative end of L ($y < 0$) going into the negative end ($v < 0$) of Γ (how?). Similarly, from the same equations, if a were negative, Γ would be the left branch of (2). Also from (1) the imaginary axis ($x = 0$) goes onto the imaginary axis ($u = 0$).

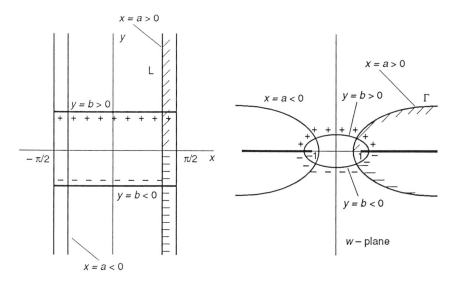

The mapping $w = \sin z$

Further, again from (1), it is clear that as $a \to \pi/2 - 0$, L collapses onto the infinite segment $v = 0$, $1 \le u < \infty$, which we take as a cut in the w-plane (because *both* ends of the line $x = \pi/2$ map onto this segment). Similarly the left branch of the hyperbola (2) collapses onto $v = 0$, $-\infty < u \le -1$ as $a \to -\pi/2 + 0$. This segment we also take as a cut in the w-plane.

Thus we see that the vertical strip $-\pi/2 < x < \pi/2$ maps conformally (since $(\sin z)' = \cos z \ne 0$) onto the w-plane cut from 1 to ∞ and from -1 to $-\infty$ on the real axis.

We now look at the images of horizontal segments. First the x-axis from $-\pi/2$ to $\pi/2$ maps into the u-axis between -1 and 1 (how?). For a segment $y = b > 0$, $-\pi/2 < x < \pi/2$, we have

$$u = \sin x \cosh b, \qquad v = \cos x \sinh b, \tag{3}$$

from which (how?)

$$\frac{u^2}{\cosh^2 b} + \frac{v^2}{\sinh^2 b} = 1, \tag{4}$$

an ellipse with foci at $(\pm 1, 0)$. But from (3) we must have $v > 0$, so the image of the segment is the upper half of (4). Similarly if b were negative, the image would be the lower half of (4).

Compositions of Mappings. We can follow one mapping by another as we did in some cases in discussing bilinear transformations. We note that if $f(z)$ and $g(w)$ are both analytic with the range of $f(z)$ overlapping the domain of $g(w)$, then $g(f(z))$ has its derivative given by

$$\frac{d}{dz} g(f(z)) = g'(f(z))f'(z).$$

Hence if neither $g'(w)$ not $f'(z)$ vanishes, then the composition produces a conformal map.

As an example we map an angle conformally onto the unit disk. Let A be the angle described by $a < \arg z < b$ with $c \equiv b - a < 2\pi$.

(i) Let $z_1 = z\, e^{-ia}$. Then A maps conformally onto $A_1 : 0 < \arg z_1 < c = b - a$.

(ii) Let $z_2 = z_1^{\pi/c}$. Then A_1 maps conformally onto the upper half-plane in the z_2-plane since

$$\arg z_2 = (\pi/c) \arg z_1$$

and $\arg z_2$ goes from 0 to π as $\arg z_1$ goes from 0 to c.

(iii) We map the upper half-plane of the z_2-plane onto the w-plane by Example 1 of Section 4.3. Thus

$$w = \frac{z_2 - \alpha_2}{z_2 - \bar{\alpha}_2}$$

where $\operatorname{Im} \alpha_2 > 0$, and α_2 maps into $w = 0$.

If we choose a point α in A to map into $w = 0$, we find its image in the z_1-plane (in A_1) to be $\alpha_1 = \alpha\, e6-ia$, and this in turn is mapped onto $\alpha_2 = \alpha_1^{\pi/2} e^{-ia\pi/c}$ which occurs in the bilinear mapping.

We could, of course, collapse all these into a single formula, but it seems simpler to leave the description of the mapping as a series of easy steps.

Significant Topics   **Page**

B Exercises

1. Under the mapping $w = z^2$ describe the image of
   (a) the strip $0 < x < 1$;
   (b) the strip $0 < y < 1'$
   (c) the rectangle $0 < x < a$, $0 < y < b$.

2. Describe the image of (a) the right half-plane $(x > 0)$; (b) the upper half-plane under $w = z^2$.

3. Describe the image of the rectangle $a < x < b$, $c < y < d$ where $d - c < 2\pi$ under $w = e^z$.

4. Map the strip $|y| < \pi/2$ conformally onto $|w| < 1$.

5. (a) Discuss the mapping of vertical strips by $w = e^{iz}$.
   (b) Map $0 < x < \pi$ conformally onto $|w| < 1$ with $z = 1$ mapping onto $w = 0$.

6. Describe the image of (a) the right half-plane (b) the upper half-plane by $w = \text{Log } z$.

7. Describe the image of $\pi/2 < x < 3\pi/2$ under the mapping $w = \sin z$. Hint: $\sin(\pi + z) = -\sin z$.

8. Describe the image of $0 < x < \pi$ under the mapping $w = \cos z$. Hint: $\sin(z + \pi/2) = \cos z$.

9. Describe the pre-image of the first quadrant under the mapping $w = \sin z$.

10. Map a semi-infinite strip $0 < x < a$, $y > 0$ conformally onto $|w| < 1$.

11. Map the region of Exercise C1 of Section 4.3 onto $|w| < 1$.

4.5 Transplanting Harmonic Functions and the Schwarz–Christoffel Formula

The Chain Rule ensures that an analytic function of an analytic function is itself analytic. That is, if $g(z)$ is analytic at $z = \alpha$, and $f(w)$ is analytic at $w = \beta$ and $g(\alpha) = \beta$, then $F(z) = f \circ g(z) = f(g(z))$ is analytic at $z = \alpha$ (and $F'(\alpha) = f'(\beta)g'(\alpha)$). If $f(w) = u(w) + iv(w)$ we get

$$F(z) = u(g(z)) + iv(g(z))$$

and so both $u(g(z))$ and $v(g(z))$ are harmonic at $z = \alpha$. We look at a simple example.

Example 1. We take $w = \xi + i\eta$, $z = x + iy$ and examine

$$f(w) = \frac{1}{w-1} + \frac{1}{w+1}$$

with real part

$$u(w) = \frac{\xi - 1}{(\xi - 1)^2 + \eta^2} + \frac{\xi + 1}{(\xi + 1)^2 + \eta^2}. \tag{1}$$

Thus $u(w)$ is harmonic in the half-plane $\eta > 0$. (And elsewhere, but we restrict out attention to this region.) It will have certain values there determined of course by its formula (1).

Now we set $w = g(z) = z^2$, so that

$$\xi = x^2 - y^2, \qquad \eta = 2xy$$

and observe that $w = z^2$ maps the first quadrant $(x > 0, y > 0)$ onto $\eta > 0$. Thus

$$u(z^2) = \frac{x^2 - y^2 - 1}{(x^2 - y^2 - 1)^2 + 4x^2y^2} + \frac{x^2 - y^2 + 1}{(x^2 - y^2 + 1)^2 + 4x^2y^2} \tag{2}$$

will take the same values when $x > 0$, $y > 0$ as $u(w)$ does when $\eta > 0$, because of the mapping property. Thus the harmonic function $u(w)$, given by (1), gets its values transplanted from $\eta > 0$ in the w-plane to the first quadrant in the z-plane and becomes the harmonic function $u(z^2)$ there, given by (2).

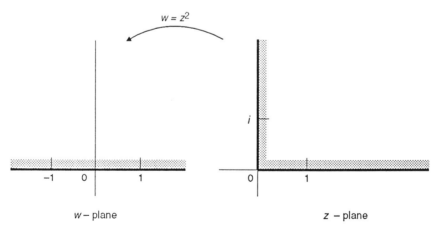

Example 1

More generally, if $u(w)$ is harmonic in a region $\mathcal{R}$ of the w-plane, and $w = g(z)$ maps R in the z-plane onto $\mathcal{R}$ in a one-to-one and conformal way,

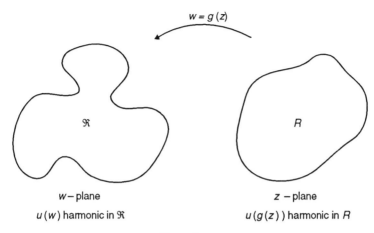

$w = g(z)$

$\mathfrak{R}$

R

w – plane z – plane

$u(w)$ harmonic in $\mathfrak{R}$ $u(g(z))$ harmonic in R

General case

then $u(g(z))$ is harmonic in R. And we can use this idea, in some cases, to transplant Dirichlet problems from one region to another where they may be easier to solve. We will have use of the following auxiliary function.

We take that branch of $\log w$ defined by

$$\log w = \ln r + i\theta, \qquad -\frac{\pi}{2} < \theta < \frac{3\pi}{2}.$$

Then $\theta = \theta(w)$ is harmonic, as the imaginary part of an analytic function, in the cut plane, so it is certainly harmonic in the closed upper half-plane $\operatorname{Im} w \geq 0$ except at the origin. Further, with $w = u + iv$, it is 0 on the positive u-axis, π on the negative u-axis, and bounded in the half-plane: $0 \leq \theta \leq \pi$. Then in the open half-plane $v > 0$ we have

$$\theta(w) = \text{arc cot } u/v = \cot^{-1} u/v. \tag{3}$$

This formula is not valid on the u-axis where $v = 0$, but from the previous remarks we know the values of $\theta(w)$ there (except of course at $w = 0$). You should graph in a $t\tau$-plane the equation $t = \cot^{-1} \tau$ to recall its properties (Exercise A1).

Example 2. A heat conductor is in the form of the first quadrant: $x \geq 0$, $y \geq 0$. The positive x-axis is held to $0°$ and the y-axis to $\pi°$. Find the steady state temperature $h(z)$ in the conductor. (We observe that this approximates the temperature near the origin in a large square $0 \leq x \leq a$, $0 \leq y \leq a$, where a is very big, and the temperaature of the far edges is bounded.)

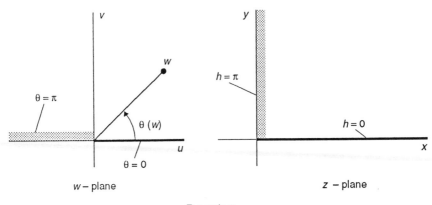

Example 2

Fig. 4.9

Solution. $w = z^2$ maps the first quadrant of the z-plane onto the upper half-plane in the w-plane; the positive x-axis goes onto the positive u-axis, and the positive y-axis onto the negative u-axis. Thus the boundary values get transplanted to the u-axis. Our auxiliary function $\theta(w)$ solves the problem in the w-plane, so the solution we seek in the z-plane is

$$h(z) = \theta(z^2) = \cot^{-1} \frac{x^2 - y^2}{2xy}.$$

Example 3. Find the two-dimensional (logarithmic) potential in $D : |z| \leq 1$, when the lower semicircle of $|z| = 1$ is at potential 0 (grounded) and the upper at potential 1. (We assume "infinitesimal" insulators separate the two semicircles at $z = \pm 1$. This problem is of course the same as finding the steady state temperature in the disk with the temperatures of the semicircles at 0 and 1, respectively.)

Solution. By Example 1, Section 4.3,

$$z = \frac{w - i}{w + 1}$$

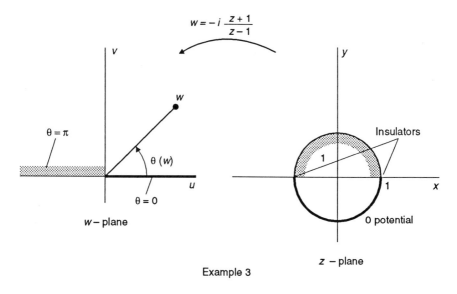

Example 3

maps the half-plane $v \geq 0$ onto the disk $|z| \leq 1$ with the right half of the u-axis going onto the lower semicircle, and the left onto the upper semicircle. Thus the inverse

$$w = -i\,\frac{z+1}{z-1}$$

maps the disk onto the half-plane. And we get

$$w = u + iv = [2y + i(1 - x^2 - y^2)]/[(x-1)^2 + y^2].$$

Then, using our auxiliary function again, we have the solution

$$\frac{1}{\pi}\,\cot^{-1}\frac{u}{v} = \frac{1}{\pi}\,\cot^{-1}[2y/(1 - x^2 - y^2)].$$

Example 4. A heat conductor in the form of an "infinite rectangle" (as illustrated) has its left side held to $10°$ and the bottom and right side to $5°$. Find the steady state temperature in the conductor.

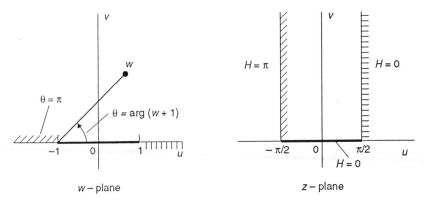

Example 4

Solution. Suppose our temperature function in $h(z)$. Then we take $H(z) = \pi[h(z) - 5]/5$, so that $H(z)$ has boundary values π and 0 as illustrated.

From Section 4.4 we know that

$$w = \sin z$$

maps the rectangle onto the upper half-plane with the bottom segment going onto the interval $|u| < 1$ of the y-axis, the right side onto the infinite segment $u > 1$, and the left side onto the segment $u < -1$ of the u-axis. Since the break between the boundary values of 0 and 1 now occurs at $w = -1$ rather than $w = 0$, we need to modify our auxiliary function. We take now

$$\theta = \theta(w) = \arg(w + 1) = \cot^{-1}\frac{u + 1}{v}$$

which has the proper boundary values. And $w = \sin z$ becomes

$$u + iv = \sin x \cosh y + i \cos x \sinh y$$

so

$$H(z) = \cot^{-1}[(\sin x \cosh y + 1)]/(\cos x \sinh y),$$

from which

$$h(z) = 5 + \left(\frac{5}{\pi}\right) H(z).$$

It is quite clear that all these examples have discontinuous boundary values, and so the solutions we calculated were also discontinuous. Our uniqueness theorem (2.8c) says that if a region R is bounded and the boundary values are continuous, then a Dirichlet problem for R can have no more than one solution. But now we see that for Example 2 we can add xy to the solution we got, and the sum is again a solution, albeit an unbounded one. There is an extension of Theorem 2.8c which asserts that if our boundary function is bounded and has only a finite number of discontinuities, then there can be no more than one bounded solution. We outline the proof in the C Exercises.

Our ability to solve problems of the sort just illustrated depends, among other things, on our ability to find appropriate mapping functions. The mapping function for mapping a half-plane $y \geq 0$ in the z-plane onto a closed polygonal region in the w-plane is given by the **Schwarz–Christoffel** formula which we now describe.

Let P be a polygon — a contour composed of a finite number, n, of straight sides — and R its interior: $R = I(P)$. The mapping function, whose existence is of course guaranteed by Riemann's theorem, will then take points $\{a_j\}_1^n$ on the x-axis onto the corners $\{w_j\}_1^n$ of P. We number these so that

$$a_1 < a_2 < \cdots < a_n,$$

and then the w_j's are in counterclockwise order on P. The interior angle of P at w_j we denote by $\pi\alpha_j$ measured counterclockwise, with corresponding exterior angle $\pi\beta_j$ as illustrated for $n = 5$ (β_j is negative at a re-entrant corner such as w_2). The angles $\pi\beta_j$ describe the changes in direction as w moves around a corner on P. Then the mapping function is given by

$$w = AF(z) + B, \tag{4}$$

A and B constants and

$$F(z) = \int_0^z (\zeta - a_1)^{-\beta_1}(\zeta - a_2)^{-\beta_2} \cdots (\zeta - a_n)^{-\beta_n} d\zeta \tag{5}$$

where (i) the path of integration is in $y \geq 0$, and (ii) the powers are principal values. We will not prove this result, but we will comment about it and illustrate its use.

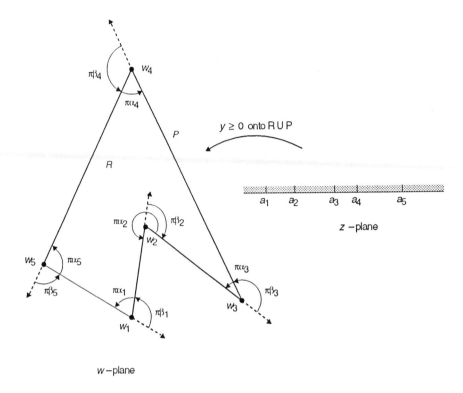

w−plane

Remark 1. Any z_0 with Im $z_0 \geq 0$ can replace 0 as the lower limit in (5) for this merely alters the value of the constant B.

Remark 2. Given a polygon P, part of the problem is to determine the a_j's. Since we can map the half-plane onto itself taking 3 points into 3 others, we can choose 3 of the a_j's arbitrarily. This is of course all of them for the case where P is a triangle.

Remark 3. $F(z)$ maps the half-plane $y \geq 0$ onto a polygon Q and its interior where Q is geometrically similar to P. Then A changes its size (by $|A|$) and its orientation (by arg A), and B translates it to coincide with P.

Remark 4. We can choose $a_n = \infty$, in which case the factor $(\zeta - a_n)$ does not occur in (5).

Remark 5. Only rarely can the integral $F(z)$ be evaluated in terms of elementary functions. Numerical values must, in general, be determined by quadratures.

Example 5. Let P be the triangle cornering at $w_1 = 1+2i$, $w_2 = 1+i$, and $w_3 = 2+i$ with exterior angles $\pi\beta_1 = 3\pi/4$, $\pi\beta_2 = \pi/2$, and $\pi\beta_3 = 3\pi/4$ as shown in the figure. Find the mapping function (4).

Solution 1. We take $a_1 = -1$, $a_2 = 0$, $a_3 = 1$ and consider

$$F_1(z) = \int_0^z (\zeta + 1)^{-3/4} \zeta^{-1/2} (\zeta - 1)^{-3/4} d\zeta$$

and determine the triangle Q of Remark 3. The corners of Q we denote by

$$\gamma_1 = F(-1), \quad \gamma_2 = F(0), \quad \gamma_3 = F(1),$$

and determine them: Clearly $\gamma_2 = 0$, and

$$\gamma_3 = \int_0^1 (\zeta + 1)^{-3/4} \zeta^{-1/2} (\zeta - 1)^{-3/4} d\zeta.$$

We set $\zeta = t$, then $\zeta - 1 = (1-t)e^{i\pi}$ since we are using principal values, and so

$$\gamma_3 = e^{-3\pi i/4} \int_0^1 t^{-1/2} (1 - t^2)^{-3/4} dt$$
$$= b\,e^{-3\pi i/4}$$

where b is the value of the integral. Next

$$\gamma_1 = \int_0^{-1} (\zeta + 1)^{-3/4} \zeta^{-1/2} (\zeta - 1)^{-3/4} d\zeta.$$

Now we set $\zeta = -t = t\,e^{i\pi/2}$, $\zeta + 1 = 1 - t$, and $\zeta - 1 = 1 - t = (1+t)e^{i\pi/2}$ and get

$$\gamma_1 = e^{-i\pi/4} \int_0^1 t^{-1/2} (11 - t^2)^{-3/4} dt$$
$$= b\,e^{-i\pi/4}.$$

Then our mapping function is

$$w = \frac{1}{b}\, e^{3\pi i/4} F_1(z) + (1 + i),$$

that is, $A = (1/b)e^{3\pi i/4}$, $B = 1 + i$.

Solution 2. This time we take $a_1 = -1$, $a_2 = 0$ and $a_3 = \infty$. Then

$$F_2(z) = \int_0^z \zeta^{-1/2} (\zeta + 1)^{-3/4} d\zeta,$$

and again $\gamma_2 = 0$ and now

$$\gamma_1 = \int_0^{-1} \zeta^{-1/2}(\zeta + 1)^{-3/4} d\zeta$$

$$= i \int_0^1 t^{-1/2}(1 - t)^{-3/4} dt$$

$$= ib'$$

This time

$$\gamma_3 = \int_0^\infty t^{-1/2}(1 + t)^{-3/4} dt,$$

and setting $t = \tau/(1 - \tau)$ we get

$$\gamma_3 = \int_0^1 \tau^{-1/2}(1 - \tau)^{-3/4} d\tau = b'.$$

Thus this time Q has the right orientation so our mapping function is

$$w = \left(\frac{1}{b'}\right) F_2(z) + (1 + i)$$

with of course our new $F(z)$.

Example 6. Map a half-plane onto a rectangle.

Solution. We suppose the rectangle has its corners at $w_1 = -b + ic$, $w_2 = -b$, $w_3 = b$, $w_4 = b + ic$. By Remark 1, we can choose $a_1 = -a$, $a_2 = 1$, $a_3 = -1$, and by symmetry we can expect to take $a_4 = a$, where of course $a > 1$. Then all the β_j's are $1/2$ and we examine

$$F(z) = \int_0^z (\zeta + a)^{-1/2}(\zeta + 1)^{-1/2}(\zeta - 1)^{-1/2}(\zeta - a)^{-1/2} d\zeta,$$

$$= \int_0^z [(1 - \zeta^2)(a^2 - \zeta^2)]^{-1/2} d\zeta$$

From here we see that

$$b = \int_0^1 [(1 - t^2)(a^2 - t^2)]^{-1/2} dt$$

and

$$ic = \int_1^a [(1 - t^2)(t^2 - a^2)]^{-1/2} dt.$$

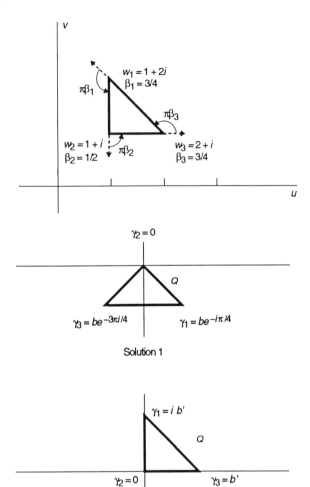

Solution 1

Solution 2

The Schwarz–Christoffel formula extends to certain "degenerate polygons." We illustrate this in our next example.

Example 7. Map the half-plane onto $|u| \leq \pi/2$, $v \geq 0$ in w-plane.

Solution. We consider this as an "infinite triangle" as illustrated. The angles at $\pi\pi/2$ are $\pi/2$ so we consider

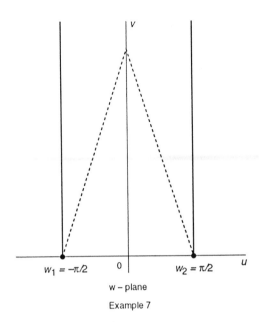

w – plane

Example 7

$$w = F(z) = \int_0^z (\zeta - 1)^{-1/2}(\zeta + 1)^{-1/2} d\zeta$$

$$= \int_0^z \frac{d\zeta}{\sqrt{\zeta^2 - 1}} = i \int_0^z \frac{d\zeta}{\sqrt{1 - \zeta^2}}$$

$$= i \sin^{-1} z$$

We know from Section 4.3 and Example 4 that $z = \sin w$ maps our "triangle" onto the half-plane, so $w = \sin^{-1} z$ is our mapping. Thus in (4) we take $A = -i$, $B = 0$ and get

$$w = AF(z) + B = -i(i \sin^{-1} z) + 0$$

$$= \sin^{-1} z.$$

Significant Topics

Temperatures in an infinite rectangle
Polygon onto a half-plane
Schwarz–Christoffel formula

A Exercises

1. Draw a graph of $t = \cot^{-1}\tau$ for $-\infty < \tau < \infty$, and show that $\lim_{\tau \to \infty} \cot^{-1}\tau = 0$, $\lim_{\tau \to -\infty} \cot^{-1}\tau = \pi$.

2. If P is a polygon as described in this section, show that $\sum_1^n \beta_j = 2$ and
$$\sum_1^n \alpha_j = n - 2.$$

3. Show geometrically that $0 < \alpha_j < 2$ and $-1 < \beta_j < 1$ for each corner of a polygon P.

4. (a) Use the Schwarz–Christoffel formula to map the half-plane onto the first quadrant;
   (b) onto the w-plane cut along the positive real axis.

B Exercises

1. (a) Suppose that $u(z)$ is a solution to a Dirichlet problem in the "rectangle" of Example 4. Show that

$$u(z) + \cos x \sinh y$$

is also a solution.
   (b) Suppose that $u(z)$ is a solution to a Dirichlet problem for the half-plane $y \geq 0$. Show that

$$u(z) + ay + bxy$$

is also a solution for any constants a and b.

2. In Example 3 consider the function v in the w-plane. What does it become when transplanted to the z-plane. Show that it is positive for $|z| < 1$, and is 0 on $|z| = 1$ except at $z = 1$.

3. Modify Example 4 by taking temperature $5°$ along the left and bottom of the "rectangle," and $10°$ on the right.

4. Modify Example 3 by taking the potential 0 on the quarter circle of $|z| = 1$ in the fourth quadrant, and 1 elsewhere.

5. A heat conducting lamina L lies in the region between (i) $x^2+y^2-y/2 = 0$ and (ii) $x^2 + y^2 - y/4 = 0$. If the temperature on (i) is 0 and on (ii) is 1, find the steady state temperature in L. Hint: First perform an inversion.

6. An infinite heat conducting lamina L lies in the right half-plane $x > 0$ outside C: $(x - 4)^2 + y^2 = 1$. The temperature along the y-axis is 0, and on C is 1. Find the steady state temperature in L. Hint: An appropriate bilinear transformation will map L onto an annulus.

7. A heat conductor L is in the form of the upper half-plane $y > 0$ with the semidisk $|z| < 1$, $y > 0$ deleted. If the temperature on the semicircle $|z| = 1$, $y > 0$ and the negative end of the x-axis from $-\infty$ to -1 is 1, and on the positive end from 1 to ∞ is 0, find the steady state temperature in L. Hint: Examine the mapping

$$w = \frac{1}{2}\left(z + \frac{1}{z}\right).$$

8. Show that

$$w = F(z) = \int_0^z \frac{d\zeta}{\sqrt{\zeta(1 - \zeta^2)}}$$

maps $y \geq 0$ onto a square in the w-plane. Express the length of its side as a definite integral.

9. Show that

$$w = F(z) = \int_0^z \frac{d\zeta}{(1 - \zeta^2)^{2/3}}$$

maps the half-plane onto an equilateral triangle.

10. By taking $z = x$ on the real axis, and $\zeta = t$ as integration variable we have that Q of Remark 2 is given by

$$Q : w = \int_0^x (t - a_1)^{-\beta_1} \cdots (t - a_n)^{-\beta_n} dt, \quad -\infty < x < \infty.$$

By examining the tangent to Q given by dw/dx show that the image of a segment $a_{j-1} < x < a_j$ of the x-axis is itself a straight line segment. Show also that as x crosses a_j the direction changes by an angle $\pi\beta_j$.

11. Evaluate the constants b and b' of Example 5 in terms of the Beta function.

C Exercises

These Exercises are concerned with the uniqueness questions raised after Example 4. We restrict our attention to the unit disk $D : |z| < 1$ with boundary $C : |z| = 1$. More general regions are discussed in Section 6.3

1. Set $h(z) = (1 + z)/(1 - z)$ and show
   (a) $u(z) = \operatorname{Re} h(z) = (1 - |z|^2)/|1 - z|^2$;
   (b) $u(z) > 0$ in D;
   (c) $u(z) \equiv 0$ on C except at $z = 1$;
   (d) $\operatorname{Im} h(z) = \cot(\theta/2)$ on C, i.e., for $z = e^{i\theta}$.
2. Similarly discuss $(1 - z)/(1 + z)$.
3. Similarly discuss $(e^{ia} + z)/(e^{ia} - z)$ where a is real.
4. Set $h(z) = 1/(1 - z)$ and show
   (a) $u(z) = \operatorname{Re} h(z) = (1 - x)/[(1 - x)^2 + y^2]$;
   (b) $u(z) > \frac{1}{2}$ in D;
   (c) $u(z) = \frac{1}{2}$ on C except at $z = 1$.
   (d) Discuss $u(z) - \frac{1}{2}$.
5. Suppose $U(z)$ is harmonic in D and continuous in $\bar{D}$ except (possibly) at one point $z_0 \in C$, and $U(z) \equiv 0$ on C except (possibly) at z_0. Suppose further that there is a constant κ for which $|U(z)| \le \kappa$. Show that $U(z) \equiv 0$ in D by arguing as follows. Set $\rho = |z - z_0|$ and $u(z) = \ln(2/\rho)$ and show
   (a) $u(z)$ is harmonic in D;
   (b) $u(z) \ge 0$ in D and on C (except at z_0).
   (c) $\lim\limits_{z \to z_0} u(z) = +\infty$.

   Now choose $\alpha \in D$ and keep it fixed for the rest of the argument. Let $\varepsilon > 0$ be given. Choose $a > 0$ so small that (i) $a < |\alpha - z_0|$ and (ii) $\varepsilon u(z) > \kappa$ for $|z - z_0| = a$ (i.e., if $\rho = a$). Set $D' = \{|z| < 1\} \cap \{|z - z_0| > a\}$ and $C' = \partial D'$. (Draw a figure!) Show
   (d) $\alpha \in D'$;
   (e) $U(z) + \varepsilon u(z)$ is harmonic in D';
   (f) $U(z) + \varepsilon u(z) \ge 0$ on C';
   (g) $U(\alpha) + \varepsilon(\alpha) \ge 0$ for each $\varepsilon > 0$:
   (h) $U(\alpha) \ge 0$;
   (i) Examine $U(z) - \varepsilon u(z)$ and conclude $U(\alpha) \le 0$.
   (j) Conclude $U(z) \equiv 0$ in D.
6. Extend C5 to the case where there is a finite number of points of discontinuity on C.

Chapter 5

Residues

5.1 Isolated Singularities

A point α is an **isolated singularity** of an analytic function $f(z)$ means that $f(z)$ is analytic in a deleted neighborhood of α, but not at α itself. In this case the Laurent theorem applies *with inner radius* 0 and outer radius some positive number, say $r > 0$. Thus

$$f(z) = \sum_{-\infty}^{\infty} a_k(z - \alpha)^k = \sum_{1}^{\infty} \frac{a_{-k}}{(z - \alpha)^k} + \sum_{0}^{\infty} a_k(z - \alpha)^k \tag{1}$$

for $0 < |z - \alpha| < r$. The second series on the right side of (1) converges, for $|z - \alpha| < r$, and the first one for $|z - \alpha| > 0$. This first one, namely,

$$P_\alpha(f; z) \equiv \sum_{1}^{\infty} \frac{a_{-k}}{(z - \alpha)^k}, \quad |z - \alpha| > 0, \tag{2}$$

is called the **principal part** of $f(z)$ at α. As noted above, it is analytic for $|z - \alpha| > 0$, that is, for $z \neq \alpha$. We sometimes simplify the notation by writing

$$P_\alpha(f; z) = P_\alpha(z) = P(z)$$

Whenever the abbreviated forms are unambiguous.

If $f(z)$ has an isolated singularity at α we distinguish three mutually exclusive cases:

(A) $f(z)$ has an **essential singularity** at α means that there are infinitely many non-zero a's of negative index in (1), so that (2) is in fact an infinite series.

(B) $f(z)$ has a **pole of order** n at α means that there is an n for which $a_{-k} = 0$ for all $k > n$ and $a_{-n} \neq 0$ so that $P_\alpha(f; z)$ has the form

$$P_\alpha(f; z) = \sum_{k=1}^{n} \frac{a_{-k}}{(z - \alpha)^k}, \quad a_{-n} \neq 0. \tag{3}$$

A pole of order 1 is called a **simple pole**, in which case $P_\alpha(f; z) \equiv a_{-1}/(z - \alpha)$. You should also observe that in the case (3) we can write

$$P_\alpha(f, z) = \frac{Q(z)}{(z - \alpha)^n} \tag{4}$$

where $Q(z)$ is a polynomial of degree $< n$, and $Q(\alpha) \neq 0$.

(C) $f(z)$ has a **removable singularity** at α means that $a_{-k} = 0$ for all $k = 1, 2, 3, \ldots$, so that $P_\alpha(f, z) \equiv 0$, and so from (1),

$$f(z) = \sum_{k=0}^{\infty} a_k(z - \alpha)^k = a_0 + a_1(z - \alpha) + \cdots . \tag{5}$$

We define $f(\alpha) = a_0$: then $f(z)$ becomes (by (5)) analytic for $|z - \alpha| < r$, and so removes the singularity.

Poles form by far the most important class (for this book, at least, and probably for the subject) of isolated singularities. We will be coping with them a great deal in this chapter.

Example 1. Discuss the singularities of $e^{1/z}$ and $\sin 1/z$.

Solution. e^w and $\sin w$ are entire functions and $w = 1/z$ has an isolated singularity at $z = 0$. Thus $e^{1/z}$ and $\sin 1/z$ each has an isolated singularity at $z = 0$. Also from the definition of e^w and $\sin w$, and the substitution $w = 1/z$ we get

$$e^{1/z} = \sum_{0}^{\infty} \frac{1}{k!} \frac{1}{z^k} = 1 + \frac{1}{z} + \frac{1}{2!z^2} + \cdots + \left(\frac{1}{k!}\right) \frac{1}{z^k} + \cdots$$

and

$$\sin 1/z = \sum_0 \frac{(-1)^k}{(2k+1)! z^{2k+1}} = \frac{1}{z} + \frac{-1}{3!} \frac{1}{z^3} + \cdots + \frac{(-1)^k}{(2k+1)!} \frac{1}{z^{2k+1}} + \cdots .$$

Thus each has an essential singularity at 0.

Example 2. Discuss the singularities of $(1 - \cos z)/z^2$.

Solution. The numerator is entire; the denominator is entire, so the only singularity is at zero where the denominator is 0, Using the definition of $\cos z$, we have

$$(1 - \cos z)/z^2 = (1/z^2)\left[1 - \left(1 - \frac{z^2}{2!} + \frac{z^4}{4!} \mp \cdots\right)\right]$$

$$= \frac{z^2/2! - z^4/4! \pm \cdots}{z^2}$$

$$= \frac{1}{2!} - \frac{z^2}{4!} \pm \cdots .$$

Thus $(1 - \cos z)/z^2$ has a removable singularity at $z = 0$.

Example 3. Discuss $(\sin z)/z$ at $z = 0$.

Solution.

$$\frac{\sin z}{z} = \frac{1}{z}\sum_0^\infty \frac{(-1)^k z^{2k+1}}{(2k+1)!}$$

$$= \frac{1}{z}\left(z - \frac{z^3}{3!} + \frac{z^5}{5!} \mp \cdots\right)$$

$$= 1 - \frac{z^2}{3!} + \frac{z^4}{5!} \mp \cdots$$

so $(\sin z)/z$ has a removable singularity at $z = 0$.

Certainly we would not consider the z in the denominator of

$$\frac{z^3 - 2z^2 + 3z}{z}$$

as causing difficulty at the origin, for we can divide by it and reduce the fraction to $z^2 - 2z + 3$, which is analytic at the origin. This shows the essence of the notion of removal of singularities. Though some cases may be technically more complicated because the cancellations are not always as obvious as in the last two examples. But the basic idea is always this simple: we cancel out common factors.

It is common practice to assume that removable singularities have been removed: thus we would say that $(\sin z)/z$ is an entire function, or that $(1 - \cos z)/z^2$ is entire. Clearly these functions are analytic in the whole plane except at 0. After removal of the singularities, as done in the two preceding examples, they are seen to be analytic at 0 as well. So these functions, modified by the removal, are entire.

There are a number of criteria for identifying removable singularities. Some of these are explored in the problems. We give now that which is probably the simplest and most useful.

5.1a Theorem. *If $f(z)$ is analytic for $0 < |z - \alpha| < r$, that is, $f(z)$ has an isolated singularity at α, then a necessary and sufficient condition that α be a removable singularity is that $f(z)$ is bounded for $0 < |z - \alpha| \leq r/2$.*

Proof. (Sufficiency) We show that all coefficients of negative index in the Laurent expansion (1) are zero. By equation (11), Section 3.3 we have

$$a_{-n} = \frac{1}{2\pi i} \int_{|z-\alpha|=\rho} f(\zeta)(\zeta - \alpha)^{n-1} d\zeta, \quad n \geq 1, \; 0 < \rho < r.$$

Then for $\rho \leq r/2$ we get

$$|a_{-n}| \leq \frac{1}{2\pi} \int_0^{2\pi} M\rho^n d\theta = M\rho^n$$

But, for each n, a_{-n} is a constant and $M\rho^n \to 0$ as $\rho \to 0$. Thus $a_{-n} = 0$. The necessity we leave as an exercise. $\square$

Suppose $f(z)$ has a pole or order n at $z = \alpha$. Then we can write, with $a_{-n} \neq 0$,

$$f(z) = \frac{a_{-n}}{(z - \alpha)^n} + \cdots + \frac{a_{-1}}{(z - \alpha)} + a_0 + a_1(z - \alpha) + \cdots$$

$$= \frac{1}{(z - \alpha)^n}[a_{-n} + a_{-n+1}(z - \alpha) + \cdots + a_0(z - \alpha)^n + \cdots]$$

$$= \frac{1}{(z - \alpha)^n}[b_0 + b_1(z - \alpha) + \cdots + b_k(z - \alpha)^k + \cdots]$$

$$= g(z)/(z - \alpha)^n$$

where $b_k = a_{-n+k}$. The series in brackets, which we have denoted by $g(z)$, converges for $0 \leq |z - \alpha| < r$ for some $r > 0$, and so is an analytic function at α, and $g(\alpha) \neq 0$. Thus there is an r', $0 < r' \leq r$ for which $g(z) \neq 0$ in the disk $|z - \alpha| < r'$. In this disk $1/g(z)$ is analytic, and so can be expanded in a power series. We get

$$\frac{1}{f(z)} = (z - \alpha)^n \cdot \left(\frac{1}{g(z)} \right)$$
$$= (z - \alpha)^n [c_0 + c_1(z - \alpha) + \cdots + c_n(z - \alpha)^n + \cdots]$$

with $c_0 = 1/g(\alpha) = 1/b_0 = 1/a_{-n} \neq 0$ from which it is clear that $1/f(z)$ has a removable singularity at α, and after removal of the singularity that $1/f(z)$ has a zero of order n.

Conversely it is easy to see, by essentially reversing the steps in the preceding calculations, that if $f(z)$ has a zero of order n at α, then $1/f(z)$ has a pole of order n there. The details are left to the exercises. Thus we have proved the following.

5.1b Theorem. *If $f(z)$ has a pole of order n at $z = \alpha$, then $1/f(z)$ has a zero of order n there. If $f(z)$ has a zero of order n at $z = \alpha$, then $1/f(z)$ has a pole of order n there.*

5.1c Corollary. *If $f(z)$ has a pole of order n at $z = \alpha$, then there is an $r > 0$ for which $f(z)$ is neither zero, nor has a singularity in $0 < |z - \alpha| < r$.*

5.1d Corollary. *For $f(z)$ to have a pole at $z = \alpha$ it is necessary and sufficient that $f(z) \to \infty$ as $z \to \alpha$.*

Proof. For $f(z) \to \infty$ is equivalent to $1/f(z) \to 0$, which is clear from Theorem 5.1b. $\square$

Thus we have established criteria for an isolated singularity being removable or a pole. It follows from these that if α is an isolated *essential* singularity of $f(z)$, then $f(z) \not\to \infty$ as $z \to \alpha$ (by 5.1d), nor is there a deleted neighborhood of α in which $f(z)$ is bounded (by 5.1a). Clearly then the behavior of $f(z)$ near an essential singularity is somewhat strange. In fact, there is a theorem (**Picard's Theorem**) which asserts that in each deleted neighborhood of an essential singularity $f(z)$ actually achieves every (complex) value with at most one exception. This is a difficult and deep result. We will content ourselves with a weaker result which still shows that functions behave very strangely near essential singularities.

5.1e Theorem. (**Casorati-Weierstrass**) *Suppose $f(z)$ has an isolated essential singularity at α. Then for each complex number, γ, for each*

$\rho > 0$, *for each $\varepsilon > 0$ there is a ζ with $|\zeta - \alpha| < \rho$ for which $|f(\zeta) - \gamma| < \varepsilon$. And (corresponding to $\gamma = \infty$) there is a ζ' with $|\zeta' - \alpha| < \rho$ for which $|f(\zeta')| > 1/\varepsilon$. That is, in each deleted neighborhood of α, $f(z)$ comes arbitrarily near to each extended complex number.*

Proof. Given a complex number, γ, if it were not true that a ζ exists as asserted, then there would be an $\varepsilon_0 > 0$ and a $\rho_0 > 0$ such that $|f(z) - \alpha| \geq \varepsilon_0$ for all $|z - \alpha| < \rho_0$. Thus $1/|f(z) - \alpha| \leq 1/\varepsilon_0$ and so $\phi(z) \equiv 1/(f(z) - \alpha)$ would be bounded for $|z - \alpha| < \rho_0$ and so $\phi(z)$ would have a removable singularity (by 5.1a). Then $f(z) - \alpha$ would either have a pole (if $\phi(\alpha) = 0$ after removal of the singularity) or would itself have a removable singularity. The case $\gamma = \infty$ we leave as an exercise. $\square$

We include the point at ∞ in our classification of isolated singularities of a function $f(z)$: If there is a constant R such that $f(z)$ is analytic for $|z| > R$, then

(i) $f(z)$ is analytic at ∞ means that $f(1/\zeta)$ is analytic or has a removable singularity at $\zeta = 0$.

(ii) $f(z)$ has a pole of order n at ∞ means that $f(1/\zeta)$ has a pole of order n at $\zeta = 0$.

(iii) $f(z)$ has an essential singularity at ∞ means that $f(1/\zeta)$ has an essential singularity at $\zeta = 0$.

Example 4. Discuss the behavior at ∞ of

(a) $\dfrac{z+1}{z^2+2z} = f_1(z)$ (b) $\dfrac{z^2+1}{z+1} = f_2(z)$ (c) $e^z = f_3(z)$.

Solution.

(a) $f_1\left(\dfrac{1}{\zeta}\right) = \dfrac{\frac{1}{\zeta}+1}{\frac{1}{\zeta^2}+2\cdot\frac{1}{\zeta}} = \dfrac{\zeta+\zeta^2}{1+2\zeta}$ which is analytic at $\zeta = 0$, so $f_1(z)$ is analytic at ∞.

(b) $f_2\left(\dfrac{1}{\zeta}\right) = \dfrac{\frac{1}{\zeta^2}+1}{\frac{1}{\zeta}+1} = \dfrac{1+\zeta^2}{\zeta+\zeta^2} = \dfrac{1}{\zeta}\cdot\left(\dfrac{1+\zeta^2}{1+\zeta}\right)$ so $f_2(z)$ has a pole or order 1 at ∞.

(c) $f_3\left(\dfrac{1}{\zeta}\right) = e^{1/\zeta} = 1 + \dfrac{1}{\zeta} + \dfrac{1}{2!}\dfrac{1}{\zeta^2} + \cdots + \dfrac{1}{n!}\dfrac{1}{\zeta^n} + \cdots$, so $f_3(z)$ has an essential singularity at ∞.

We are now in a position to establish the partial fraction expansion (Theorem 1.5h) used in Chapter 1. So suppose $f(z) = p(z)/q(z)$ is a proper rational function in lowest terms with

$$q(z) = C(z - \beta_1)^{p_1}(z - \beta_2)^{p_2}\cdots(z - \beta_k)^{p_k}$$

where the β's are the distinct zeros of $q(z)$ and p_j is the order of β_j. Then $f(z)$ has a pole of order p_j at β_j with principal part in the form

$$P_j(z) = \sum_{n=1}^{p_j} \frac{a_{jn}}{(z - \beta_j)^n}.$$

5.1f Theorem. *In the notation above*

$$f(z) = \sum_{j=1}^{k} P_j(z) = \sum_{j=1}^{k} \sum_{n=1}^{p_j} \frac{a_{jn}}{(z - \beta)^n}.$$

Proof. Set

$$g(z) = f(z) - \sum_{j=1}^{k} P_j(z),$$

and observe that $g(z)$ is analytic in $\mathbb{C}$ except at the β's. We examine it near β_m:

$$g(z) = f(z) - P_m(z) - \sum_{j=1}^{m-1} P_j(z) - \sum_{j=m+1}^{k} P_j(z).$$

Each P_j in the two sums is analytic at β_m, and $f(z) - P_m(z)$ has a removable singularity at β_m. Thus $g(z)$ is analytic at β_m. Since m is arbitrary, $1 \le m \le k$, $g(z)$ is entire. And also (how?)

$$g(z) \to 0 \quad \text{as} \quad z \to \infty. \tag{6}$$

From this we can conclude that $g(z)$ is bounded, for choose R so large that

$$|g(z)| < 1 \quad \text{for} \quad |z| > R. \tag{7}$$

Then, by continuity, there is an M for which

$$|g(z)| \le M \quad \text{for} \quad |z| \le R. \tag{8}$$

Thus (7) and (8) together show that $g(z)$ is bounded for all z:

$$|g(z)| \le \max[M, 1].$$

Liouville's theorem then implies that $g(z)$ is a constant, and by (6) that constant is 0. That is,

$$g(z) = f(z) - \sum_1^k P_j(z) = 0. \quad \square$$

We may note in closing that we have discussed only *isolated* singularities. Let us examine

$$f(z) = \csc(1/z) = 1/\sin(1/z).$$

This function clearly has a singularity at $z = 0$. It also has simple poles at $z = 1/n\pi$, $n = \pm 1, \pm 2, \ldots$. And $\lim_{n \to \pm \infty} 1/n\pi = 0$, so that 0 is a limit of poles—each neighborhood of 0 has poles of $f(z)$ in it. Thus 0 is *not an isolated singularity*. This is merely one simple example to illustrate the fact that there are more complicated singularities than we discuss here.

Significant Topics page

A Exercises

1. Locate and classify all finite singularities of the following:

   (a) $\dfrac{1 - \cos z}{z}$;   (b) $\dfrac{1 - \cos z}{z^2}$;   (c) $\dfrac{1 - \cos z}{z^3}$;

   (d) $\dfrac{z}{1 - \cos z}$;   (e) $\dfrac{z^2}{1 - \cos z}$;   (f) $\dfrac{z^3}{1 - \cos z}$;

   (g) $\tan z$;   (h) $\cot^2 z$;   (i) $1/(e^z - 1)$.

2. Classify the behavior at ∞ of

   (a) z;   (b) $(z - 1)/z$;   (c) 1;

   (d) $\left(\sum_0^n a_k z^k\right) \Big/ \left(\sum_0^m b_k z^k\right)$;   (e) $\sin z$;   (f) $\csc z$;

3. Prove formula (4) for $P_\alpha(f, z)$.

B Exercises

1. Show that in each neighborhood of zero, $e^{1/z} = \alpha$ has infinitely many distinct solutions for any $\alpha \neq 0$.
2. Complete the proof of Theorem 5.1b by showing that if $f(z)$ has a zero of order n at α, then $1/f(z)$ has a pole of order n at α.
3. Prove Corollary 5.1c.
4. Complete the proof of Theorem 5.1e by examining the case $\gamma = \infty$.
5. Suppose $g(z)$ and $h(z)$ are both analytic at α, and $f(z) \equiv g(z)/h(z)$.
   (a) If each of $f(z)$ and $g(z)$ has a zero of order $n \geq 1$ at α, then show that $f(z)$ has a removable singularity at α and, after removal, $f(\alpha) = g^{(n)}(\alpha)/h^{(n)}(\alpha)$.
   (b) If $g(z)$ has a zero of order n at α and $h(z)$ has a zero of order $m \neq n$, classify the behavior of $f(z)$ at α.
   (c) Examine the character of $f(z)$ if $g(z)$ and $h(z)$ each has a pole at α.
6. Take the negative real axis as a branch cut for $\log z$. Show
   (a) $1/\text{Log } z$ has a simple pole at $z = 1$;
   (b) $1/\log z$ is analytic at $z = 1$ for each other branch of $\log z$;
   (c) $(\text{Log } z)/(z - 1)$ has a removable singularity at $z = 1$;
   (d) $(\log z)/(z - 1)$ has a simple pole at $z = 1$ for each other branch of $\log z$.

C Exercise

1. Suppose $f(z)$ has an isolated singularity at α and $|f(z)| \leq M|z - \alpha|^a$, $a > -1$. Show that the singularity is removable. If $-1 \geq a > n - 1$, then $f(z)$ has a pole of order at most n.

5.2 The Residue Theorem

If $f(z)$ has an isolated singularity—an "honest" one, not a removable one—at α, then the **residue** of $f(z)$ at α is denoted by Res(f, α) or, if the f is clear, simply by Res(α), and is defined to be

$$\text{Res}(f; \alpha) = \frac{1}{2\pi i} \int_{|z-\alpha|=r} f(z)dz \tag{1}$$

where r is so small that the only singularity of $f(z)$ inside $|z - \alpha| = r$ is the one at α. This is the same as

$$\text{Res}(f, \alpha) = a_{-1} \tag{2}$$

in the Laurent expansion

$$f(z) = \sum_{-\infty}^{\infty} a_n(z - \alpha)^n = \sum_{1}^{\infty} \frac{a_{-n}}{(z-\alpha)^n} + \sum_{0}^{\infty} a_n(z-\alpha)^n$$

with inner radius zero. That is, Res(f, α) is the coefficient of $(z - \alpha)^{-1} = \dfrac{1}{z - \alpha}$ in this expansion.

5.2a Theorem. (Cauchy's Residue Theorem) *Suppose $f(z)$ is analytic in and on a contour C except for isolated singularities in $I(C)$ which occur at points $\{\alpha_k\}$. Then*

$$\int_C f(z)dz = 2\pi i \sum_k \text{Res}(f, \alpha_k). \tag{3}$$

$$= 2\pi i \text{ (sum of residues inside } C\text{)}.$$

Before we prove this result we make two comments. (A) By hypothesis no singularity of $f(z)$ can occur on C itself. Thus for a given function $f(z)$ they must occur inside C where they contribute to the formula, or outside C and are ignored by the formula. (B) The sum on the right side of (3) is always a finite sum. In any particular example this is usually easy to see. As a general result—a lemma for the theorem, really—it follows from theorems established in real calculus.

Proof. Let us look at one of the α_k's; we have

$$f(z) = \sum_{n=0}^{\infty} a_n^{(k)}(z - \alpha_k)^n + P_{\alpha_k}(z)$$

where $P_{\alpha_k}(z)$ is the principal part of $f(z)$ at α_k. Then

$$f(z) - P_{\alpha_k}(z) = \sum_{n=0}^{\infty} a_n^{(k)}(z - \alpha_k)^n,$$

and so $f(z) - P_{\alpha_k}(z)$ is analytic (i.e., has a removable singularity) at α_k.
Then

$$f(z) - \sum_k P_{\alpha_k}(z),$$

where the sum is taken over all $\alpha_k \in I(C)$, is analytic in and on C since each
$P_{\alpha_k}(z)$ is analytic at α_j when $\alpha_j \neq \alpha_k$. Thus Cauchy's theorem applies
and we have

$$\int_C \left[f(z) - \sum_k P_{\alpha_k}(z) \right] dz = 0$$

so

$$\int_C f(z)dz = \sum_k \int_C P_{\alpha_k}(z)dz \tag{4}$$

But, by Cauchy's theorem for two contours,

$$\int_C P_{\alpha_k}(z)dz = \int_{|z-\alpha|=\rho} P_{\alpha_k}(z)dz$$

$$= \int_{|z-\alpha_k|=\rho} f(z)dz + \int_{|z-\alpha_k|=\rho} [P_{\alpha_k}(z) - f(z)]dz$$

$$= 2\pi i R(f, \alpha_k) + 0$$

if ρ is sufficiently small that all α_j's except α_k are outside $|\rho - \alpha_k| = \rho$.
Then, from (4) we get

$$\int_C f(z)dz = 2\pi i \sum_k R(f; \alpha_k). \quad \square$$

It seems clear that the residue theorem will be useful for the evaluation
of integrals. (It turns out that it is quite useful for other purposes as well.)
Before we get seriously into the utility of that theorem we concern ourselves
with the evaluation of residues so that we can compute the right hand side
of equation (3). In this we limit ourselves to the residues at poles because
there are no residues at removable singularities, and we will in general not
be concerned with essential singularities.

5.2b Theorem. *If $f(z)$ has simple pole (order 1) at α, then*

$$\text{Res}(f, \alpha) = \lim_{z \to \infty} (z - \alpha) f(z).$$

Proof. We have

$$f(z) = \frac{a_{-1}}{(z - \alpha)} + a_0 + a_1(z - \alpha) + \cdots,$$

so

$$(z - \alpha) f(z) = a_{-1} + a_0(z - \alpha) + a_1(z - \alpha)^2 + \cdots$$

from which the result is clear. $\square$

You should observe that really $(z - \alpha) f(z)$ is analytic at α (i.e., has a removable singularity there) and we simply evaluate this analytic function at α. Similarly, if $f(z)$ has a pole of order $n > 1$ at α, then $(z - \alpha)^n f(z)$ is analytic at α, for

$$(z - \alpha)^n f(z) = (z - \alpha)^n \left[\frac{a_{-n}}{(z - \alpha)^n} + \cdots + \frac{a_{-1}}{z - \alpha} + a_0 + \cdots \right]$$

$$= a_{-n} + a_{-n+1}(z - \alpha) + \cdots + a_{-1}(z - \alpha)^{n-1} + a_0(z - \alpha)^n + \cdots$$

Thus it is clear that if we differentiate $(n - 1)$ times and evaluate at $z = \alpha$, we get

$$\left(\frac{d}{dz} \right)^{n-1} [(z - \alpha)^n f(z)] \bigg|_{z=\alpha} = (n - 1)! a_{-1},$$

so that we have proved the following.

5.2c Theorem. *If $f(z)$ has a pole of order n at α, then*

$$\text{Res}(f; \alpha) = \frac{1}{(n - 1)!} \frac{d}{dz} [(z - \alpha)^n f(z)] \bigg|_{z=\alpha}.$$

We consider a special case of Theorem 5.2a.

5.2d Theorem. *If $f(z) = h(z)/g(z)$ where $h(z)$ is analytic at α with $h(\alpha) \neq 0$, and $g(z)$ is analytic at α with a simple (first order) zero there, then $\text{Res}(f, \alpha) = h(\alpha)/g'(\alpha)$.*

Proof. By Theorem 4.2a the residue is

$$\lim_{z \to \alpha} (z - \alpha) h(z)/g(z)$$

Since $g(\alpha) = 0$, this is

$$\lim_{z \to \alpha} \frac{h(z)}{\frac{g(z)-g(\alpha)}{z-\alpha}} = \lim_{z \to \infty} \frac{h(\alpha)}{\frac{g(z)-g(\alpha)}{z-\alpha}} = \frac{h(\alpha)}{g'(\alpha)}. \quad \square$$

Example 1. locate and classify all singular points and find their residues for $f(z) = z/\sin z$.

Solution. The denominator is 0 for $z = k\pi$, $k = 0, \pm 1, \pm 2, \ldots$. For $k = 0$ we have $f(z)$ is analytic at $z = 0$ (after removal of a removable singularity), and for $k \neq 0$ we have simple poles with residue $k\pi/\cos k\pi = (-1)^k k\pi$ at $z = k\pi$ by Theorem 5.2c.

Example 2. Locate and classify all singular points and find their residues for $f(z) = \tan z$.

Solution. Again by Theorem 5.2d with $h(z) = \sin z$, $g(z) = \cos z$, we have simple poles at $z = (k+\frac{1}{2})\pi$ with residue $\sin(k+\frac{1}{2})\pi/(-\sin(k+\frac{1}{2})\pi) = -1$.

Example 3. Locate and classify all singularities and find the residues for $f(z) = ze^{iz}/(z+\pi)^2$.

Solution. The only finite singularity is at $z = -\pi$. It is a pole of order 2 and, by Theorem 5.2c,

$$\begin{aligned} \operatorname{Res}(f, -\pi) &= \frac{d}{dz}\left(ze^{iz}\right)\Big|_{z=-\pi} \\ &= e^{iz} + ize^{iz}\Big|_{z=-\pi} = e^{-i\pi} - i\pi e^{-i\pi} \\ &= -1 + i\pi. \end{aligned}$$

A Exercises

1. For each function in Exercise A1 of Section 5.1 which has poles, find the residue at each pole.
2. If $g(z)$ is analytic and not zero at α, then show that the residue at α of
   (a) $g(z)/(z-\alpha)$ is $g(\alpha)$;
   (b) $g(z)/(z-\alpha)^2$ is $g'(\alpha)$.
   (c) Find the residue at α of $g(z)/(z-\alpha)^k$.
3. (a) Show that $\Gamma(z)$ has a simple pole with residue $(-1)^n/n!$ at $z = -n$, $n = 0, 1, 2, \ldots$.
   (b) Use Exercise C6(d), Section 3.6 to show that $1/\Gamma(z)$ is an entire funciton with simple zeros at $-n$, $n = 0, 1, 2, \ldots$.

B Exercises

1. Find the residues at the poles of the functions in Exercise B6 of Section 5.1.
2. (a) If $g(z)$ is analytic and not zero at α, and $h(z)$ is analytic with a second order zero at α, then show that $g(z)/h(z)$ has a second order pole at α with residue $[6g'(\alpha)h''(\alpha)-2g(\alpha)h'''(\alpha)]/3[h''(\alpha)]^2$.
   (b) What if $g(\alpha) = 0$?
3. (a) Suppose $f(z) = p(z)/q(z)$ where p and q are polynomials of degree m and n respectively, with $n > m+1$. Show that the sum of the residues of $f(z)$ is zero.
   (b) If C is a contour which does not go through a zero of $q(z)$, show that the sum of the residues of $f(z)$ at poles in $I(C)$ is equal to the negative of the sum at poles in $E(C)$.
   (c) Show that the sum of the residues of $f(z)$ is a_{n-1}/b_n if
   $$p(z) = \sum_0^{n-1} a_k z^k \quad \text{and} \quad q(z) = \sum_0^n b_k z^k.$$
4. (a) If $f(z)$ is analytic at α with a zero of order n there, show that $f'(z)/f(z)$ has a simple pole at α with residue n.
   (b) If $f(z)$ has a pole of order m at α, show that $f'(z)/f(z)$ has a simple pole at α with residue $-m$.
5. (Continuation) Suppose $f(z)$ is analytic in and on a contour C except for poles in $I(C)$ and that $f(z)$ does not vanish on C itself. Let Z be

the number of zeros of $f(z)$ and P the number of poles of $f(z)$ in $I(C)$, each zero and pole being counted according to its multiplicity. Show

$$\frac{1}{2\pi i} \int_C \frac{f'(z)}{f(z)} \, dz = Z - P.$$

6. (Continuation) Under the conditions of B5 above show

$$\frac{1}{2\pi i} \int_C \frac{z f'(z)}{f(z)} \, dz = \sum_k n_k \alpha_k - \sum_j m_j \beta_j$$

where the α_k's are the zeros of $f(z)$ of order n_k, respectively, and the β_j's are the poles of $f(z)$ of order m_j, respectively.

7. If $g(z)$ is analytic in and on C, and $f(z)$ satisfies the conditions of B5 above, show

$$\frac{1}{2\pi i} \int_C \frac{g(z) f'(z)}{f(z)} \, dz = \sum_k n_k g(\alpha_k) - \sum_j m_j g(\beta_j).$$

8. (a) Suppose $f(z)$ is analytic for $|z| \leq 2$ except for a simple pole with residue R at $z = 1$, and $f(z) = \sum_0^\infty a_n z^n$ is the Taylor series of $f(z)$ about $z = 0$. Show that

$$a_n = -R + b_n/2^n$$

where $b_n \to 0$ as $n \to \infty$.

   (b) Generalize to $f(z)$ analytic for $|z| \leq c$, where $c > 1$.

C Exercises

1. (Continuation of B5 above) Under the conditions of B5, evaluate the integral to establish the formula

$$Z - P = \frac{1}{2\pi} \Delta_C \arg f(z)$$
$$= \text{the (net) number of times } f(z) \text{ goes around}$$
$$\text{the origin as } z \text{ traces out } C.$$

This result is called the **principle of the argument**.

2. **(Rouché's Theorem)** Suppose both $f(z)$ and $g(z)$ are analytic in and on a contour C and that $|g(z)| < |f(z)|$ on C. Then $f(z)$ and $f(z) + g(z)$ have the same number of zeros in $I(C)$.

   Hint: Let m be the number of zeros of $f(z)$ in $I(C)$ and n be the number of zeros of $f(z) + g(z)$ in $I(C)$. Apply the principle of the argument to

   $$F(z) = \frac{f(z) + g(z)}{f(z)} = 1 + \frac{g(z)}{f(z)}$$

   and note that

   $$Re(F(z)) = 1 + Re(g(z)/g(z))$$
   $$> 1 - |g(z)/f(z)| > 0,$$

   and conclude that $\Delta_C \arg F(z) = 0$.

3. Give an alternative proof of the fundamental theorem of algebra (3.2h). Apply Rouché's Theorem to $f(z) = a_n z^n$, $g(z) = \sum_0^{n-1} a_k z^k$ on a large circle. $C : |z| = R$, and conclude that $p(z) \equiv \sum_0^n a_k z^k$ has n zeros if $a_n \neq 0$.

4. (Continuation) Let the n zeros of $p(z)$ be $z_1, z_2, \ldots, z_n$ with multiple zeros repeated according to their multiplicity. Use Liouville's Theorem (3.2g) to conclude that

   $$p(z)/(z - z_1)(z - z_2) \cdots (z - z_n)$$

   is a constant and deduce

   $$p(z) = a_n(z - z_1)(z - z_2) \cdots (z - z_n).$$

5.3 Improper Integrals of Rational Functions

If $f(x)$ is continuous for all real x, and if

$$\lim_{b \to \infty} \int_0^b f(x)dx$$

exists, it is called an **improper** integral and is denoted by

$$\int_0^\infty f(x)dx = \lim_{b \to \infty} \int_0^b f(x)dx \tag{1}$$

Similarly

$$\int_{-\infty}^{0} f(x)dx = \lim_{a \to -\infty} \int_{a}^{0} f(x)dx. \tag{2}$$

If *both* (1) and (2) exist, then

$$\int_{-\infty}^{\infty} f(x)dx = \int_{-\infty}^{0} f(x)dx + \int_{0}^{\infty} f(x)dx \tag{3}$$

is the definition of the left side of (3). However, if (1) and (2) fail to exist, it is still possible that

$$\lim_{b \to \infty} \left[\int_{-b}^{0} f(x)dx + \int_{0}^{b} f(x)dx \right] = \lim_{b \to \infty} \int_{-b}^{b} f(x)dx \tag{4}$$

exists. If (4) exists, it is called the **Cauchy principal value** of (3) at ∞ and is denoted by $P_\infty \int_{-\infty}^{\infty} f(x)dx$.

Similarly, if $f(x)$ is continuous for $a \le x \le b$, except at c where it becomes unbounded, then the improper integral

$$\int_{a}^{b} f(x)dx \equiv \lim_{\varepsilon \to 0+} \int_{a}^{c-\varepsilon} f(x)dx + \lim_{\varepsilon \to 0+} \int_{c+\varepsilon}^{b} f(x)dx$$

where the left side is defined by the right provided *both* limits exist. Again if (4) fails to exist, but

$$\lim_{\varepsilon \to 0+} \left[\int_{a}^{c-\varepsilon} f(x)dx + \int_{c+\varepsilon}^{b} f(x)dx \right]$$

exists, then it is called the **Cauchy principal value** of (4) at c and denoted by

$$P_c \int_{a}^{b} f(x)dx.$$

You should note the meaning of the symbols P_c, P_∞ here: they are warning flags calling your attention to the principal value character of the integrals, and giving the location of the singular point involved.

Example 1. Compute $\int_{0}^{\infty} \dfrac{dx}{1+x^2}$.

Solution. $\displaystyle\int_{0}^{b} \frac{dx}{1+x^2} = \arctan x \Big|_{0}^{b} = \arctan b \to \pi/2$ as $b \to \infty$, and

$$\int_{a}^{0} \frac{dx}{1+x^2} = -\arctan a \to -\left(-\frac{\pi}{2}\right) = \frac{\pi}{2} \quad \text{as} \quad a \to -\infty$$

so
$$\int_{-\infty}^{\infty} \frac{dx}{1+x^2} = \int_0^{\infty} \frac{dx}{1+x^2} + \int_{-\infty}^0 \frac{dx}{1+x^2} = \frac{\pi}{2} + \frac{\pi}{2} = \pi.$$

Example 2. Compute $P_{\infty} \int_{-\infty}^{\infty} x\,dx$.

Solution. First we observe that
$$\int_0^{\infty} x\,dx = \lim_{b\to\infty} \int_0^b x\,dx = \lim_{b\to\infty} \frac{b^2}{2},$$

which does not exist, so the improper integral $\int_{-\infty}^{\infty} x\,dx$ does not exist. However,
$$\int_{-b}^b x\,dx = \left(\frac{b^2}{2}\right) - \left(\frac{b^2}{2}\right) = 0 \to 0 \quad \text{as} \quad b \to \infty$$

so
$$P_{\infty} \int_{-\infty}^{\infty} x\,dx = 0.$$

You should observe that the Cauchy principal value computed in Example 2 exists because two "infinities" cancel: $b^2/2 - b^2/2 = 0$. On the other hand in Example 1, each end of the integral has an independent existence. *This is basically the difference between ordinary improper integrals and principal value integrals.* This remark also applies when the singularity is at a finite point.

There is a class of integrals of rational functions which can be evaluated by elementary calculus methods, though the calculations in general get very complicated. These integrals can frequently be evaluated more easily by use of the residue theorem and contour integration.

Example 3. Use the residue theorem to prove
$$\int_{-\infty}^{\infty} \frac{dx}{1+x^2} = \pi.$$

Solution. Clearly, as we did a few paragraphs back, one can evaluate this integral simply and easily by elementary calculus. *The point here is to illustrate a new method in a simple case.*

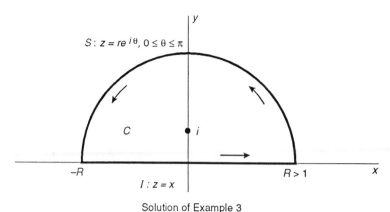

Solution of Example 3

Let C be the semicircular contour as illustrated, consisting of the interval I from $-R$ to $+R$ along the real axis and the semicircle S from R to $-R$ in the upper half plane. We consider

$$\int_C \frac{dz}{1+z^2} = \int_C \frac{dx}{(z+i)(z-i)}.$$

The function $f(z) = 1/(1+z^2) = 1/(z+i)(z-i)$ has one pole in the upper half plane, a simple pole at $z = i$, with residue

$$\text{Res}\,(i) = \frac{1}{2i}.$$

So if we take $R > 1$ we have this one pole in $I(C)$ so

$$\int_C \frac{dz}{1+z^2} = 2\pi i \left(\frac{1}{2i}\right) = \pi.$$

And

$$\int_C \frac{dz}{1+z^2} = \int_I \frac{dz}{1+z^2} + \int_S \frac{dz}{1+z^2},$$

where

$$\int_I \frac{dz}{1+z^2} = \int_{-R}^{R} \frac{dx}{1+x^2} \to \int_{-\infty}^{\infty} \frac{dx}{1+x^2} \quad \text{as} \quad R \to \infty$$

which is the integral we want. If we can now show that the other integral goes to zero in the limit as $R \to \infty$, we've got it made.

But

$$\left| \int_S \frac{dz}{1+z^2} \right| = \left| \int_0^\pi \frac{iRe^{i\theta}\,d\theta}{1+R^2 e^{2i\theta}} \right|$$
$$\leq \int_0^\pi \frac{R\,d\theta}{R^2-1} = \frac{\pi R}{R^2-1} \to 0 \text{ as } R \to \infty.$$

This completes the evaluation.

We discuss the estimate in the last integral: First we use $\left| \int_a^b f(\theta)\,d\theta \right| \leq \int_a^b |f(\theta)|\,d\theta$ to get

$$\left| \int_0^\pi \frac{iRe^{i\theta}\,d\theta}{1+R^2 e^{2i\theta}} \right| \leq \int_0^\pi \frac{R}{|1+R^2 e^{2i\theta}|}\,d\theta.$$

Then we use the *lower* end of the triangle inequality (Corollary 1.2c)

$$|\alpha + \beta| \geq ||\alpha| - |\beta||$$

which gives

$$|1+R^2 e^{2i\theta}| \geq |1-R^2| = R^2 - 1$$

since $R > 1$. Then

$$\frac{1}{|1+R^2 e^{2i\theta}|} \leq \frac{1}{R^2-1}$$

and the estimate is complete.

We now consider an extension of this method to the case where we permit *first order* poles on the x-axis and the resulting integral is a principal value integral.

Example 4. Evaluate $P_1 \int_{-\infty}^{\infty} \frac{dx}{(x-1)(x^2+1)}$ by the residue theorem.

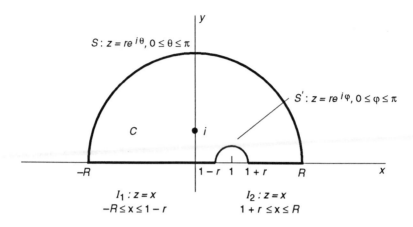

Example 4; Exercise A2

Solution. We consider a contour C as illustrated consisting of a semicircular contour as before but indented by S' at $z = 1$. Then for $R > 1$ and $0 < r < 1$ we have, as before

$$\int_C \frac{dz}{(z-1)(z^2+1)} = 2\pi i \operatorname{Res}(i) = 2\pi i \frac{1}{i-1} \cdot \frac{1}{2i} \quad \text{(how?)}$$

$$= \frac{\pi}{i-1} = -\frac{\pi(i+1)}{2}$$

As before

$$\int_S \frac{dz}{(z-1)(z^2+1)} \to 0 \quad \text{as} \quad R \to \infty$$

(see Exercise A2) and we now look at the simple pole at 1. There we can write, since $\operatorname{Res}(1) = \frac{1}{2}$ (how?),

$$\frac{1}{(z-1)(z^2+1)} = \frac{1}{2(z-1)} + g(z)$$

where $g(z)$ is analytic at $z = 1$, and consequently is bounded, say $|g(z)| \le M$, for $|z-1| < 1$. Then

$$\int_{S'} \frac{dz}{(z-1)(z^2+1)} = \frac{1}{2} \int_{S'} \frac{dz}{(z-1)} + \int_{S'} g(z)dz$$

Now

$$\left| \int_{S'} g(z)dz \right| \le M \cdot \pi r \to 0 \quad \text{as} \quad r \to 0,$$

and setting $z - 1 = re^{i\phi}$ we get

$$\frac{1}{2} \int_{S'} \frac{dz}{z - 1} = \frac{1}{2} \int_\pi^0 \frac{ire^{i\phi} d\phi}{re^{i\phi}} = i \left(\frac{1}{2}\right) \int_\pi^0 d\phi$$
$$= -\pi i \left(\frac{1}{2}\right)$$

so that

$$\lim_{r \to 0} \int_{S'} \frac{dz}{(z - 1)(z^2 + 1)} = -\pi i/2.$$

Now putting this all together we have

$$-\frac{\pi}{2}(i + 1) = \int_C \frac{dz}{(z - 1)(z^2 + 1)}$$
$$= \int_{-R}^{1-r} \frac{dx}{(x - 1)(x^2 + 1)} + \int_{S'} \frac{dz}{(z - 1)(z^2 + 1)}$$
$$+ \int_{1+r}^R \frac{dx}{(x - 1)(x^2 + 1)} + \int_S \frac{dz}{(z - 1)(z^2 + 1)}$$

Then we let $R \to \infty$ and we get

$$-\frac{\pi}{2}(i + 1) = \int_{-\infty}^{1-r} \frac{dx}{(x - 1)(x^2 + 1)} + \int_{1+r}^\infty \frac{dx}{(x - 1)(x^2 + 1)}$$
$$+ \int_{S'} \frac{dz}{(z - 1)(z^2 + 1)}$$

Now we let $r \to 0$ and get

$$-\frac{\pi}{2}i - \frac{\pi}{2} = P_1 \int_{-\infty}^\infty \frac{dx}{(x - 1)(x^2 + 1)} - \frac{\pi i}{2}$$

so that

$$P_1 \int_{-\infty}^\infty \frac{dx}{(x - 1)(x^2 + 1)} = -\frac{\pi}{2}.$$

The methods we used on the last two examples enable us to establish the following theorems.

5.3a Theorem. *If $f(z) = p(z)/q(z)$ where p and q are polynomials with $\deg q \geq \deg p + 2$, and if $q(z) = 0$ has no real solutions, then*

$$\int_{-\infty}^\infty f(x)dx = 2\pi i \sum_{\text{Im}\,\alpha_k > 0} \text{Res}\,(f, \alpha_k).$$

5.3b Theorem. *If $f(z)$ is the same as in 5.3a, except that it may have simple poles on the real axis, then*

$$P \int_{-\infty}^{\infty} f(x)dx = 2\pi i \sum_{\text{Im } \alpha_k > 0} \text{Res}\,(f; \alpha_k) + \pi i \sum_{\text{Im } \alpha_k = 0} \text{Res}\,(f; \alpha_k).$$

Proof of Theorem 5.3a.

All that is required here is to show that, in the previous notation,

$$\lim_{R \to \infty} \int_S f(z)dz = 0$$

where S is the large semicircle of radius R, for if R is large enough, then all poles in the upper half-plane will be contained in $I(C)$. We estimate $q(z)$, the denominator in $f(z) = p(z)/q(z)$:

$$q(z) = \sum_0^n a_k z^k = a_n z^n + a_{n-1} z^{n-1} + \cdots + a_0$$

where $a_n \neq 0$. Then (recall the estimate in the proof of the fundamental theorem of algebra, 3.2f)

$$|q(z)| \geq |a_n z^n| - \left| \sum_0^{n-1} a_k z^k \right|$$

$$\geq |a_n||z|^n \left\{ 1 - \sum_0^{n-1} \left| \frac{a_k}{a_n} \right| \frac{1}{|z|^{n-k}} \right\}.$$

Then there is an R_0 for which the sum in the braces is $< 1/2$ if $|z| = R > R_0$. Then for $|z| = R > R_0$ we have

$$|q(z)| \geq |a_n||z|^n/2.$$

And

$$p(z) = \sum_0^{n-2} b_k z^k,$$

where b_{n-2} may or may not be zero, but for $|z| > 1$,

$$|p(z)| \leq \sum_0^{n-2} |b_k||z|^k \leq \sum_0^{n-2} |b_k||z|^{n-2}$$

$$= |z|^{n-2} \sum_0^{n-2} |b_k| = K_1 |z|^{n-2}$$

where K is a constant. Then for z on S, the semicircle of radius R, we have

$$|f(z)| = |p(z)/q(z)| \leq K|z|^{n-2}/[|a_n||z|^n/2]$$
$$= [2K/|a_n|/|z|^2 = K_1/|z|^2.$$

And, finally,

$$\left| \int_S f(z)dz \right| \leq \frac{K_1}{R^2} \cdot \pi R = \frac{K_1 \pi}{R} \to 0 \quad \text{as} \quad R \to \infty. \quad \square$$

Proof of Theorem 5.3b.

The only additional argument required here is to show that for a *simple pole* at α on the real axis, the contribution is $\pi i \operatorname{Res}(\alpha)$. But for such a pole we have

$$f(z) = \frac{a_{-1}}{z - \alpha} + g(z)$$

where $g(z)$ is analytic, and hence bounded in a neighborhood of α. Then, exactly as in the example,

$$\int_{S'} g(z)dz \to 0 \quad \text{as} \quad r \to 0$$

if r is the radius of S'. And

$$\int_{S'} \frac{a_{-1}dz}{z - \alpha} = a_{-1} \int_\pi^0 \frac{ire^{i\theta}\,d\theta}{re^{i\phi}} = -\pi i a_{-1}.$$

Thus

$$\int_{S'} f(z)dz \to -\pi i a_{-1}$$

at each simple pole on the axis. $\square$

There is a result analogous to the case where $f(z)$ has a simple pole on the real axis. That is the case where $f(z)$ has a simple zero at ∞. We get the following result.

5.3c Theorem. *If* $q(z) = \sum_0^n a_k z^k$ *and* $p(z) = \sum_0^{n-1} b_k z^k$ *where* $a_n \neq 0$ *and* $b_{n-1} \neq 0$, *and if* $f(z) = p(z)/q(z)$ *has simple poles, if any, on the real axis, then*

$$P\int_{-\infty}^\infty f(z)dz = 2\pi i \sum_{\operatorname{Im} \alpha_k > 0} \operatorname{Res}(f, \alpha_k) + \pi i \sum_{\operatorname{Im} \alpha_k = 0} \operatorname{Res}(f, \alpha_k) - \pi i a_n/b_{n-1}.$$

The proof goes the same as in the previous cases except that

$$\lim_{R\to\infty}\int_S f(z)dz = \pi i a_n/b_{n-1}.$$

The details are left to the exercises.

We make two remarks concerning the integrals we deal with in this section.

A. Though the calculations in general are complicated all the integrals can be evaluated by the *methods* of elementary calculus.

B. Further, this section could come immediately after Section 1.5. We would merely have needed to replace $e^{i\theta}$ by $E(\theta) = \cos\theta + i\sin\theta$.

Significant Topics   Page

A Exercises

1. What is wrong here? Since $1/x^2 > 0$, then $\int_I dx/x^2$ must be positive for any interval I. But $\int_{-1}^{1} dx/x^2 = -1/x\Big|_{-1}^{1} = -1 -1 = -2.$

2. In Example 4 show that the integral over $S \to 0$ as $R \to \infty$.

3. Complete the proof of Theorem 5.3c by showing that the integral over $S \to \pi i b_{n-1}/a_n$ as $R \to \infty$.

4. Evaluate

(a) $P_0 \int_{-1}^{2} \dfrac{dx}{x}$;

(b) $P_0 \int_{-a}^{b} \dfrac{dx}{x}$, $a > b > 0$;

(c) $\int_{-1}^{2} \dfrac{dx}{\sqrt{|x|}}$;

(d) $P_{0,\infty} \int_{-\infty}^{\infty} \dfrac{dx}{x}$;

(e) $P_1 \int_{0}^{2} \dfrac{dx}{1 - x^2}$;

(f) $P_0 \int_{-\infty}^{\infty} \dfrac{dx}{x(x^2 - 1)}$;

5. Show that the following *do not exist*:

(a) $P_0 \displaystyle\int_{-1}^{2} \frac{dx}{x^2}$;

(b) $P_0 \displaystyle\int_{-a}^{b} \frac{dx}{x^4}$, $a > 0$, $b > 0$;

(c) $P_\infty \displaystyle\int_{-\infty}^{\infty} \sqrt{|x|}\, dx$;

(d) $P_\infty \displaystyle\int_{-\infty}^{\infty} |x|^a dx$, $a > -1$.

6. Evaluate $\displaystyle\int_{-\infty}^{\infty} \frac{dx}{x^2 - 2x + 5}$

   (a) by elementary calculus;   (b) by residues.

   (c) Evaluate $\displaystyle\int_{0}^{\infty} x^2 dx/(1+x^6)$.

7. Reconsider Example 3 by integrating over the illustrated contour.

8. Reconsider Example 4 by integrating over the illustrated contour.

9. Reconsider again Example 4 by integrating over the illustrated contour.

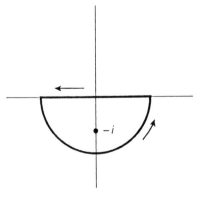

Exercise A7

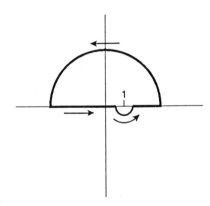

Exercise A8

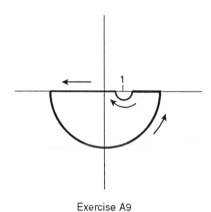

Exercise A9

B Exercises

Evaluate by the residue theorem.

1. $\displaystyle\int_{-\infty}^{\infty} \frac{dx}{x^2 + a^2}$.

2. $\displaystyle\int_{0}^{\infty} \frac{dx}{(x^2 + a^2)(x^2 + b^2)}$.

3. $\displaystyle\int_{0}^{\infty} \frac{dx}{(x^2 + z^2)^2}$.

4. $\displaystyle P_b \int_{-\infty}^{\infty} \frac{dx}{(x - b)(x^2 + a^2)}$.

5. Show that the answer to B2 tends to that of B3 as $b \to a$.

6. In Theorem 5.3a show that

$$\int_{-\infty}^{\infty} f(x)dx = -2\pi i \sum_{\text{Im}\,\alpha_k < 0} \text{Res}\,(f, \alpha_k).$$

5.4 Infinite Trigonometric Integrals.

In this section we discuss integrals of the form

$$\int_{-\infty}^{\infty} f(x) \cos ax\, dx \quad \text{and} \quad \int_{-\infty}^{\infty} f(x) \sin ax\, dx \tag{1}$$

where $f(x)$ is a rational function. As in the previous section the semicircular contours (with indentations where appropriate) would seem to be useful. However, we must modify the integrands, because both $\cos ax$ and $\sin ax$ grow exponentially off the real axis. We examine

$$\int_{-\infty}^{\infty} f(x) e^{iax}\, dx \tag{2}$$

and if (2) is known we can recapture the integrals (1) as the real and imaginary parts of (2).

Example 1. Evaluate $\int_{-\infty}^{\infty} \dfrac{\cos ax}{1+x^2}\,dx$, $a > 0$.

Solution. We consider

$$\int_C \frac{e^{iaz}dz}{1+z^2}$$

where C is the standard semicircular contour. Then, by the residue theorem, we have

$$2\pi i\left(\frac{e^{-a}}{2i}\right) = \int_C \frac{e^{iaz}}{1+z^2}\,dz = \int_{-R}^{R} \frac{e^{iax}}{1+x^2}\,dx + \int_S \frac{e^{iaz}}{1+z^2}\,dz. \qquad (3)$$

We examine

$$\int_S \frac{e^{iaz}dz}{1+z^2} = \int_S \frac{e^{iax}e^{-ay}dz}{1+z^2}$$

Now $|e^{iax}| = 1$ and $|e^{-ay}| \le 1$ on S since $y > 0$ and $a > 0$. Thus, as before

$$\left|\int_S \frac{e^{iaz}dz}{1+z^2}\right| \le \frac{\pi R}{R^2 - 1} \to 0 \quad \text{as} \quad R \to \infty$$

Then from (3) we get, as $R \to \infty$,

$$\int_{-\infty}^{\infty} \frac{e^{iax}dx}{1+x^2} = \pi e^{-a} = \pi/e^a.$$

Equating real and imaginary parts we get

$$\int_{-\infty}^{\infty} \frac{\cos ax\,dx}{1+x^2} = \frac{\pi}{e^a}; \qquad \int_{-\infty}^{\infty} \frac{\sin ax\,dx}{1+x^2} = 0.$$

The second integral is clearly zero since the integrand is an odd function, and from the first we also have

$$\int_0^{\infty} \frac{\cos ax\,dx}{1+x^2} = \frac{\pi}{2e^a}$$

since the integrand is even.

If $f(x) = p(x)/q(x)$ where p and q are polynomials with $\deg q \ge \deg p + 2$, then the integrals (1), converge at ∞. So the method illustrated by the examples of the previous section are appropriate here. However if $\deg q = \deg p + 1$, then things are a little more difficult.

Example 2. Evaluate $\displaystyle\int_0^\infty \frac{\sin x}{x}\,dx$.

Solution. We consider

$$\int_C \frac{e^{iz}}{z}\,dz$$

where C is the usual semicircular contour, but indented at the origin. Then, since there are no singularities in $I(C)$, we have

$$
\begin{aligned}
0 &= \int_C \frac{e^{iz}}{z}\,dz \\
&= \int_S \frac{e^{iz}}{z}\,dz + \int_{-R}^{-r} \frac{e^{ix}}{x}\,dx + \int_r^R \frac{e^{ix}}{x}\,dx + \int_{S'} \frac{e^{iz}}{z}\,dz \\
&= I_1 + I_2 + I_3 + I_4
\end{aligned}
\tag{4}
$$

respectively. Setting $x = -t$ in I_2 we get

$$I_2 = \int_R^r \frac{e^{-it}}{t}\,dt = -\int_r^R \frac{e^{-it}}{t}\,dt = -\int_r^R \frac{e^{ix}}{x}\,dx$$

so that

$$I_2 + I_3 = \int_r^R [e^{ix} - e^{-ix}]\frac{dx}{x} = 2i \int_r^R \frac{\sin x}{x}\,dx.$$

For S' we observe that

$$\frac{e^{iz}}{z} = \frac{1}{z} + \left(\frac{e^{iz}-1}{z}\right) = \frac{1}{z} + g(z)$$

where $g(z) = (e^{iz}-1)/z$ is analytic (has a removable singularity) at the origin and so is bounded near $z = 0$. Then as before

$$\int_{S'} \frac{e^{iz}}{z}\,dz \to -\pi i \quad \text{as} \quad r \to 0.$$

So we have, from (4),

$$0 = 2i \int_0^R \frac{\sin x}{x}\,dx - \pi i + I_1. \tag{5}$$

To consider the limit of I_1 as $R \to \infty$ we introduce polar coordinates:

$$I_1 = \int_S \frac{e^{iz}}{z}\, dz = \int_0^\pi \frac{e^{iR\cos\theta} e^{-R\sin\theta} Rie^{i\theta}}{Re^{i\theta}}\, d\theta$$

from which

$$|I_1| \leq \int_0^\pi e^{-R\sin\theta}\, d\theta$$

We will show that this integral has limit 0 as $R \to \infty$. Then from (5) we get

$$\int_0^\infty \frac{\sin x}{x}\, dx = \frac{\pi}{2}$$

Rather than showing directly that $I_1 \to 0$ as $R \to \infty$, we will establish a more general result which implies that $I_1 \to 0$.

It seems clear from the graph of $y = \sin x$, that between 0 and $\pi/2$ the curve lies above its secant line. But the secant line from $(0,0)$ to $(\pi/2, 1)$ is

$$y = \frac{2x}{\pi}.$$

This then suggests that

$$\sin x \geq 2x/\pi, \quad 0 \leq \theta \leq \pi/2 \tag{6}$$

The formula (6) is known as **Jordan's inequality**. We outline in the C Exercise how this can be rigorously established. Then using (6) we can prove the following, which is called **Jordan's Lemma**.

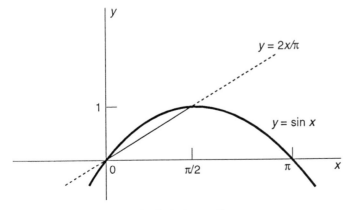

Jordan's Inequality

5.4a Lemma. *Suppose S denotes the semicircle $|z| = R$ in the upper half plane and $a > 0$. Then*

$$\lim_{R \to \infty} \int_S e^{iaz} f(z) dz = 0$$

provided that $f(Re^{i\theta}) \to 0$ uniformly for $0 \le \theta \le \pi$ as $R \to \infty$.

Proof. Denote the integral by I. Then

$$I = \int_0^\pi e^{iaR\cos\theta} e^{-aR\sin\theta} f(Re^{i\theta}) i Re^{i\theta} d\theta$$

so that

$$|I| < R \int_0^\pi e^{-aR\sin\theta} |f(Re^{i\theta})| d\theta.$$

Then given $\varepsilon > 0$ there is an R_0 for which $|f(Re^{i\theta})| < \varepsilon$ if $R > R_0$. Thus, for $R > R_0$,

$$|I| \le \varepsilon R \int_0^\pi e^{-aR\sin\theta} d\theta$$

$$= 2\varepsilon R \int_0^{\pi/2} e^{-aR\sin\theta} d\theta \quad \text{(how?)}.$$

Applying Jordan's inequality (6) we get

$$|I| \le 2\varepsilon R \int_0^{\pi/2} e^{-aR2\theta/\pi} d\theta$$

$$= (2\varepsilon R)(\pi/2aR) \int_0^{\pi/2} e^{-2aR\theta/\pi} (2aR/\pi) d\theta$$

$$= (\pi\varepsilon/a)[-e^{-2aR\theta/\pi}]_0^{\pi/2}$$

$$= \pi\varepsilon a[1 - e^{-aR}] < \pi a\varepsilon$$

Since ε is arbitrary, this proves the lemma and so completes Example 2. □

There is a somewhat more tedious, but basically simpler, approach which avoids the use of Jordan's Lemma. We illustrate that by the following

Example 3. Evaluate $\displaystyle\int_0^\infty \frac{x \sin x}{x^2 + a^2} \, dx$, $a > 0$.

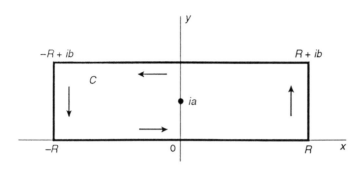

Example 3

Solution. Let C be the rectangle cornering at $\pm R$ and $\pm R + ib$. Then we have

$$\int_C \frac{ze^{iz}}{z^2 + z^2}\, dz = 2\pi i\left(\frac{aie^{-a}}{2ai}\right) = \pi ie^{-a}$$

so

$$\pi ie^{-a} = \int_{-R}^{R} \frac{xe^{ix}}{x^2 + a^2}\, dx + \int_{R}^{R+ib} \frac{ze^{iz}}{z^2 + a^2}\, dz + \int_{R+ib}^{-R+ib} \frac{ze^{ia}}{z^2 + a^2}\, dz$$

$$+ \int_{-R+ib}^{-R} \frac{ze^{iz}dz}{z^2 + a^2} = I_1 + I_2 + I_3 + I_4$$

respectively. Clearly

$$I_1 = 2i\int_0^R \frac{x\sin x}{x^2 + a^2}\, dx$$

and so its limit, as $R \to \infty$, involves the integral we wish to evaluate. We consider next I_3. In I_3, $z = x + ib$, $dz = dx$, so

$$I_3 = -\int_{-R}^{R} \frac{(x + ib)e^{ix} \cdot e^{-b}dx}{(x + ib)^2 + a^2}$$

so

$$|I_3| \le \frac{(b + R)e^{-b} \cdot 2R}{a^2}.$$

And in I_2, $z = R + iy$, $dz = idy$, so

$$I_2 = \int_0^b \frac{(R + iy)e^{-y}idy}{(R + iy)^2 + a^2}$$

so

$$|I_2| \leq \frac{1}{R^2 + a^2} \int_0^b (R + y)e^{-y}\,dy \leq \frac{1}{R^2 + a^2} \int_0^\infty (R + y)e^{-y}\,dy$$
$$= \frac{R}{R^2 + a^2} + \frac{1}{R^2 + a^2}$$

We get the same estimate for I_4. We first let $b \to \infty$, then $R \to \infty$, and we have all of $I_2, I_3, I_4 \to 0$. Thus we get

$$2i \int_0^\infty \frac{x \sin x}{x^2 + a^2}\,dx = i\pi e^{-a}$$

or

$$\int_0^\infty \frac{x \sin x}{x^2 + a^2}\,dx = \frac{\pi e^{-a}}{2}.$$

Significant Topics	Page

A Exercises

1. Reconsider Example 3 using a semicircular contour and Jordan's lemma. Evaluate the following with $a > 0$

2. $\int_0^\infty \frac{x \sin ax}{1 + x^2}\,dx,$

3. $\int_0^\infty \frac{\cos x}{a^2 + x^2}\,dx$

4. $\int_0^\infty \frac{x \sin x}{(1 + x^2)^2}\,dx,$

5. $\int_0^\infty \frac{\cos x}{(a^2 + x^2)^2}\,dx$

6. $\int_0^\infty \frac{\cos x}{(1 + x^2)^2}\,dx,$

7. $\int_0^\infty \frac{x \sin x}{(a^2 + x^2)^2}\,dx.$

8. $\int_0^\infty \frac{x^2 \cos x}{(1 + x^2)^2}\,dx.$

9. Reconsider Example 3 by integrating $ze^{iz}/(z^2 + a^2)$ over a semicircular contour in the *lower half-plane*.

B Exercises

1. $\displaystyle\int_0^\infty \frac{\sin^2 x}{x^2}\,dx$, Hint: $P_0 \displaystyle\int_{-\infty}^\infty \frac{1 - e^{2ix}}{x^2}\,dx.$

2. $\displaystyle\int_0^\infty \frac{\cos x}{(x^2 + a^2)(x^2 + b^2)}\,dx.$

3. $\displaystyle\int_0^\infty \frac{x \sin x}{x^4 + 16}\,dx.$

4. $\displaystyle\int_{-\infty}^\infty \frac{\sin x}{x^2 - 2x + 2}\,dx.$

5. $\displaystyle\int_{-\infty}^\infty \frac{\cos x}{x^2 + 4x + 5}\,dx.$

6. $\displaystyle\int_0^\infty \frac{dx}{x^6 + 1}.$

7. $P \displaystyle\int_{-\infty}^\infty \frac{\cos x}{a^2 - x^2}\,dx.$

8. $\displaystyle\int_0^\infty \frac{dx}{x^3 + a^3}.$

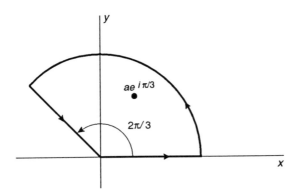

C Exercise

1. (Jordan's inequality) (a) Suppose $f(x)$ is strictly decreasing on $I = [0, c]$. Take $g(0) = f(0)$ and $g(x) = (1/x) \displaystyle\int_0^x f(t)dt$ and show that $g(x)$ is strictly decreasing on I. Hint: (i) Show $g(x) < g(0)$ if $0 < x \leq c$ and (ii) take $0 < a < b \leq c$ and examine

$$\frac{1}{a}\int_0^a f(t)dt - \frac{1}{b}\int_0^b f(t)dt = \left(\frac{1}{a} - \frac{a}{b}\right)\int_a^a f(t)dt - \frac{1}{b}\int_a^b f(t)dt$$

$$> \left(\frac{1}{a} - \frac{1}{b}\right)af(a) - \frac{a}{b}(b - a)f(a) = 0.$$

(b) Apply part (a) to $f(x) = \cos x$ and conclude

$$1 \geq \frac{\sin x}{x} \geq \frac{1}{(\pi/2)} = \frac{2}{\pi}.$$

5.5 Integrals of the form $\int_0^\infty x^{a-1} f(x) dx$

Example 1. Evaluate

$$I = \int_0^\infty \frac{t^{a-1}}{1+t^2} \, dt$$

and determine the set of real a's for which this evaluation is valid.

We find that our experience over the last few sections persuades us to examine

$$\int_C \frac{z^{a-1}}{1+z^2} \, dz,$$

but in so doing we are dealing with a multivalued integrand. Thus we must be judicious in our choice of C, and clear in our minds about the choice of the branch of z^{a-1} we use, and about the limitations of that choice.

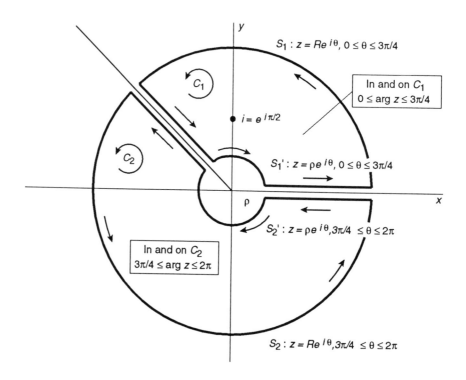

Solution. We first take C_1 as illustrated, and then C_2 as illustrated. The segments, drawn in the figure as double lines for the purpose of illustration, are along the real axis, and along the ray $\theta = 3\pi/4$. Then taking $z^{a-1} = z^a/z$ where z^a is the principal value in and on C_1, the integrand is analytic except for a simple pole at $z = i$. There we have the residue $(i)^{a-1}/2i = -(i)^a/2$ where i^a is the principal value of i^a which is

$$i^a = (e^{i\pi/2})^a = e^{i\pi a/2}.$$

Thus

$$\int_{C_1} \frac{z^{a-1}}{1+z^2}\, dz = 2\pi i(-e^{i\pi a/2}/2) = -\pi i e^{i\pi a/2} \tag{1}$$

Now we take C_2, and determine z^a in and on C_2 so that it has the same value on $\theta = 3\pi/4$ as it had on C_1, and so that it is single valued and analytic in and on C_2; therefore we must have z^a determined by $3\pi/4 \leq \theta \leq 2\pi$ in and on C_2. Then the integrand is single valued and analytic in and on C_2 except for a simple pole at $-i = e^{3\pi i/2}$. Thus the residue there is

$$\frac{(e^{3\pi i/2})^a/(-i)}{(-2i)} = -\frac{1}{2}\, e^{3\pi i a/2}$$

so that

$$\int_{C_2} \frac{z^{a-1}}{1+z^2}\, dz = 2\pi i\left(-\frac{1}{2}\, e^{3\pi i a/2}\right) = -\pi i e^{3\pi i a/2}. \tag{2}$$

If we add (1) and (2) we see that along the ray $\theta = 3\pi/4$, the integrand is the same in the two integrals *by the way we chose the branch of* z^a. Thus the integrals on this ray cancel since they are in opposite directions. However, along the segment of the real axis the integrands are different since in C_1, $\arg z = 0$ there, while in C_2, $\arg z = 2\pi$, so that the integrals over this segment *do not cancel.*

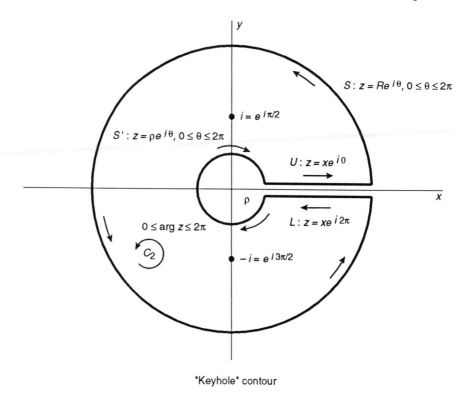

"Keyhole" contour

Thus the result of adding formulas (1) and (2) is an integral over the curve C in the second figure in which the segment from ρ to R of the real axis is covered twice. It is convenient to note that the positive real axis is a branch cut, and the two integrals over the segment are on different branches of z^a. We take the branch on which $\arg z = 0$ and call it (for convenience) the "upper branch" and label it U. The other branch we call the "lower branch" and label it L. Thus the results of adding the formulas (1) and (2) amount to an extension of the residue theorem to the curve C which is *not* simple. Thus we have

$$\int_C \frac{z^{a-1}\,dz}{1+z^2} = -\pi i e^{i\pi a/2} - \pi i e^{3\pi i a/2}$$
$$= -\pi i e^{i\pi a}[e^{-i\pi a/2} + e^{i\pi a/2}]$$
$$= -2\pi i e^{i\pi a}\cos(\pi a/2)$$

and

$$\int_C \frac{z^{a-1}}{1+z^2} = \left[\int_U + \int_S + \int_L + \int_{S'} \right] \frac{z^{a-1}\,dz}{1+z^2}$$
$$= I_1 + I_2 + I_3 + I_4, \quad \text{respectively.}$$

Let us first consider I_4 in polar coordinates:

$$I_4 = \int_{2\pi}^{0} \frac{\rho^{a-1}e^{i\theta(a-1)} \cdot \rho e^{i\theta} i\,d\theta}{1+\rho^2 e^{2i\theta}}.$$

Then, for $\rho < 1$,

$$|I_4| \le \frac{2\pi\rho^a}{1-\rho^2} \to 0 \quad \text{as} \quad \rho \to 0 \quad \text{if} \quad a > 0.$$

We estimate I_2 the same way, but $R > 1$ so that

$$|I_2| \le \frac{2\pi R^a}{R^2 - 1} \to 0 \quad \text{as} \quad R \to \infty \quad \text{if} \quad a < 2.$$

Then we have, after letting $\rho \to 0$ and $R \to \infty$,

$$I_1 + I_3 = -2\pi i e^{i\pi a} \cos(\pi a/2),$$

which is

$$\int_0^\infty \frac{x^{a-1}\,dx}{1+x^2} + \int_\infty^0 \frac{x^{a-1}e^{i2\pi a}\,dx}{1+x^2} = -2\pi i e^{i\pi a} \cos(\pi a/2)$$

or

$$(1 - e^{i2\pi a}) \int_0^\infty \frac{x^{a-1}\,dx}{1+x^2} = -2\pi i e^{i\pi a} \cos(\pi a/2).$$

Thus

$$\int_0^\infty \frac{x^{a-1}\,dx}{1+x^2} = \frac{-2\pi i e^{i\pi a} \cos(\pi a/2)}{1 - e^{i2\pi a}}$$
$$= \pi \cos(\pi a/2) \Big/ \left(\frac{e^{-\pi i a} - e^{i\pi a}}{-2i} \right)$$
$$= \frac{\pi \cos(\pi a/2)}{\sin \pi a} = \frac{\pi}{2\sin(\pi a/2)}.$$

Thus we have finally,

$$\int_0^\infty \frac{x^{a-1}\,dx}{1+x^2} = \frac{\pi}{2} \csc\left(\frac{\pi a}{2} \right), \quad 0 < a < 2.$$

If, in this integral we make the substitution $x^2 = t$ we get

$$\int_0^\infty \frac{x^{a/2-1}\,dx}{1+x} = \pi\csc(\pi a/2), \quad 0 < a < 2$$

or, equivalently

$$\int_0^\infty \frac{x^{a-1}\,dx}{1+x} = \pi\csc\pi a, \quad 0 < a < 1.$$

We can evaluate this latter integral directly from the residue theorem, which we do to illustrate a variation of the method we used in Example 1.

Example 2. Evaluate $\displaystyle\int_0^\infty \frac{x^{a-1}}{1+x}\,dt$, a real, and determine the values of a for which the evaluation holds.

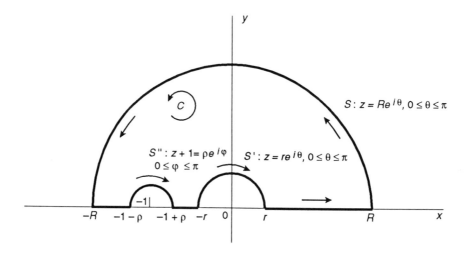

Solution 1. Let us choose C to be our standard semicircular contour indented at -1 *and indented at* 0, because 0 is not a point of analyticity of the integrand. Then we choose z^{a-1} to be z^a/z where z^a means the *principal branch*. Then z^a is real along the segment r to R of the positive real axis.

With these choices of C and of the branch of z^a, the integrand is analytic in and on C, so Cauchy's theorem gives

$$0 = \int_C \frac{z^{a-1}dz}{1+z}$$

$$= \left\{ \int_S + \int_{-R}^{-1-\rho} + \int_{S''} + \int_{-1+\rho}^{-r} + \int_{S'} + \int_r^R \right\} \frac{z^{a-1}}{1+z}\, dz$$

$$= I_1 + I_2 + I_3 + I_4 + I_4 + I_6, \text{ respectively,}$$

and we proceed to examine these integrals individually. Let us take I_3 first.

Near $z = -1$, we have

$$\frac{z^{a-1}}{z - (-1)} = \frac{e^{i\pi(a-1)}}{z - e^{i\pi}} + \frac{z^{a-1} - e^{i\pi(a-1)}}{z - e^{i\pi}}$$

$$= \frac{-e^{i\pi a}}{z+1} + g(z)$$

where $g(z)$ is analytic and bounded near $z = -1$. Thus, as before we have

$$I_3 = \int_{S''} \frac{z^{a-1}dz}{z+1} \to \pi i e^{i\pi a} \quad \text{as} \quad \rho \to \infty$$

We next consider I_5. Estimating as we did in Example 1 we get

$$|I_5| = \left| \int_\pi^0 \frac{r^{a-1}e^{i\theta(a-1)} \cdot re^{i\theta}id\theta}{1+re^{i\theta}} \right|$$

$$\leq \frac{\pi\rho^a}{1-\rho} \to 0 \quad \text{as} \quad \rho \to 0 \quad \text{if} \quad a > 0.$$

Similarly

$$|I_1| \leq \frac{\pi R^a}{R-1} \to 0 \quad \text{as} \quad R \to \infty \quad \text{if} \quad a < 1.$$

We have then, after $\rho \to 0$ and $R \to \infty$

$$0 = P_{-1} \int_{-\infty}^0 \frac{z^{a-1}dz}{1+z} + \int_0^\infty \frac{x^{a-1}dx}{1+x} + i\pi e^{i\pi a}$$

But on the negative x axis we have $z = xe^{i\pi}$, so that

$$P_{-1} \int_{-\infty}^0 \frac{z^{a-1}dz}{1+z} = P_1 \int_\infty^0 \frac{x^{a-1}e^{i\pi(a-1)}e^{i\pi}dx}{1+xe^{i\pi}}$$

$$= e^{i\pi a}\left(P_1 \int_0^\infty \frac{x^{a-1}dx}{1-x} \right)$$

Thus

$$e^{i\pi a} \left(P_1 \int_0^\infty \frac{x^{a-1} dx}{1-x} \right) + \int_0^\infty \frac{x^{a-1} dx}{1+x} = -i\pi e^{i\pi a} \tag{3}$$

Equating imaginary parts of (3) we get

$$\sin \pi a \left(P_1 \int_0^\infty \frac{x^{a-1} dx}{1-x} \right) = -\pi \cos \pi a$$

or

$$P_1 \int_0^\infty \frac{x^{a-1} dx}{1-x} = -\pi \cot \pi a, \quad 0 < a < 1$$

Then, multiplying (3) by $e^{-i\pi a}$, and again equating imaginary parts we get

$$\sin \pi a \int_0^\infty \frac{x^{a-1} dx}{1+x} = \pi$$

or, as in Example 1,

$$\int_0^\infty \frac{x^{a-1} dx}{1+x} = \pi \csc \pi a, \quad 0 < a < 1.$$

Solution 2. We consider

$$\int_C \frac{z^{a-1} dz}{1+z}$$

where C is the "keyhole contour" obtained after combining C_1 and C_2 in the solution of Example 1. More precisely, the same procedure used in example 1 applies here. Since we now have a single pole at $-1 = e^{i\pi}$ with residue $e^{i\pi(a-1)} = -e^{i\pi a}$, we have

$$\int_C \frac{z^{a-1} dz}{1+z} = -2\pi i e^{i\pi a}.$$

Then, as before,

$$\int_S \frac{z^{a-1} dz}{1+z} \to 0 \quad \text{as} \quad R \to \infty \quad \text{if} \quad a < 1$$

and

$$\int_{S'} \frac{z^{a-1} dz}{1+z} \to 0 \quad \text{as} \quad \rho \to 0 \quad \text{if} \quad a > 0.$$

and we have left the contributions from U and L. Thus we get

$$\int_0^\infty \frac{x^{a-1}dx}{1+x}[1 - e^{2\pi i a}] = -2\pi i e^{i\pi a}$$

or

$$\int_0^\infty \frac{x^{a-1}dx}{1+x} = \frac{-2\pi i e^{i\pi a}}{1 - e^{2\pi i a}}$$

$$= \frac{\pi}{(e^{-i\pi a} - e^{i\pi a})/2i}$$

$$= \frac{\pi}{\sin \pi a} = \pi \csc \pi a.$$

The methods of this section apply to other multivalued functions. Thus some integrals involving logarithms occur in the exercises.

The "keyhole" contour is not, strictly speaking, a contour at all since it is not a simple curve. However, if we view it as a curve on the Riemann surface of the function z^a, it is a simple curve, and hence, on that surface, is a contour. This suggests, and it is true, that the Cauchy integral theorem, the Cauchy formula, and the residue theorem are valid for contours on Riemann surfaces.

Significant Topics **Page**

B Exercises

Evaluate and determine the real values of a for which the evaluation is correct.

1. $\displaystyle\int_0^\infty \frac{x^{a-1}}{(1+x^2)^2}\,dx.$ 2. $\displaystyle\int_0^\infty \frac{x^{a-1}}{(1+x)^2}\,dx.$

3. For $|a| < 1$ and $0 < \alpha < \pi$ show

$$\int_0^\infty \frac{x^a\,dx}{1 + 2x\cos\alpha + x^2} = \frac{\pi\sin a\alpha}{\sin a\pi \sin\alpha}$$

Hint: The denominator factors into $(x + e^{i\alpha})(x + e^{-i\alpha})$.

4. (a) By integrating $(\text{Log } z)^2/(z^2 + 1)$ around the usual semicircular contour indented at 0 deduce

$$\int_0^\infty \frac{(\ln x)^2}{1 + x^2}\, dx = \pi^3/8; \qquad \int_0^\infty \frac{\ln x}{1 + x^2}\, dx = 0$$

(b) By writing $\int_0^\infty = \int_0^1 + \int_1^\infty$ and substituting $x = 1/t$ in $\int_1^\infty$

show that the second integral of part (a) is elementary.

5. Integrate $z^{1/2} \log z$ around the "keyhole contour" and deduce the values of

$$\int_0^\infty \frac{t^{1/2} \ln t}{(1 + t)^2}\, dt \quad \text{and} \quad \int_0^\infty \frac{t^{1/2}}{(1 + t)^2}\, dt.$$

6. Evaluate

(a) $\displaystyle\int_0^\infty \frac{\ln x}{(1 + x^2)^2}\, dx;$ (b) $\displaystyle\int_0^\infty \frac{\ln x}{(a^2 + x^2)^2}\, dx.$

C Exercises

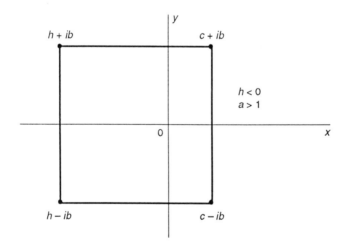

Exercise C1

1. For $c > 0$ and $a > 0$ integrate a^z/z^2 around a rectangle cornering at $c \pm ib$, and $h \pm ib$ and deduce

$$\frac{1}{2\pi i} \int_{c-i\infty}^{c+i\infty} \frac{a^z}{z^2}\, dz = \begin{cases} \ln a, & a > 1, \\ 0, & 0 < a \le 1. \end{cases}$$

Hint: For $a > 1$ take $h < 0$, for $0 < a \le 1$ Take $h > 0$.

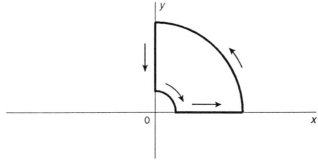

Exercise C2

2. Integrate $e^{-\zeta}\zeta^{z-1}$ with respect to ζ around the illustrated contour and deduce, for $0 < \mathrm{Re}\, z < 1$,

$$\int_0^\infty t^{z-1} \cos t \, dt = \Gamma(z) \cos(\pi z/2),$$

$$\int_0^\infty t^{z-1} \sin t \, dt = \Gamma(z) \sin(\pi z/2).$$

Extend the second formula to $|\mathrm{Re}\, z| < 1$ by analytic continuation.

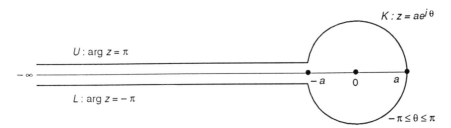

Exercise C3

3. **(Hankel's Integral)** Let C be the illustrated "keyhole" contour consisting of the lower branch L of a cut from $-\infty$ to $-a$ along the negative x-axis where $\arg z = -\pi$, the circle K given b y $z = ae^{i\theta}$, $-\pi \le \theta \le \pi$, and the upper branch U of the cut where $\arg z = \pi$ from $-a$ to $-\infty$. Thus $C = L \cup K \cup U$. Define

$$F(\alpha) = \frac{1}{2\pi i} \int_C e^z z^{-\alpha} dz.$$

(a) By writing $\displaystyle\int_C = \int_L + \int_K + \int_U$, show that $F(\alpha)$ is an entire function of α.

(b) Show that

$$F(-n) = 0, \qquad n = 0, 1, 2, \ldots,$$
$$F(n) = \frac{1}{(n-1)!}, \qquad n = 1, 2, 3, \ldots,$$

(c) Take α real, $0 < \alpha < 1$ and show that the integral over $K \to 0$ as $a \to 0$ and deduce that $F(\alpha) = \Gamma(1-\alpha)/\pi \csc \pi\alpha$.

(d) Use Exercise C6(b) of Section 3.6 to get

$$\frac{1}{\Gamma()\alpha} = \frac{1}{2\pi i} \int_C e^z z^{-\alpha} dz.$$

(e) Use part (a) above to show that $\Gamma(\alpha) = 0$ has no solutions. (See Exercise C6(e) of Section 3.6.)

(f) Show that

$$B(\alpha, \beta) = \frac{\Gamma(\alpha)\Gamma(\beta)}{\Gamma(\alpha + \beta)}$$

for all $\alpha \neq -n$, $n = 0, 1, 2, \ldots$, and all $\beta \neq -m$, $m = 0, 1, 2, \ldots$.

4. **(Integrals and derivatives of arbitrary order.)** Suppose $f(t)$ is continuous for $0 \leq t < a \leq \infty$, and set

$$I^1 f(t) = \int_0^t f(\tau) d\tau, \quad I^2 f(t) = \int_0^t I^1 f(\tau) d\tau, \ldots .$$

(a) Show

$$I^2 f(t) = \int_0^t (t - \tau) f(\tau) d\tau$$

and, in general,

$$I^n f(t) = \frac{1}{(n-1)!} \int_0^t (t - \tau)^{n-1} f(\tau) d\tau.$$

(b) For $\operatorname{Re} \alpha > 0$, define $I^\alpha f(t)$ by

$$I^\alpha f(t) = \frac{1}{\Gamma(\alpha)} \int_0^t (t - \tau)^{\alpha - 1} f(\tau) d\tau,$$

and show that

(i) $I^\alpha f(t)$ is analytic in α;

(ii) $I^\alpha(I^\beta f(t)) = I^{\alpha+\beta} f(t)$ (if also $\operatorname{Re}\beta > 0$);

(iii) $\dfrac{d}{dt} I^{\alpha+1} f(t) = I^\alpha f(t).$

Thus $I^\alpha f(t)$ defines an "iterated" integral of order α for $\operatorname{Re}\alpha > 0$. It interpolates between $n, n + 1$, and extrapolates to complex orders. For example,

$$I^{1/2} f(t) = \frac{1}{\Gamma(\frac{1}{2})} \int_0^t (t - \tau)^{-1/2} f(\tau)d\tau.$$

We now suppose that $f(t)$ is a differentiable as we please and continue:

(c) Show, by integrating by parts

$$I^\alpha f(t) = I^{\alpha+1} f'(t) + \frac{1}{\Gamma(\alpha + 1)} t^\alpha f(0)$$

and further

$$I^\alpha f(t) = I^{\alpha+2} f''(t) + \frac{1}{\Gamma(\alpha + 2)} t^{\alpha+1} f'(0) + \frac{1}{\Gamma(\alpha)} t^\alpha f(0),$$

and

$$I^\alpha f(t) = I^{\alpha+n} f^{(n)}(t) + \cdots + \frac{1}{\Gamma(\alpha)} t^\alpha f(0).$$

(d) Use the fact that $1/\Gamma(\alpha)$ is entire and is zero for $\alpha = 0, -1, -2, \ldots,$ to show that $I^\alpha f(t)$ can be continued analytically to $\operatorname{Re}\alpha > -n$ and that

$$I^{-k} f(t) = f^{(k)}(t), \quad k = 0, 1, 2, \ldots, n - 1.$$

and (i), (ii), (iii) of part (b) continue to hold. Thus we can interpret $I^\alpha f(t)$ to be a derivative of order $-\alpha$ for $\operatorname{Re}\alpha < 0$.

(e) If $f(t)$ is infinitely differentiable, show $I^\alpha f(t)$ is an entire function of α.

(f) Find $I^{-1/2} f(t) = \left(\dfrac{d}{dt}\right)^{1/2} f(t).$

5.6 Laplace Transforms and Differential Equations

In Section 3.5 we defined the Laplace transform and calculated transforms in a few simple cases. Here we will develop some operational properties and show how transforms can be applied to the solution of differential equations.

If $f(t)$ is defined on $[0, \infty)$, then

$$F(z) = \int_0^\infty f(t)e^{zt} dt$$

is the Laplace transform of f whenever the integral exists. We will also use the notation $\mathcal{L}(f)$ for $F(z)$. We showed that if $f(t)$ is continuous and there is a real constant a, and a positive constant M for which

$$|f(t)| \le Me^{at}, \quad 0 \le t < \infty \tag{1}$$

then $F(z) = \mathcal{L}(f)$ exists and is analytic for $\text{Re } z > a$. It is an easy observation (Exercise A1) that

$$\mathcal{L}(af + bg) = a\mathcal{L}(f) + b\mathcal{L}(g)$$

if both transforms exist.

A function which satisfies an inequality of the form (1) for real a and positive M on the interval $[0, \infty)$ is said to be of **exponential order** there. We will denote the class of such functions by E. Thus $f \in E$ means that (i) $f(t)$ is continuous on $[0, \infty)$, (ii) for some positive M and some real a, $f(t)$ satisfies (1). The assumption of continuity can be weakened, and in some of the exercises we consider transforms of discontinuous functions. But the inequality (1) is very near the fastest growth as $t \to \infty$ that can be permitted. You should be aware that if $f \in E$ and $g \in E$, then in general the inequality (1) uses different constants a and M for the two functions. We also showed in Section 3.5 that if $f \in E$, then

$$F'(z) = -\int_0^\infty e^{-zt} t f(t) dt,$$

which can be written as

$$\mathcal{L}(t f(t)) = -F'(z). \tag{2}$$

By iteration we get

$$\mathcal{L}(t^n f(t)) = (-1)^n F^{(n)}(z). \tag{3}$$

If both $f(t)$ and $f'(t) \in E$ (see Exercise B3), then $\mathcal{L}(f')$ exists and we compute

$$\mathcal{L}(f') = \int_0^\infty e^{-zt} f'(t) dt$$
$$= e^{-zt} f(t) \Big|_0^\infty + z \int_0^\infty e^{-zt} f(t) dt$$
$$= z\mathcal{L}(f) - f(0) + \lim_{c \to \infty} e^{-zc} f(c).$$

To compute the limit we know that there is an a for which

$$|f(t)| \leq Me^{at},$$

and we have $x = \operatorname{Re} z > a$, so

$$|e^{-zc} f(c)| \leq Me^{-xc} e^{ac} = Me^{-(x-a)c} \to 0$$

as $c \to \infty$. We get

$$\mathcal{L}(f') = z\mathcal{L}(f) - f(0)$$

and by iteration

$$\mathcal{L}(f'') = z^2 \mathcal{L}(f) - zf(0) - f'(0),$$

provided, of course, that f'', f', and f all are in E. Assuming higher derivatives are also in E we can iterate further. (See Exercise B1.)

The notion of the **convolution** of two functions will be useful. If f and g are continuous on $[0, \infty)$, then the convolution of f with g is denoted by $f * g$ and defined by

$$f * g(t) = \int_0^t f(\tau) g(t - \tau) d\tau,$$

and it is easy to see that $f * g = g * f$, and $f * g \in E$ if f and g are. (Exercise B2)

5.6a Theorem. *If f and g are in E, then*

$$\mathcal{L}(f * g) = \mathcal{L}(f)\mathcal{L}(g) = F(z)G(z). \tag{4}$$

Proof. To see that (4) holds we compute

$$\mathcal{L}(f * g) = \int_0^\infty e^{-zt} \left[\int_0^t f(\tau)g(t - \tau)d\tau \right] dt$$

$$= \int_0^\infty \int_0^t e^{-zt} f(\tau)g(t - \tau)d\tau dt$$

$$= \int_0^\infty \int_\tau^\infty e^{-zt} f(\tau)g(t - \tau)dt d\tau$$

$$= \int_0^\infty e^{-z\tau} f(\tau) \left[\int_\tau^\infty e^{-z(t-\tau)} g(t - \tau)dt \right] d\tau.$$

The inner integral, with the substitution $t - \tau = \sigma$, becomes

$$\int_0^\infty e^{-z\sigma} g(\sigma)d\sigma = G(z)$$

and can be factored out of the outer integral which is then $F(z)$:

$$\mathcal{L}(f * g) = \int_0^\infty e^{-z\tau} f(\tau)G(z)d\tau = \left[\int_0^\infty e^{-z\tau} f(\tau)d\tau \right] G(z)$$

$$= F(z)G(z). \quad \square$$

Example 1. If $\alpha = a + ib$, compute the transform of $e^{\alpha t}$.

Solution.

$$\mathcal{L}(e^{\alpha t}) = \int_0^\infty e^{\alpha t} e^{-zt} dt = \int_0^\infty e^{-(z-\alpha)t} dt$$

$$= -\frac{1}{z - \alpha} e^{-(z-\alpha)t} \Big|_0^\infty$$

$$= \frac{1}{z - \alpha} - \lim_{t \to \infty} \frac{e^{-(z-\alpha)t}}{z - \alpha}$$

$$= \frac{1}{z - \alpha} \quad \text{if} \quad \operatorname{Re} z > a = \operatorname{Re} \alpha.$$

By (2) we get

$$\mathcal{L}(te^{\alpha t}) = \frac{1}{(z - \alpha)^2}$$

and by (3)

$$\mathcal{L}(t^n e^{\alpha t}) = \frac{n!}{(z - \alpha)^{n+1}}. \tag{5}$$

Example 2. Find a function $h(t)$ for which

$$H(z) = \mathcal{L}(h) = \frac{1}{z(z-\alpha)}.$$

Solution. From Section 3.5 we have $\mathcal{L}(1) = \frac{1}{z}$ and from above $\mathcal{L}(e^{\alpha t}) = \frac{1}{z-\alpha}$. So by Theorem 5.6a

$$h(t) = 1 * e^{\alpha t} = e^{\alpha t} * 1$$
$$= \int_0^t e^{\alpha \tau} \ell\tau = \frac{1}{\alpha} e^{\alpha \tau} \Big|_0^t$$
$$= \frac{1}{\alpha}(e^{\alpha t} - 1).$$

We can verify this by transforming:

$$\mathcal{L}(h) = \frac{1}{\alpha} \mathcal{L}(e^{\alpha t}) - \frac{1}{\alpha} \mathcal{L}(1)$$
$$= \frac{1}{\alpha}\frac{1}{z-\alpha} - \frac{1}{\alpha}\frac{1}{z} = \frac{1}{\alpha}\frac{z-(z-\alpha)}{z(z-\alpha)}$$
$$= \frac{1}{z(z-\alpha)}.$$

This last example raises the question of how to invert the transform, and whether the answer is unique. That is, can $f(t) \neq g(t)$ be such that $\mathcal{L}(f) = \mathcal{L}(g)$ so that when we try to invert, two (or more) answers are possible? The answer is that if $f(t)$ and $g(t)$ are in E and $F(z) \equiv G(z)$, then $f(t) \equiv g(t)$, and if $F(z)$ is known, we can recapture $f(t)$ by

$$f(t) = \frac{a}{2\pi i} \int_{c-i\infty}^{c+i\infty} F(z) e^{zt} dz \tag{6}$$

where the integral is over a vertical line in the region of analyticity of $F(z)$. This formula is called the **inverse Laplace transform** and is denoted by $\mathcal{L}^{-1}(F)$. Thus, symbolically, we have

$$f(t) = \mathcal{L}^{-1}(F(z)).$$

In this example we have found one function, namely $(e^{\alpha t} - 1)/\alpha$, whose transform is $1/[z(z-\alpha)]$. But, since not all problems have just one solution (e.g., quadratic equations) there might be others. However, we do have only one solution here. We state a uniqueness theorem without proof.

5.6b Theorem. *If $f \in E$ and $g \in E$ and $\mathcal{L}(f) = \mathcal{L}(g)$, then $f(t) \equiv g(t)$.*

Thus we know that if $F(z)$ is a transform of a function $f(t)$ in E, there is only one such $f(t)$ which we call the **inverse Laplace transform** of $F(z)$ and write

$$f(t) = \mathcal{L}^{-1}(F(z)).$$

We can now seek a formula for $\mathcal{L}^{-1}(F(z))$, that is, a formula which recaptures $f(t)$ from a knowledge of $F(z)$. This formula is

$$f(t) = \frac{1}{2\pi i} \int_{c-i\infty}^{c+i\infty} e^{zt} F(z) dz \tag{7}$$

where the integral is over any vertical line $x = c$ in the half-plane $\operatorname{Re} z > a$ when $|f(t)| \leq Me^{at}$. We will not prove this general formula, but will establish a simple formula valid for a restricted class of functions.

5.6c Theorem. *If $f \in E$ and $\mathcal{L}(f) = F(z)$ where $F(z)$ is a rational fraction, then*

$$f(t) = \frac{1}{2\pi i} \int_C F(z) e^{zt} dz \tag{8}$$

where C is any contour with all poles of $F(z)$ in $I(C)$.

Proof. Since $F(z)$ is a rational function, it must be a proper one (Exercise A7). So, by Theorem 5.1f, we can write it in the form

$$F(z) = \sum_{j=1}^{k} \sum_{n=1}^{p_j} \frac{a_{jn}}{(z - \beta_j)^n}$$

where $\{\beta_j\}_1^k$ are the distinct poles of $F(z)$ and p_j is the order of β_j. Then we define $g(t)$ by

$$g(t) = \frac{1}{2\pi i} \int_C F(z) e^{zt} dt = \sum_{j=1}^{k} \sum_{n=1}^{p_j} \frac{a_{jn}}{2\pi i} \int_C \frac{e^{zt} dz}{(z - \beta_j)^n}.$$

We calculate

$$\frac{1}{2\pi i} \int_C \frac{e^{zt} dz}{(z - \beta_j)^n} = \frac{e^{\beta_j t}}{2\pi i} \int_C \frac{e^{(z - \beta_j)t} dz}{(z - \beta_j)^n}$$

$$= \frac{e^{\beta_j t}}{2\pi i} \int_C \sum_{m=0}^{\infty} \frac{1}{m!} (z - \beta_j)^{m-n} t^m dz = \frac{e^{\beta_j t} t^{n-1}}{(n-1)!},$$

from which

$$g(t) = \sum_{j=1}^{k} \sum_{n=1}^{p_j} a_{jn} e^{\beta_j t} t^{n-1}/(n-1)!$$

Then we compute $\mathcal{L}(g)$:

$$\mathcal{L}(g) = \sum_{j=1}^{k} \sum_{n=1}^{p_j} \frac{a_{jn}}{(n-1)!} \mathcal{L}(e^{\beta_j t} t^{n-1})$$

$$= \sum_{j=1}^{k} \sum_{n=1}^{p_j} \frac{a_{jn}}{(n-1)!} \frac{(n-1)!}{(z-\beta_j)^n} = \sum_{j=1}^{k} \sum_{n=1}^{p_j} \frac{a_{jn}}{(z-\beta_j)^n}$$

$$= F(z)$$

where the calculation of $\mathcal{L}(e^{\beta_j} t^{n-1})$ is by equation (5). Thus $\mathcal{L}(f) = \mathcal{L}(g) = F(z)$ and by the uniqueness of the inverse transform (unproven here, though true) we conclude that $g(t) \equiv f(t)$. This establishes (8). □

We now want to apply the transform to the solution of differential equations. Since $\mathcal{L}$ converts differentiation into multiplication by z (with a contribution from $f(0)$), we expect a differential equation to transform into an algebraic equation which is easier to solve. The pattern is shown in the diagram where DE means differential equation and AE means algebraic equation.

$$DE \xrightarrow[IV]{Want} \text{Soln. } DE$$
$$\mathcal{L} \downarrow I \qquad III \uparrow \mathcal{L}^{-1}$$
$$AE \xrightarrow{II} \text{Soln. } AE$$
$$I + II + III \to IV.$$

Let us now come to a differential equations problem. We seek to solve the initial value problem (IVP)

$$u'(t) + \alpha u(t) = f(t), \quad u(0) = a, \quad 0 \le t < \infty.$$

Since we are going to use the transform, we need to know that $f(t)$ is transformable, so we consider only the case where $f(t)$ is a known (given) function in E. We further assume a and α are known constants, and we seek a solution $u(t)$ for which both u and u' are in E.

Though this is a particularly simple case, it (with $\alpha = R/L$) describes the current response to an e.m.f. of Lf of an electrical circuit with an inductance L in series with a resistance R. Furthermore it illustrates in a simple way the steps outlined in our solution diagram above.

I. We transform both sides of the DE:

$$zU(z) - a + \alpha U(z) = F(z).$$

II. We solve for $U(z)$:

$$U(z) = F(z) \cdot \frac{1}{z + \alpha} + \frac{a}{z + \alpha}.$$

III. We take inverse transforms:

$$u(t) = \mathcal{L}^{-1}(U) = \mathcal{L}^{-1}\left(F(z) \cdot \frac{1}{a + \alpha}\right) + \mathcal{L}^{-1}\left(\frac{a}{z + \alpha}\right)$$

$$= f * e^{-\alpha t} + ae^{-\alpha t}$$

$$= \int_0^t f(\tau)e^{-\alpha(t-\tau)}d\tau + ae^{-\alpha t}.$$

IV. Thus we arrive at our solution:

$$u(t) = e^{-\alpha t}\int_0^t f(\tau)e^{\alpha\tau}d\tau + ae^{-\alpha t}. \tag{9}$$

We verify the solution: Clearly, $u(0) = a$ since $e^{-\alpha \cdot 0} = 1$ and the integral vanishes at $t = 0$. We compute

$$u'(t) = e^{-\alpha t}f(t)e^{\alpha t} - \alpha e^{-\alpha t}\int_0^t f(\tau)e^{\alpha\tau}d\tau - \alpha ae^{-\alpha t}$$

$$= f(t) - \alpha\left[e^{\alpha t}\int_0^t f(\tau)e^{\alpha\tau}d\tau + ae^{-\alpha t}\right]$$

$$= f(t) - \alpha u(t)$$

or

$$u'(t) + \alpha u(t) = f(t),$$

so the formula we found does indeed give a solution to our I.V.P.

Now let us examine our solution. First we have the solution as a sum of two parts, say $u = v + w$ where

$$v = f * e^{\alpha t} \quad \text{and} \quad w = ae^{\alpha t}.$$

We observe

$$w' - \alpha w = 0, \qquad w(0) = a,$$

and

$$v' - \alpha v = f, \qquad v(0) = 0.$$

Thus one part satisfies the homogeneous equation and the given initial conditions, and the other part satisfies the non-homogeneous equation with zero initial conditions.

Further, since the Laplace transform operates on functions defined on the interval $[0, \infty)$, we have no reason to expect it to provide a solution for negative values of t. But the formula (9) for $u(t)$ defines a function on $(-\infty, \infty)$, provided, of course, that $f(t)$ is defined there. Or if $f(t)$ is defined say on $(-75, \infty)$, then $u(t)$ is defined there and, by our calculation, satisfies the differential equation there. (The question of the physical sense for the circuit problem is another matter.)

Still further, in order to apply the transform to the differential equation we had to assume that $f(t)$ had a transform, that it was in E, and this involves $f(t)$ for all $t \geq 0$. Thus for instance $f(t) = e^{t^2}$ would not be an admissible f, for it grows too large, nor would $f(t) = 1/(t-1)$ for this function has a non-integrable singularity at $t = 1$. But again, if we examine our formula (9), we see that for e^{t^2}, the solution is valid for all t, and for $1/(t-1)$ it is valid for $0 < t < 1$. Indeed, in this latter case the solution is valid for $-\infty < t < 1$. Those conditions of growth rate and the domain of f were imposed by the *method* and were not a part of the problem.

Thus the transform procedure has delivered into our hands a formula which was derived under certain restrictions. An examination of the solutions formula *on its own merit* has shown that the formula represents a valid solution under conditions which were not covered in its derivation, and so provides an extension of the conditions under which we can solve the original problem. This is not unusual. There are other transforms (e.g., the Fourier Transform) for which this situation occurs as well. Thus, whenever a solution is found by methods which are *not inherent in the problem itself*, it pays to examine the solution to see if its range of validity can be extended beyond those permitted under the *method of solution*.

The differential equations governing the behavior of complicated electrical networks can be very difficult to handle. One type of problem arising is of the form of the IVP

$$P\left(\frac{d}{dt}\right) u = 1, \quad u(0) = u'(0) = \cdots = u^{(n-1)}(0) = 0$$

where P is a polynomial of degree n in the differential operator d/dt. This becomes

$$P(z)U(z) = \frac{1}{z}$$

under the action of the Laplace transform. Then

$$U(z) = \frac{1}{zP(z)}$$

and the solution is

$$u(t) = \mathcal{L}^{-1}\left(\frac{1}{zP(z)}\right).$$

To factor the polynomial and get a partial fraction expansion so that the inverse transform may be computed can be difficult. There is an alternative approach, due in substance to the British physicist Oliver Heaviside, which may sometimes be useful.

Since $zP(z)$ is of degree $n+1$, then $1/(zP(z))$ has, counting multiplicities, $n+1$ poles. Thus there is an r for which $1/(zP(z))$ is analytic for $|z| > r$, and so $1/(zP(z))$ has a Laurent expansion

$$\frac{1}{zP(z)} = \sum_{k=n}^{\infty} \frac{a_k}{z^{k+1}},$$

which converges uniformly for $|z| \geq R$, for any $R > r$. The series begins with the $1/z^{n+1}$ term since $zP(z)$ has degree $n+1$. By Theorem 5.6c

$$u(t) = \mathcal{L}^{-1}\left(\frac{1}{zP(z)}\right) = \frac{1}{2\pi i}\int_{|z|=R}\frac{e^{zt}dz}{zP(z)}$$

$$= \frac{1}{2\pi i}\int_{|z|=R}e^{zt}\left[\sum_{n}^{\infty}\frac{a_k}{z^{k+1}}\right]dz.$$

By uniform convergence

$$u(t) = \sum_{n}^{\infty}a_k\frac{1}{2\pi i}\int_{|z|=R}e^{zt}\frac{dz}{z^{k+1}}$$

$$= \sum_{n}^{\infty}a_k\mathcal{L}^{-1}\left(\frac{1}{z^{k+1}}\right)$$

$$u(t) = \sum_{n}^{\infty}a_k\frac{t^k}{k!}.$$

Thus we have a power series solution for our problem, and have proved the following.

4.6d Theorem. (Heaviside) *The IVP*

$$P\left(\frac{d}{dt}\right)u = 1, \quad u(0) = u'(0) = \cdots = u^{(n-1)}(0) = 0,$$

where P is a polynomial of degree n, has a power series solution of the form

$$u(t) = \sum_{n}^{\infty} a_k \frac{t^k}{k!}$$

where the a_k's are the Laurent coefficients as described above.

This does not answer all difficulties, for the a_k's may be difficult to compute, and the series difficult to sum. It replaces the difficulty of getting a partial fraction development for $1/(zP(z))$ by the difficulty of coping with the series.

We give below a short table of the operational properties of $\mathcal{L}$ and those few functions whose transforms we have established. There are books containing copious tables available for those who use the transform extensively.

$f(t)$	$F(z) = \mathcal{L}(f) = \displaystyle\int_0^\infty e^{-zt} f(t)dt$
$af + bg$	$aF(z) + bG(z)$
$tf(t)$	$-F'(z)$
$t^n f(t)$	$(-1)^n F^{(n)}(z)$
$f'(t)$	$zF(z) - f(0)$
$f''(t)$	$z^2 F(z) - zf(0) - f'(0)$
$f * g$	$F(z)G(z)$
1	$1/z$
e^{at}	$1/(z - a)$
$t^n e^{at}$	$n!/(z - a)^{n+1}$
$\cos at$	$z/(z^2 + a^2)$
$\sin at$	$a/(z^2 + z^2)$
$t \cos at$	$(z^2 - a^2)/(z^2 + a^2)^2$
$t \sin at$	$2az/(z^2 + a^2)^2$
$\dfrac{1}{2\pi i}\displaystyle\int_C F(z)e^{zt}dz$	$F(z)$

A Exercises

1. Show that $\mathcal{L}(af + bg) = a\mathcal{L}(f) + b\mathcal{L}(g)$ for a, b constant and f, g in E.
2. If f and g are continuous on $[0, \infty)$, show that $f * g = g * f$.
3. If f and g are in E, show that $f * g \in E$.
4. Show that if $f \in E$ and $\mathcal{L}(f(t)) = F(z)$, then $\mathcal{L}(e^{\beta t} f(t)) = F(z - \beta)$.
5. Solve the following initial value problems by Laplace transforms and verify your solutions.
   (a) $u' - 3u = t$, $u(0) = 0$;
   (b) $u' + 3u = e^{-3t}$, $u(0) = 0$;
   (c) $u'' + 3u' + 2u = 1$, $u'(0) = u(0) = 0$;
   (d) $u'' - 3u' + 2u = 1$, $u'(0) = u(0) = 0$;
   (e) $u'' - 3u' + 2u = e^{-t}$, $u'(0) = u(0) = 0$.

6. If $f \in E$ show $\mathcal{L}\left(\int_0^t f(\tau)d\tau \right) = F(z)/z$.

7. If $f \in E$ with $|f(t)| \le Me^{at}$, show that $F(z) \to 0$ as $\operatorname{Re} z \to \infty$, and even more, that $|F(z)| \le \dfrac{M}{x - a}$, $x = \operatorname{Re} z > a$.

B Exercises

1. If $f^{(k)}(t) \in E$ for $k = 0, 1, 2, \ldots, n$, then show $\mathcal{L}(f^{(n)}) = z^n \mathcal{L}(f) + p(z)$ where $p(z)$ is a polynomial of degree $n - 1$. Find $p(z)$.
2. If $f \in E$, show that $tf(t) \in E$.

3. To see that if f is differentiable and is in E, then f' may not be, examine $f(t) = \sin(e^{t^2})$.

4. Though h_c defined by

$$h_c(t) = \begin{cases} 0 & 0 \le t < c \\ 1 & t \ge c \end{cases}$$

   is not continuous, show that $\mathcal{L}(h_c)$ exists and calculate it.

5. Use Heaviside's method to solve

$$u'' + a^2 u = 1, \qquad u'(0) = u(0) = 0.$$

Show that this leads to

$$u(t) = \frac{1}{a^2} \left[1 - \cos at\right]$$

and verify that this is correct.

Chapter 6

Potential Theory

6.1 Poisson and Schwarz Integrals

We have seen how Laplace's equation arises in applied problems and have had some examples of solutions. We now want to give an elementary systematic account of the theory of this equation in two dimensions. That is, we study logarithmic potential theory. Some of the early parts of this chapter will overlap with some previous work.

The study of Laplace's equation

$$u_{xx} + u_{yy} = 0, \tag{1}$$

which we write symbolically as

$$\Delta u = 0,$$

is called **potential theory**. The subject takes its name from the fact that certain potentials, e.g., gravitational or electrostatic, satisfy this equation in three dimensions, and modified forms of them satisfy it in two. The equation also arises in fluid dynamics, in diffusion theory, and in the theory of heat conduction. One of the main reasons why complex variables is an important subject for applied people is that two-dimensional solutions of (1) are characterized by being the real parts of analytic functions. In this section we develop certain basic formulas involving harmonic functions (solutions of (1)) in a half-plane and in a disk.

324

We begin with the half-plane which we take to be $y > 0$, that is, the upper half-plane. Since this is an unbounded region we are going to have to impose conditions at ∞ on our functions. This can involve a certain unfamiliar manner of expression. For example if we have a continuous, or even piecewise continuous, function f on the real axis, and we want to say that both

$$\int_1^\infty |f(t)/t|dt \quad \text{and} \quad \int_{-\infty}^{-1} |f(t)/t|dt \tag{2}$$

converge, we can combine these two conditions into one by saying

$$\int_{-\infty}^\infty \frac{|f(t)|}{1+|t|} \, dt \tag{3}$$

converges. (See Exercise A1.)

Also by saying that $f(z)$ **converges uniformly to zero** ($f(z) \to 0$ uniformly) as $z \to \infty$ for $y \geq 0$ we mean that for each $\varepsilon > 0$ there is an $R = R(\varepsilon)$ for which $|f(z)| < \varepsilon$ if $y \geq 0$ and $|z| > R$.

6.1a Theorem. *Suppose*

   (i) $f(z)$ *is analytic for* $y \geq 0$,

   (ii) $f(z) \to 0$ *uniformly as* $z \to \infty$ *for* $y \geq 0$,

   (iii) $\displaystyle\int_{-\infty}^\infty \frac{|f(t)|dt}{1+|t|}$ *converges.*

Then

$$\frac{1}{2\pi i} \int_{-\infty}^\infty \frac{f(t)dt}{t-z} = \begin{cases} f(z) & y > 0 \\ 0 & y < 0. \end{cases} \tag{4}$$

Proof. We take the usual semicircular contour C consisting of an interval $[-b, b]$ on the real axis and the semicircle S described by $z = be^{i\theta}, 0 \leq \theta \leq \pi$. We choose z with $y \neq 0$, and choose $b > |z|$. Then by the Cauchy integral theorem we have

$$\frac{1}{2\pi i} \int_C \frac{f(t)dt}{t-z} = \begin{cases} f(z), & y > 0 \\ 0, & y < 0. \end{cases}$$

Condition (iii) ensures that the integral in formula (4) converges (how?). It remains to show that the integral over S goes to zero as $b \to \infty$. So, given $\varepsilon > 0$, we choose $b > R(\varepsilon/2)$ (condition (ii)) and $|b| > 2|z|$:

$$\left| \frac{1}{2\pi i} \int_S \frac{f(t)dt}{t-z} \right| \leq \frac{1}{2\pi} \cdot \frac{\varepsilon/2}{b-|b|} \cdot 2\pi b$$

$$\leq \frac{\varepsilon}{2} \cdot \frac{b}{b-|z|} < \frac{\varepsilon}{2} \cdot \frac{b}{b-b/2} = \varepsilon. \quad \square$$

In the following theorem we use a symmetry property which involves certain judicious cancellations. The procedure is called the **method of electric images** from its meaning in electrostatics.

6.1b Theorem. *If $f(z) = u + iv$ satisfies the conditions of Theorem 6.1a, then for $y > 0$,*

$$u(x, y) = \frac{y}{\pi} \int_{-\infty}^{\infty} \frac{u(t, 0)dt}{(x - t)^2 + y^2} \tag{5}$$

and

$$v(x, y) = \frac{1}{\pi} \int_{-\infty}^{\infty} \frac{(x - t)u(t, 0)dt}{(x - t)^2 + y^2} \tag{6}$$

Proof. Choose z with $y > 0$. Then by (4)

$$f(z) = \frac{1}{2\pi i} \int_{-\infty}^{\infty} \frac{f(t)}{t - z} \, dt$$

$$0 = \frac{1}{2\pi i} \int_{-\infty}^{\infty} \frac{f(t)}{t - \bar{z}} \, dt.$$

We subtract and get

$$\begin{aligned}
f(z) &= \frac{1}{2\pi i} \int_{-\infty}^{\infty} \left(\frac{1}{t - z} - \frac{1}{t - \bar{z}} \right) f(t)dt \\
&= \frac{1}{2\pi i} \int_{-\infty}^{\infty} \frac{(z - \bar{z})f(t)dt}{(t - z)(t - \bar{z})} \\
&= \frac{y}{\pi} \int_{-\infty}^{\infty} \frac{f(t)dt}{|t - z|^2} \\
&= \frac{y}{\pi} \int_{-\infty}^{\infty} \frac{[u(t, 0) + iv(t, 0)]dt}{(x - t)^2 + y^2}.
\end{aligned}$$

Taking real parts gets (5). If we add rather than subtract, we get (6); the details are left to the exercises (A5). □

There is an extension of formula (5) which admits a larger class of functions and extends the validity of this formula to the case where $f(z)$ may grow like $|z|^a$, $a < 1$ as $|z| \to \infty$ in the upper half-plane, and where the analyticity may break down on the real axis. See Exercise C2.

Formula (5) is called the **Poisson integral** for the half-plane, and (6) is called the **conjugate Poisson integral** for the half-plane. The function

$$p(t; x, y) = p(t; z) = \frac{y}{\pi} \cdot \frac{1}{(x - t)^2 + y^2} = \frac{\operatorname{Im}(z - t)}{\pi |t - z|^2}$$

is called the **Poisson kernel** for the half-plane. The function

$$q(t; x, y) = q(t, z) = \frac{1}{\pi} \cdot \frac{x - t}{(x - t)^2 + y^2} = \frac{\operatorname{Re}(z - t)}{\pi |t - z|^2}$$

is called the **conjugate Poisson kernel** for the half-plane.

Let us see what information we have: Suppose we know that $f(z)$ satisfies the conditions of Theorem 6.1a, but we know only the values of its real part on the real axis. Then (5) reconstructs the whole real part. From the real part we could solve the Cauchy–Riemann equations to get the imaginary part. But (6) does that for us: from u on the real axis we reconstruct v. Then knowing v we can form $u + iv$ to reconstruct all of $f(z)$. That is the next theorem. But first we do a calculation on the kernels.

$$
\begin{aligned}
p(t, z) + iq(t, z) &= \frac{1}{\pi} \frac{y + i(x - t)}{(x - t)^2 + y^2} \\
&= \frac{1}{\pi} \frac{i(\bar{z} - t)}{|z - t|^2} = \frac{i}{\pi} \frac{(\bar{z} - t)}{(z - t)(\bar{z} - t)} \\
&= \frac{i}{\pi} \frac{1}{z - t} = \frac{1}{\pi i} \frac{1}{t - z}.
\end{aligned} \tag{7}
$$

6.1c Corollary. *If $f(z)$ satisfies the conditions of Theorem 6.1a, then*

$$f(z) = \frac{1}{\pi i} \int_{-\infty}^{\infty} \frac{u(t, 0) dt}{t - z} \tag{8}$$

The proof is immediate from (7) and 6.1b.

We now derive similar formulas for the disk. The method of electric images works here as well, but we must use the analogue, for this case, of the conjugate $\bar{z}$ in Theorem 6.1b. If one maps a half-plane onto a disk by a bilinear transformation, then points symmetric to the bounding line of the half-plane (as z and $\bar{z}$ are for $y = 0$) go into so called **inverse points** with respect to the bounding circle of the disk. Points z and z^* are **inverse** with respect to the circle $|\zeta| = a$ if $\bar{z} z^* = a^2$. (See Example 4, Section 4.3).

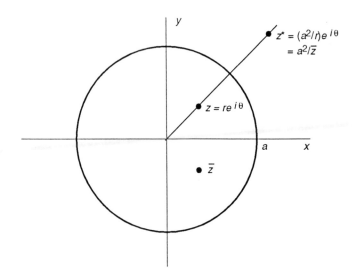

Theorem 6.1d

6.1d Theorem. *Suppose $f(z) = u(z) + iv(z)$ is analytic in and on the circle $C : |z| = a$. Then for $|z| < a$*

$$u(z) = \frac{1}{2\pi} \int_C \frac{(|\zeta|^2 - |z|^2)u(\zeta)}{|\zeta - z|^2} \frac{d\zeta}{i\zeta}, \tag{9}$$

or, in polar coordinates with $z = re^{i\theta}$, $\zeta = ae^{it}$,

$$u(t,\theta) = \frac{1}{2\pi} \int_0^{2\pi} \frac{(a^2 - r^2)u(a,t)dt}{a^2 - 2ar\cos(\theta - t) + r^2}, \quad 0 \le r < a. \tag{10}$$

Proof. If $|z| < a$, then $|z^*| = |a^2/\bar{z}| > a$ and so by Cauchy's formula

$$f(z) = \frac{1}{2\pi i} \int_C \frac{f(\zeta)d\zeta}{\zeta - z} \tag{11}$$

and

$$0 = \frac{1}{2\pi i} \int_C \frac{f(\zeta)d\zeta}{\zeta - a^2/\bar{z}}. \tag{12}$$

We subtract (12) from (11) and get

$$f(z) = \frac{1}{2\pi i} \int_C \left[\frac{1}{\zeta - z} - \frac{1}{\zeta - a^2/\bar{z}} \right] f(\zeta)d\zeta$$

$$= \frac{1}{2\pi i} \int_C \frac{(z - a^2/\bar{z})f(\zeta)d\zeta}{(\zeta - z)(\zeta - a^2/\bar{z})}$$

$$= \frac{1}{2\pi i} \int_C \frac{(|z|^2 - a^2)f(\zeta)d\zeta}{(\zeta - z)(\zeta\bar{z} - a^2)}$$

But $a^2/\zeta = \bar{\zeta}$ (how?) so

$$f(z) = \frac{1}{2\pi} \int_C \frac{(|z|^2 - |\zeta|^2)f(\zeta)}{(\zeta - z)(\bar{z} - \bar{\zeta})} \frac{d\zeta}{i\zeta}$$

$$= \frac{1}{2\pi} \int_C \frac{(|\zeta|^2 - |z|^2)f(\zeta)}{|\zeta - z|^2} \frac{d\zeta}{i\zeta}$$

But $d\zeta/i\zeta$ is real ($= dt$), so taking real parts gives (9). Then in (9) setting $z = re^{i\theta}$, $\zeta = ae^{it}$ gives, as remarked, $d\zeta/i\zeta = dt$, and also

$$|\zeta - z|^2 = (\zeta - z)(\bar{\zeta} - \bar{z})$$

$$= (ae^{it} - re^{i\theta})(ae^{-it} - re^{-i\theta})$$

$$= a^2 - are^{i(\theta - t)} - are^{-i(\theta - t)} + r^2$$

$$= a^2 - 2ar\,\cos(\theta - t) + r^2.$$

Substituting these into (9) gives (10). □

The formula (10), is the **Poisson integral** for the disk, and the function

$$P(a, t; r, \theta) = P(\zeta, z) = \frac{1}{2\pi} \frac{a^2 - r^2}{a^2 - 2ar\,\cos(\theta - t) + r^2}$$

$$= \frac{1}{2\pi} \frac{|\zeta|^2 - |z|^2}{|\zeta - z|^2}$$

is the **Poisson kernel** for the disk.

6.1e Corollary. $\int_0^{2\pi} P(a, t; r, \theta)dt = 1.$

Proof. Take $f(z) \equiv 1$ in Theorem 6.1d. □

There is a conjugate integral and kernel which can be found by adding equations (11) and (12). (Exercise A6) But we take a computationally simpler, if less direct, way to the remaining formulas analogous to those for the half-plane.

We define the **Schwarz kernel** $S(\zeta, z)$ by

$$S(\zeta; z) = \frac{1}{2\pi} \frac{\zeta + z}{\zeta - z} \tag{13}$$

We first note that

$$S(\zeta; z) = \frac{1}{\pi} \frac{\zeta}{\zeta - z} - \frac{1}{2\pi}. \tag{14}$$

Then with $\zeta = ae^{it}$, $z = re^{i\theta}$ again, we get

$$
\begin{aligned}
S(\zeta, z) &= \frac{1}{2\pi} \frac{(\zeta + z)(\bar{\zeta} - \bar{z})}{(\zeta - z)(\bar{\zeta} - \bar{z})} \\
&= \frac{1}{2\pi} \frac{a^2 - r^2}{a^2 - 2ar\,\cos(\theta - t) + r^2} \\
&\quad + \frac{i}{\pi} \frac{ar\,\sin(\theta - t)}{a^2 - 2ar\,\cos(\theta - t) + r^2}
\end{aligned}
$$

so that

$$
\begin{aligned}
S(\zeta, z) &= P(\zeta, z) + iQ(\zeta, z) \\
&= P(a, t; r, \theta) + iQ(a, t; r, \theta)
\end{aligned} \tag{15}
$$

where

$$Q(\zeta, z) = Q(a, t; r, \theta) = \frac{1}{\pi} \frac{ar\,\sin(\theta - t)}{a^2 - 2ar\,\cos(\theta - t) + r^2} \tag{16}$$

is called the **conjugate Poisson kernel** for the disk.

6.1f Theorem. *If $f(z) = u(z) + iv(z)$ is analytic in and on the circle $C : |\zeta| = a$, then for $|z| < a$*

$$f(z) = \frac{1}{2\pi} \int_C \frac{\zeta + z}{\zeta - z} u(\zeta) \frac{d\zeta}{i\zeta} + iv(0). \tag{17}$$

Proof. Let us, for now, denote the right side by $g(z)$: we must show that $f(z) = g(z)$. By (14)

$$g(z) = \frac{1}{\pi i} \int_C \frac{u(\zeta)d\zeta}{\zeta - z} - \frac{1}{2\pi i} \int_C u(\zeta) \frac{d\zeta}{\zeta}.$$

The first integral here is analytic for $|z| < a$ by Theorem 3.2b, and the second term is a constant—and equal to $-u(0)$ (how?). Thus both $f(z)$ and $g(z)$ are analytic for $|z| < a$. From (15) we get

$$g(z) = \int_0^{2\pi} P(a, t; r, \theta)u(a, t)dt + i\int_0^{2\pi} Q(a, t; r, \theta)u(a, t)dt + iv(0)$$

$$= u(z) + i\int_0^{2\pi} Q(a, t; r, \theta)u(a, t)dt + iv(0).$$

Thus $\operatorname{Re} g(z) = u(z) = \operatorname{Re} f(z)$, and so (how?)

$$f(z) = g(z) + ik,$$

where k is a constant. Then setting $z = 0$ in this last equation we get

$$f(0) = u(0) + iv(0) = g(0) + ik = u(0) + iv(0) + ik,$$

from which $k = 0$.   □

6.1g Corollary. *If $f(z) = u(z) + iv(z)$ is analytic in and on $C : |z| = a$, then for $|z| < a$*

$$v(z) - v(0) = \int_C Q(\zeta, z)u(\zeta)d\zeta/i\zeta \tag{18}$$

or, in polar coordinates,

$$v(r, \theta) - v(0) = \int_0^{2\pi} Q(a, t; r, \theta)u(a, t)dt. \tag{19}$$

Significant Topics   **Page**

A Exercises

1. Show that conditions (2) and (3) are equivalent.
2. Show that condition (iii) of Theorem 6.1a ensures that the integral in (4) converges.
3. Compute $\partial^2 p/\partial x^2$ and $\partial^2 p/\partial y^2$ and verify that $\Delta p = 0$ for each t if $(x, y) \neq (t, 0)$.
4. Show that $\displaystyle\int_{-\infty}^{\infty} p(t; x, y)dt = 1$ for $y \neq 0$.
5. Add the first two equations in the proof of Theorem 6.1b and so establish formula (6).
6. Add equations (11) and (12) and thus establish Corollary 6.1g.
7. How should Theorems 6.1a, b, c be modified to apply to (a) the half-plane $y \geq a$; (b) the half-plane $x \geq 0$; (c) the half-plane $x \geq a$?
8. How should Theorems 6.1d, f, g be modified to apply to the circle $|z - z_0| = a$?
9. Explain, in intuitive terms, why the term $v(0)$ appears in (18) and (19).

B Exercises

1. Use the methods of Section 4.4 to show that

$$\frac{1}{2\pi i} \int_{-\infty}^{\infty} \frac{e^{it}\,dt}{t - z} = \begin{cases} e^{iz}, & y > 0, \\ 0, & y < 0, \end{cases}$$

   where the integral is a Cauchy principal value.
2. Show that
   (a) $\displaystyle\frac{1}{2\pi}\frac{a - r}{a + r} \leq P(a, t; r, \theta) \leq \frac{1}{2\pi}\frac{a + r}{a - r}$ if $r < a$;
   (b) if $u(z)$ is harmonic and positive for $|z| \leq a$, then for $|z| = r < a$,

$$\frac{a - r}{a + r}\, u(0) \leq u(z) \leq \frac{a + r}{a - r}\, u(0);$$

   (c) if $\{u_n(z)\}_1^\infty$ is an increasing sequence of harmonic functions in $|z| < a$ (i.e., $u_n(z) \leq u_{n+1}(z)$) and if $\{u_n(0)\}$ is bounded, then $\{u_n(z)\}$ converges uniformly in $|z| \leq b$ for each $b < a$. Hint: $u_n(z) - u_1(z) \geq 0$.

C Exercises

1. Extend Theorem 6.1a as follows: Suppose
    (i) $f(z)$ is analytic for $y \geq 0$;
    (ii) $f(z)$ is bounded for $y \geq 0$:
    (iii) $f(R\,e^{i\theta}) \to 0$ uniformly as $R \to \infty$ for $a \leq \theta \leq \pi - a$ for each a, $0 < a < \pi/2$;
    (iv) $\displaystyle\int_{-\infty}^{\infty} \frac{|f(t)|}{1+|t|} \, dt$ converges.

    Then
    $$\frac{1}{2\pi i} \int_{-\infty}^{\infty} \frac{f(t)}{t-z} \, dt = \begin{cases} f(z), & y < 0; \\ 0, & y < 0. \end{cases}$$

2. Suppose (i) $f(z) = u(x,y) + iv(x,y)$ is continuous for $y \geq 0$ and analytic for $y > 0$; (ii) there is an $a < 1$, an $M > 0$, and an $R > 0$ for which $|f(r\,e^{i\theta})| \leq Mr^a$ for $r > R$. Show that
    $$u(x,y) = \int_{-\infty}^{\infty} p(t; x, y) u(t, 0) \, dt, \qquad y > 0.$$

    Hints: Use Theorem 3.1e to get, in the notation of the proof of Theorem 6.1a,
    $$\frac{1}{2\pi i} \int_C \frac{f(t)\,dt}{t-z} = f(z) \quad \text{and} \quad \frac{1}{2\pi i} \int_C \frac{f(t)\,dt}{t-\bar{z}} = 0$$

    for $y > 0$. Then subtract, and show that the integral over $S \to 0$ as $b \to \infty$. Also note that e^{iz}, $\sqrt{z}$, and $\sqrt{z}\mathrm{Log}\,z$ are examples of such functions.

6.2 The Dirichlet Problem for a Half-Plane

The most important problem studied in the development of potential theory is the Dirichlet problem. We now give a general formulation of this problem.

Dirichlet Problem. *Suppose R is a region with boundary B, and $g(\zeta)$ is a continuous function on B. Find a function $u(z)$ which is continuous in $\bar{R} = R \cup B$ and*
    (i) $\Delta u(z) = 0, \qquad z \in R$;
    (ii) $u(\zeta) = g(\zeta), \qquad \zeta \in B.$

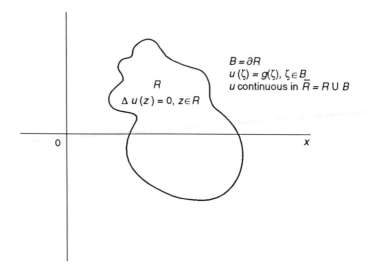

The Dirichlet Problem

Laplace's equation $u_{xx} + u_{yy} = 0$ describes, among other things, steady state temperatures in a heat conducting lamina (plate) which is insulated on its flat surfaces. Thus one (but not the only) physical interpretation of the Dirichlet problem is to seek the steady state temperature u at points z in a plate in the shape of the region R when the temperature $u = g$ is known on the uninsulated edge (boundary of R).

It would seem at first glance that we have the solution at hand in the case where R is either a half-plane or a disk. Doesn't the Poisson integral provide exactly the solution we seek? To properly answer this question you must clearly understand the distinction between our state of knowledge in Section 6.1 and the present section. In Section 6.1 we assumed that we *had at hand* analytic and harmonic functions with certain specified properties. We then concluded that certain formulas, specifically, the Poisson integrals, were valid. Now we *seek* harmonic functions having given boundary values. We don't a priori know that these functions we seek actually exist, and we don't yet know if the formulas we would like to use are valid. Our approach will be to write out the functions defined by the formulas and attempt to show that they do in fact solve the problems we want to solve.

First we consider the problem where R is the half-plane $y > 0$ with boundary $y = 0$. Then the conditions (i) and (ii) become

    (i) $\Delta u(z) = 0$, $y > 0$

(ii) $u(t) = g(t)$, $-\infty < t < \infty$.

Also we will consider a slight generalization: we will permit $g(t)$ to be piecewise continuous.

6.2a Theorem. *Suppose that $g(t)$ is piecewise continuous on $-\infty < t < \infty$ and that*

$$\int_{-\infty}^{\infty} \frac{|g(t)|dt}{1 + |t|}$$

converges. Then $u(x,y)$ is harmonic for $y > 0$ where

$$u(x,y) = \frac{y}{\pi} \int_{-\infty}^{\infty} \frac{g(t)dt}{(x-t)^2 + y^2} = \int_{-\infty}^{\infty} p(t; x, y)g(t)dt$$

Proof. The theorem states that $u_{xx} + u_{yy} = 0$. To show this we could compute the derivatives and verify that the equation is satisfied. This would involve justifying differentiation under the integral sign. To avoid this difficulty we could show that u is the real part of an analytic function. Thus, as we have seen before,

$$\frac{1}{\pi i} \frac{1}{t - z} = p(t; x, y) + i q(t; x, y)$$

and so

$$\frac{1}{\pi i} \int_{-\infty}^{\infty} \frac{g(t)dt}{t - z} = \int_{-\infty}^{\infty} p(t; x, y)g(t)dt + i \int_{-\infty}^{\infty} q(t; x, y)g(t)dt$$

The first integral on the right is $u(x,y)$, and the integral on the left is analytic by Theorem 3.5d. Thus u is the real part of an analytic function. $\square$

Remark. For (x,y) fixed, $p(t; x, y)$ goes to zero like $1/t^2$ as $t \to \infty$. Thus one might conjecture that the condition on g could be relaxed to requiring convergence of

$$\int_{0}^{\infty} \frac{|g(t)|dt}{1 + t^2}.$$

This is indeed the case; the more stringent condition was required by our *method* rather than any inherent property of harmonic functions. (See Exercise C1.)

Example 1. Find a function which is harmonic for $y > 0$ and which is 1 on the real axis from a to b, and zero on the rest of the real axis.

Solution 1. We take

$$g(t) = \begin{cases} 1, & a < t < b \\ 0, & t < a \text{ or } t > b \end{cases}$$

and evaluate

$$u(x,y) = \frac{y}{\pi} \int_{-\infty}^{\infty} \frac{g(t)dt}{(x-t)^2 + y^2} = \frac{y}{\pi} \int_a^b \frac{dt}{(x-t)^2 + y^2}$$

$$= \frac{y}{\pi} \int_{x-b}^{x-a} \frac{d\tau}{\tau^2 + y^2} = \frac{1}{\pi} \arctan \frac{x-a}{y} - \frac{1}{\pi} \arctan \frac{x-b}{y}$$

To put our answer in a more tractable form we write $z - a$ and $z - b$ in polar form:

$$z - a = \rho e^{i\phi}, \quad 0 < \phi < \pi;$$
$$z - b = r e^{i\theta}, \quad 0 < \theta < \pi.$$

Then $\arctan(x-a)/y = \pi/2 - \phi$, and $\arctan(x-b)/y = \pi/2 - \theta$. Then we get

$$u = \frac{1}{\pi} \left[\left(\frac{\pi}{2} - \phi \right) - \left(\frac{\pi}{2} - \theta \right) \right] = (\theta - \phi)/\pi$$

Solution 2. The function we seek is the real part of

$$f(z) = \frac{1}{\pi i} \int_a^b \frac{dt}{t - z} = \frac{1}{\pi i} \left[\text{Log}\,(z - b) - \text{Log}\,(z - a) \right]$$

$$= \frac{1}{\pi i} \left[(\ln r + i\theta) - (\ln \rho + i\phi) \right]$$

$$= (\theta - \phi)/\pi - (i/\pi)\ln(r/\rho).$$

So

$$u(z) = (\theta - \phi)/\pi.$$

We want to examine the solution just found for various intervals, both finite and infinite. Let us denote the solution arising from an interval I by $u_I(z)$. Then we can characterize $u_I(z)$ as follows:

$$u_I(z) = (1/\pi) \text{ times the angle subtended by } I \text{ at } z.$$

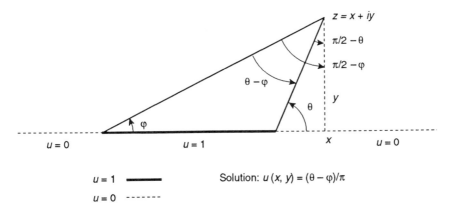

$u = 1$ ━━━ Solution: $u(x, y) = (\theta - \varphi)/\pi$

$u = 0$ ┄┄┄┄

Remark 1. If $I = (-\infty, c)$ for any finite real c, then

$$u_I = \theta/\pi, \qquad \theta = \text{Arg}\,(z - c),$$

and u_I is still harmonic for $y > 0$. (Exercise A1 (a). See also Exercise C1.)

Remark 2. If $I = (c, \infty)$ for any finite real c, then

$$u_I(z) = (\pi - \theta)/\pi, \qquad \theta = \text{Arg}\,(z - c)$$

and u_I is still harmonic for $y > 0$. (Exercise A1 (b).)

Remark 3. If x_0 is a point on the real axis and is *not* an end point of I, then

$$\lim_{z \to x_0} u_I(z) = \begin{cases} 1 & \text{if } x_0 \in I \\ 0 & \text{if } x_0 \notin I. \end{cases}$$

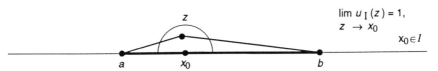

$\lim u_I(z) = 1,$
$z \to x_0$

$x_0 \in I$

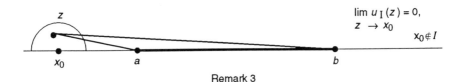

$\lim u_I(z) = 0,$
$z \to x_0$

$x_0 \notin I$

Remark 3

These remarks are geometrically clear and their proofs are consigned to the exercises. Remark 3 shows that the solution provided by the Poisson integral does indeed satisfy its boundary condition in case of Example 1.

6.2b Theorem. *Suppose $g(t)$ is bounded $(|g(t)| \leq M)$, and piecewise continuous, and $u(x, y)$ is defined to be*

$$u(x, y) = \frac{y}{\pi} \int_{-\infty}^{\infty} \frac{g(t)dt}{(x - t)^2 + y^2}$$

then $\lim_{\substack{x \to x_0 \\ y \to 0+}} u(x, y) = g(x_0)$ *at each point x_0 where g is continuous.*

Proof. Choose x_0 a point of continuity of g. Then for each $\varepsilon > 0$ there is a $\delta > 0$ for which

$$|g(t) - g(x_0)| < \frac{\varepsilon}{3} \quad \text{whenever} \quad |t - x_0| < \delta.$$

Next, choose $a < x_0$ and $b > x_0$ so that $|x_0 - a| < \delta$ and $|x_0 - b| < \delta$. Then we consider $D \equiv u(x, y) - g(x_0)$:

$$D = \frac{y}{\pi} \int_{-\infty}^{\infty} \frac{g(t)dt}{(x - t)^2 + y^2} - g(x_0)$$

$$= \frac{y}{\pi} \int_{-\infty}^{\infty} \frac{g(t)dt}{(x - t)^2 + y^2} - g(x_0)\frac{y}{\pi} \int_{-\infty}^{\infty} \frac{dt}{(x - t)^2 + y^2}$$

by Exercise A4, Section 6.1. then

$$D = \frac{y}{\pi} \int_{-\infty}^{\infty} \frac{g(t)dt}{(x - t)^2 + y^2} - \frac{y}{\pi} \int_{-\infty}^{\infty} \frac{g(x_0)dt}{(x - t)^2 + y^2}$$

$$= \frac{y}{\pi} \int_{-\infty}^{\infty} \frac{[g(t) - g(x_0)]dt}{(x - t)^2 + y^2}.$$

From which

$$|D| < \frac{y}{\pi} \int_{-\infty}^{\infty} \frac{|g(t) - g(x_0)|dt}{(x - t)^2 + y^2}$$

$$= \frac{y}{\pi} \left\{ \int_{-\infty}^{a} + \int_{a}^{b} + \int_{b}^{\infty} \right\} \frac{|g(t) - g(x_0)|}{(x - t)^2 + y^2} dt$$

$$\equiv D_1 + D_2 + D_3 \quad \text{respectively.}$$

Then, since $|g(t)| \leq M$ and $|g(x_0)| \leq M$, we have

$$D_1 \leq 2M \frac{y}{\pi} \frac{dt}{(x-t)^2 + y^2} = 2M u_{(-\infty,a)}(x,y)$$

$$D_2 \leq \frac{\varepsilon}{3} \frac{y}{\pi} \int_a^b \frac{dt}{(x-t)^2 + y^2} = \frac{\varepsilon}{3} u_{(a,b)}(x,y) \leq \frac{\varepsilon}{3}$$

$$D_3 \leq 2M \frac{y}{\pi} \int_b^\infty \frac{dt}{(x-t)^2 + y^2} = 2M u_{(b,\infty)}(x,y).$$

But by Remark 3, there is a ρ for which

$$0 \leq u_{(-\infty,a)}(x,y) < \frac{\varepsilon}{6M}$$

and

$$0 \leq u_{(b,\infty)}(x,y) < \frac{\varepsilon}{6M}$$

if $|x - x_0| < \rho$, and $0 < y < \rho$. Then under these conditions we have

$$|D| \leq D_1 + D_2 + D_2 < \frac{\varepsilon}{3} + \frac{\varepsilon}{3} + \frac{\varepsilon}{3} = \varepsilon,$$

and so the proof is complete. $\square$

We can now solve the Dirichlet problem for the half-plane, because the two previous theorems give the following.

6.2c Theorem. *If $g(t)$ is continuous for $-\infty < t < \infty$ and*

$$\int_{-\infty}^\infty \frac{|g(t)|}{1 + |t|} \, dt$$

converges, then

$$u(x,y) = \begin{cases} \dfrac{y}{\pi} \displaystyle\int_{-\infty}^\infty \dfrac{g(t)dt}{(x-t)^2 + y^2}, & y > 0 \\[4mm] g(x), & y = 0 \end{cases}$$

solves the Dirichlet problem.

Significant Topics **Page**

A Exercises

1. (a) In Remark 1 we are really taking $g(t) = 1$ for $t < c$ and $= 0$ for $t > c$. This $g(t)$ doesn't fit the conditions of Theorem 6.2a. How do we know the solution here, namely θ/π, is harmonic as asserted?

   (b) The same question for Remark 2.

2. Explain carefully the steps in the limits involved in Remark 3 and thus prove that remark.

3. Solve the Dirichlet problem

$$\Delta u(x,y) = 0, \quad y > 0$$
$$u(x,0) = 1/(1 + x^2)$$

   by evaluating the Poisson integral. Hint: The integral can be evaluated by residues.

4. Take $g(t) = t$ for $a \le t \le b$ and $= 0$ for $t < a$ and $t > b$. Evaluate

$$\int_{-\infty}^{\infty} p(t; x, y)g(t)dt$$

   and show that it is harmonic for $y > 0$.

5. Suppose $g(t)$ is continuous, or even piecewise, continuous for $-\infty < t < \infty$, and $|g(t)| \le M$. Show that

$$\left| \int_{-\infty}^{\infty} p(t; x, y)g(t)dt \right| \le M.$$

B Exercises

1. See Exercise C2 of Section 6.3.

   (a) Take

$$h_n(t) = \begin{cases} 0, & |t| > n; \\ t + n, & -n \le t \le -n + 1; \\ 1, & |t| \le n - 1; \\ n - t, & n - 1 \le t \le n \end{cases}$$

and show that each $h_n(t)$ is an admissible $g(t)$ for Theorem 6.2c.

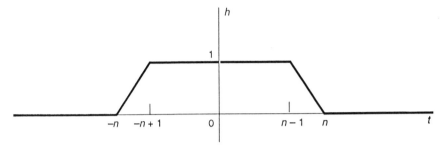

(b) Evaluate the integrals and so express

$$\int_{-\infty}^{\infty} p(t; x, y) h_n(t) dt$$

in elementary terms.

Hint: Use Exercise A4 and Example 1.

2. Suppose $u(x, y)$ is a solution to the Dirichlet problem

$$\Delta u = 0, \qquad y > 0,$$
$$u(x, 0) = g(x), \qquad -\infty < x < \infty,$$

where $g(x)$ is continuous. Show that

$$w_1(x, y) \equiv u(x, y) + ay$$

and

$$w_2(x, y) = u(x, y) + a \sin x \sinh y$$

are also solutions for each constant a. How do these examples fit with the work of Sections 6.1 and 6.2?

C Exercises

1. Suppose $g(t)$ is piecewise continuous and

$$\int_{-\infty}^{\infty} \frac{|g(t)| dt}{1 + t^2}$$

converges.

(a) Show that

$$u(x,y) = \int_{-\infty}^{\infty} p(t; x, y)g(t)dt$$

converges.

(b) Show that if differentiation under the integral sign is permissible, then $\Delta u = 0$ for $y \neq 0$.

2. Suppose $g(t)$ satisfies the conditions of Theorem 6.2c. Solve the Dirichlet problem for the *lower* half-plane:

$$\Delta u = 0, \quad y < 0$$
$$u(x,0) = g(x).$$

6.3 The Dirichlet Problem for a Disk

We turn to the Dirichlet problem for a disk. Since the boundary values are now on a circle, we state them as a function of the polar angle. Thus our conditions (i) and (ii) of Section 6.2 become

(i) $\Delta u(z) = 0$, $0 \leq r < a$;

(ii) $u(a, \theta) = g(\theta)$, $0 \leq \theta \leq 2\pi$,

with $g(0) = g(2\pi)$. We can then assume further that g is given as a periodic function: $g(\theta + 2\pi) = g(\theta)$. As in the case of the half-plane we examine the Poisson integral. Since the integral is sometimes expressed as a contour integral, it is convenient to set, as before, $\zeta = ae^{it}$, $t = \arg \zeta$, and we will denote $g(t) = g(\arg \zeta)$ by $G(\zeta)$.

6.3a Theorem. *Suppose $g(\theta)$ is periodic (period 2π) and piecewise continuous, and that $u(z) = u(r, \theta)$ is defined by*

$$u(r, \theta) = \int_0^{2\pi} P(a, t; r, \theta)g(t)dt$$

$$= \frac{1}{2\pi} \int_0^{2\pi} \frac{(a^2 - r^2)g(t)dt}{a^2 - 2ar\cos(\theta - t) + r^2}$$

or equivalently

$$u(z) = \int_{|\zeta|=a} P(\zeta, z)G(\zeta)d\zeta/i\zeta.$$

Then $\Delta u = 0$ for $0 \leq r < a$.

Proof. Again, we could compute derivatives under the integral sign, but, again, it is easier to show that u is the real part of an analytic function. By equation (15) of Section 6.1

$$S(\zeta, z) = P(a, t; r, \theta) + iQ(a, t; r, \theta).$$

We multiply by $g(t)dt = G(\zeta)d\zeta/i\zeta$ and integrate to get

$$\frac{1}{2\pi i} \int_{|\zeta|=a} \frac{\zeta+z}{\zeta-z} G(\zeta)\frac{d\zeta}{i\zeta} = \int_{|\zeta|=a} P(\zeta, z)G(\zeta)d\zeta/i\zeta +$$
$$i \int_{|\zeta|=a} Q(\zeta, z)G(\zeta)d\zeta/i\zeta \qquad (1)$$
$$= u(z) + iv(z)$$

where

$$v(z) = \int_{|\zeta|=a} Q(\zeta, z)G(\zeta)d\zeta/i\zeta.$$

But by (14) Section 6.1, the left side of (1) is

$$\frac{1}{\pi i} \int_{|\zeta|=a} \frac{G(\zeta)d\zeta}{\zeta-z} - \frac{1}{2\pi i} \int_{|\zeta|=a} G(\zeta)\,d\zeta/\zeta$$

which is analytic. By (1), $u(z)$ is the real part of an analytic function and so is harmonic. $\square$

Example. Find a function which is harmonic for $|z| < a$, and which is 1 on an arc A from α to β on $|z| = a$, and zero on the complementary arc $B = C - A$.

Solution. We take $g(t) = 1$ on A and 0 on B and calculate

$$u_A(r, \theta) = \int_\alpha^\beta P(a, t; r, \theta)dt.$$

But

$$P(a, t; r, \theta) = P(\zeta, z) = \text{Re } S(\zeta, z)$$
$$= \text{Re } \left(\frac{1}{\pi}\frac{\zeta}{\zeta-z}\right) - \frac{1}{2\pi}$$

so

$$u_A(r,\theta) = \operatorname{Re} \frac{1}{\pi} \int_\alpha^\beta \frac{\zeta}{\zeta - z} \frac{d\zeta}{i\zeta} - \frac{1}{2\pi} (\beta - \alpha)$$

$$= \operatorname{Re} \left\{ \frac{1}{\pi i} \int_\alpha^\beta \frac{d\zeta}{\zeta - z} \right\} - \frac{1}{2\pi} (\beta - \alpha)$$

$$= \operatorname{Re} \frac{1}{\pi i} [\log(\zeta - \beta) - \log(\zeta - \alpha)] - \frac{1}{2\pi} (\beta - \alpha)$$

$$= \frac{1}{\pi} (\phi(\beta) - \phi(\alpha)) - \frac{1}{2\pi} (\beta - \alpha)$$

where $\phi(t) = \arg(\zeta - z)$, and in particular $\phi(\beta) = \arg(ae^{i\beta} - z)$ and $\phi(\alpha) = \arg(ae^{i\alpha} - z)$. Thus again we can interpret our solution geometrically in terms of angles: If A is the arc on which $g = 1$, with $g = 0$ elsewhere on C, then

$$u_A(z) = (1/\pi) \text{ times the angle subtended by } A \text{ at } z$$

minus $1/2\pi$ times the angle subtended by A at the center.

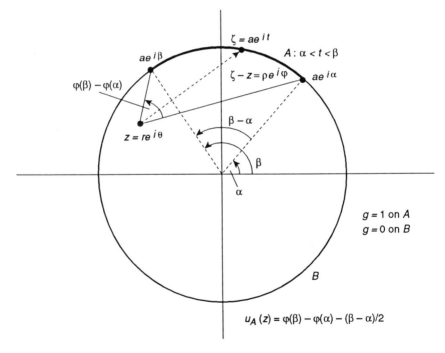

To verify that $u(z)$ achieves the proper boundary values, that is $u = g$ on C, we make use of a theorem from elementary geometry: The angle subtended by an arc A at a point ζ_0 on C but not on A, is 1/2 the angle subtended by A at the center. Thus clearly we have that if $\zeta_0 \in B$ and not an end point)

$$
\begin{cases}
\lim_{z \to \zeta_0} u_A(\zeta) = \lim_{z \to \zeta_0} (\phi(\beta) - \phi(\alpha))/\pi - (\beta - \alpha)/2\pi \\
\qquad = \dfrac{1}{2}\,(\beta - \alpha)/\pi - (\beta - \alpha)/2\pi \qquad\qquad (2) \\
\qquad = 0.
\end{cases}
$$

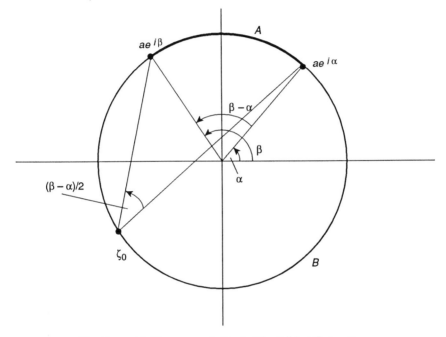

The Geometric Theorem and $\lim u_A(z) = 0$ if $\zeta_0 \in B$ $z \to \zeta_0$

And if $\zeta_0 \in A$ (and not an end point), then

$$
\begin{aligned}
\lim_{z \to \zeta_0} u_A(z) &= \lim_{z \to \zeta_0} [\phi(\beta) - \phi(\alpha)] - (\beta - \alpha)/2\pi \\
&= \frac{1}{\pi}\left[\pi + \frac{1}{2}\,(\beta - \alpha)\right] - (\beta - \alpha)/2\pi \qquad (3) \\
&= 1.
\end{aligned}
$$

Explain the steps in this limit calculation (Exercise A1).

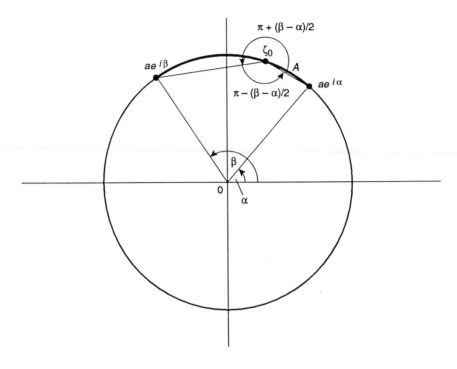

$$\lim_{z \to \zeta_0} u_A(z) = 1, \ \zeta_0 \in A,$$

6.3b Theorem. *If $g(t)$ is piecewise continuous, $0 \le t \le 2\pi$, with $g(0) = g(2\pi)$, and $u(z) = u(r, \theta)$ is defined by*

$$u(z) = u(r, \theta) = \int_0^{2\pi} P(a, t; r, \theta) g(t) dt,$$

then if $\zeta_0 = ae^{it_0}$,

$$\lim_{z \to \zeta_0} u(z) = g(t_0)$$

at each point t_0 of continuity of g.

Proof. We write $\zeta - z$ in polar form

$$\zeta - a = \rho e^{i\phi}$$

and differentiate with respect to t to get

$$\frac{d\zeta}{dt} = \rho e^{i\phi} i \frac{d\phi}{dt} + e^{i\phi} \frac{d\rho}{dt}$$

$$= \rho e^{i\phi} \left(\frac{1}{\rho} \frac{d\rho}{dt} + i \frac{d\phi}{dt} \right)$$

$$= (\zeta - z) \left(\frac{1}{\rho} \frac{d\rho}{dt} + i \frac{d\phi}{dt} \right).$$

So

$$u(z) = \mathrm{Re} \left(\frac{1}{\pi i} \int_{|\zeta|=a} \frac{g(t)d\zeta}{\zeta - a} \right) - \frac{1}{2\pi} \int_0^{2\pi} g(t)dt$$

$$= \mathrm{Re} \left(\frac{1}{\pi i} \int_0^{2\pi} g(t) \left(\frac{1}{\rho} \frac{d\rho}{dt} + i \frac{d\phi}{dt} \right) dt \right) - \frac{1}{2\pi} \int_0^{2\pi} g(t)dt$$

$$= \frac{1}{\pi} \int_0^{2\pi} g(t) \frac{d\phi}{dt} \, dt - \frac{1}{2\pi} \int_0^{2\pi} g(t)dt$$

$$= \frac{1}{\pi} \int_0^{2\pi} g(t) \left[\frac{d\phi}{dt} - \frac{1}{2} \right] dt.$$

Now choose t_0 a point of continuity of g. Then given $\varepsilon > 0$, there is a $\delta > 0$ for which $|g(t) - g(t_0)| < \varepsilon$ whenever $|t - t_0| < \delta$. Then choose $\alpha < t_0$ and $\beta > t_0$ so that $|\alpha - t_0| < \delta$. Denote the arc of C from α to β by A and the complementary arc by B. Then

$$u(z) - g(t_0) = \int_0^{2\pi} P(a,t;r,\theta)g(t) - g(t_0)$$

$$= \int_0^{2\pi} P(a,t;r,\theta)[g(t) - g(t_0)]dt$$

$$= \frac{1}{\pi} \int_A \left[\frac{d\phi}{dt} - \frac{1}{2} \right] [g(t) - g(t_0)]dt$$

$$+ \frac{1}{\pi} \int_B \left[\frac{d\phi}{dt} - \frac{1}{2} \right] [g(t) - g(t_0)]dt.$$

Then, using $|g(t)| \le M$ on B and $|g(t) - g(t_0)| \le \varepsilon/2$ on A we get

$$|u(z) - g(t_0)| \le \frac{\varepsilon}{2} \cdot \frac{1}{\pi} \int_A \left(\frac{d\phi}{dt} - \frac{1}{2} \right) dt + 2M \frac{1}{\pi} \int_B \left(\frac{d\phi}{dt} - \frac{1}{2} \right) dt$$

$$= \frac{\varepsilon}{2} \, u_A(z) + 2M u_B(z)$$

$$\le \frac{\varepsilon}{2} + 2M u_B(z)$$

By (2), there is a δ' for which $u_B(z)| < \varepsilon/4M$ if $|z - ae^{it_0}| < \delta'$. Then

$$|u(z) - g(t_0)| < \frac{\varepsilon}{2} + \frac{\varepsilon}{2} = \varepsilon. \quad \square$$

6.3c Theorem. *If $g(t)$ is continuous for $0 \leq t \leq 2\pi$, $g(0) = g(2\pi)$, then the Dirichlet problem*

$$\Delta u = 0, \qquad |z| < a$$
$$u(ae^{it}) = g(t), \qquad |z| = r < a$$

has the solution

$$u(z) = u(re^{i\theta}) = \begin{cases} \int_0^{2\pi} P(a, t; R, \theta)g(t)dt, & r < a \\ g(\theta), & r = a. \end{cases} \qquad (4)$$

We can extend this theorem to the case where we permit g to be piecewise continuous on the boundary.

6.3d Theorem. *If in 6.3c, $g(t)$ is piecewise continuous, that is, it is bounded and has a finite number of discontinuities, then (4) is valid except at the discontinuities. Furthermore it is the only bounded solution.*

Proof. The uniqueness of bounded solutions follows from the C Exercises of Section 5.5. The rest is immediate. $\square$

There is an extension of the Riemann mapping theorem which extends the mapping function to the boundary. (In fact there are many, but we will examine only one.) Suppose Γ is a Jordan curve in the w-plane and R its interior: $R = I(\Gamma)$. Let us also denote by D the unit disk $|z| < 1$ in the z-plane, and by C the unit circle $|z| = 1$. Then the theorem is the following.

6.3e Theorem. *If $z = \psi(w)$ maps R conformally and one-to-one onto D, then it can be extended to Γ so that it maps Γ one-to-one onto C. Furthermore ψ and ψ^{-1} are continuous in $\bar{R} = R \cup \Gamma$ and $\bar{D} = D \cup C$, respectively.*

We will not prove this theorem but show that it enables us to solve the Dirichlet problem for R. We will denote by ζ the w points on Γ and we let $G(\zeta)$ be a given continuous function on Γ. We seek a function $u(w)$ continuous in $\bar{R} = R \cup \Gamma$ for which

  (i) $\Delta u(w) = 0$ in R,

  (ii) $u(\zeta) = G(\zeta)$ on Γ.

6.3f Theorem. *The Dirichlet problem for R has a solution.*

Proof. We define $g(\theta) \equiv G(\psi^{-1}(e^{i\theta}))$. This transplants the boundary values onto C. Furthermore $g(\theta)$ is continuous on C and so the Poisson integral solves the Dirichlet problem for D with boundary values $g(\theta)$. Denote this function by $U(z)$. Then $U(\psi(w))$ solves the Dirichlet problem for R with boundary values $g(\zeta)$. $\square$

The difficulty with this even after we know a proof of Theorem 6.3d, is in *actually finding the mapping function* ψ. The examples we examined in Section 5.5 were limited to simple situations where we actually knew the mapping function. Furthermore, by Theorem 6.3d and Exercise C3 of Section 5.5 we can handle a finite number of discontinuities of the boundary function.

B Exercises

1. Carefully explain the steps in the limit calculation (3).
2. Carefully explain how the proof of Theorem 6.3c is in the previous work of the section.
3. Suppose $g(t)$ is continuous for $0 \le t \le 2\pi$ and $g(0) = g(2\pi)$. Solve the Dirichlet problem for the *exterior* of a circle:

$$\Delta u(z) = 0, \quad |z| > a$$

$$u(ae^{it}) = g(t).$$

4. Solve the Dirichlet problem

$$\Delta u(z) = 0, \quad |z - \alpha| < a$$

$$u(\alpha + ae^{it}) = g(t).$$

C Exercises

1. Suppose $u_n(x, y)$ is a sequence of harmonic functions in a region R, and that $u_n(x, y) \to u(x, y)$ uniformly in R. Show that $u(x, y)$ is harmonic. Hint: Use Theorems 6.3a and 6.1d.

2. See Exercises C2 of Section 6.1, C1 of Section 6.2, C1 above, and let $\{h_n(t)\}$ be the sequence of functions of Exercise B1 of Section 6.2. Suppose $g(t)$ is continuous and non-negative on R, and that there are constants $a < 1$, $M > 0$ and $T > 0$ for which

$$|g(t)| < Mt^a \quad \text{for} \quad t > T.$$

(a) Set $g_n(t) = g(t)h_n(t)$ and show that each $g_n(t)$ satisfies the conditions of Theorem 6.2c.

(b) Let $u_n(x, y)$ be the solution of the Dirichlet problem for $g_n(t)$ of part (a), and show that for each fixed (x, y) with $y > 0$, $u_n(x, y)$ is nondecreasing as $n \to \infty$.

(c) For each fixed (a, b) with $b > 0$ show that the sequence $u_n(x, y)$ is bounded in the disk $D = \{(x, y) : (x - a)^2 + (y - b)^2 \leq b^2/4\}$.

(d) Show that $u_n(x, y)$ converges uniformly in D.

(e) Show that $u_n(x, y)$ converges to a function $u(x, y)$ which solves the Dirichlet problem

$$\Delta u = 0, \quad y > 0$$

$$u(t, 0) = g(t).$$

(f) Remove the condition that $g(t) \geq 0$, and solve the problem

$$\Delta u = 0, \quad y > 0$$

$$u(t, 0) = g(t).$$

6.4 The Neumann Problem

Neumann Problem. *Let R be a region whose boundary is a contour $C : \zeta = \zeta(t)$, $a \leq t \leq b$. Let $g(\zeta)$ be a continuous function on C. Find a function u which, with its first order partial derivatives, is continuous in $\bar{R} = R \cup C$ and*

    (i) $\Delta u(z) = 0$, $z \in R$;

    (ii) $\dfrac{\partial u}{\partial n}(\zeta) = g(\zeta)$, $\zeta \in C$,

where $\partial / \partial n$ means the derivative in the direction of the normal to C at ζ.

    A physical interpretation in terms of steady state heat conduction comes from the fact that $\partial u / \partial n$ at a point ζ on C is proportional to the rate of flow (flux) of heat across C at ζ. Thus the problem is asking for the steady state temperature in an insulated lamina when the heat flux $g(\zeta)$ is known at each point ζ on the boundary.

    Clearly if $u(z)$ is one solution, then $u(z) + A$, for each constant A, is also a solution (why?). Furthermore, since the temperature is steady, we would expect the total heat flowing into the lamina to be balanced by the total heat flowing out. This means we will need the condition

$$\int_C g(\zeta) d\zeta = 0.$$

    For the disk we formulate the Neumann problem conditions as

    (i) $\Delta u = 0$, $|z| = r < 1$.

    (ii) $\lim\limits_{r \to 1} u_r(r, \theta) = g(\theta)$.

In the discussion of this problem we assume we can differentiate under the integral sign.

6.4a Theorem. *If $g(\theta)$ is piecewise continuous and periodic and*

$$\int_0^{2\pi} g(\theta) d\theta = 0$$

then

$$u(r, \theta) = -\frac{1}{2\pi} \int_0^{2\pi} \ln[1 - 2r \cos(\theta - t) + r^2] g(t) dt$$

is a solution to the Neumann problem for the disk.

Proof. That $\Delta u = 0$ for $r < 1$ we leave as an exercise (B1). To compute $u_r(r, \theta)$ we differentiate under the integral sign, and get

$$u_r(r, \theta) = -\frac{1}{2\pi} \int_0^{2\pi} \frac{-2\cos(\theta - r) + 2r}{1 - 2r\cos(\theta - t) + r^2} g(t)dt$$

$$= \frac{1}{2\pi r} \int_0^{2\pi} \frac{2r\cos(\theta - t) - 2r^2}{1 - 2r\cos(\theta - t) + r^2} g(t)dt$$

$$= \frac{1}{2\pi r} \int_0^{2\pi} \left[\frac{1 - r^2}{1 - 2r\cos(\theta - t) + r^2} \right] g(t)dt - \frac{1}{2\pi r} \int_0^{2\pi} g(t)dt$$

$$= \frac{1}{r} \int_0^{2\pi} P(1, t; r, \theta)g(t)dt - \frac{1}{2\pi r} \int_0^{2\pi} g(t)dt$$

$$= \frac{1}{r} \int_0^{2\pi} P(1, t; r, \theta)g(t)dt.$$

Then $u_r(r, \theta) \to g(\theta_0)$ at each point θ_0 where g is continuous as $(r, \theta) \to (1, \theta_0)$ by Theorem 6.3b. □

Remark. Though we do not prove it, the solution is unique to within an additive constant. That is, if u is a solution, then $w = u + C$ is also for any constant C, and *there are no other solutions*.

If R is a half-plane $y > 0$, then the boundary is $y = 0$. We examine the Neumann problem in the form
   (i) $\Delta u = 0$, $y > 0$.
   (ii) $\dfrac{\partial u}{\partial y}(x, 0) = g(x)$, $-\infty < x < \infty$.

6.4b Theorem. *If $g(t)$ is continuous for $-\infty < t < \infty$ and if*
$$\int_{-\infty}^{\infty} |g(t)| \ln(1 + |t|)dt \text{ converges, then}$$

$$u(x, y) = \frac{1}{2\pi} \int_{-\infty}^{\infty} g(t) \ln[(x - t)^2 + y^2]dt$$

solves the Neumann problems for the half plane.
   The proof is left to the exercises.

Significant Topics   **Page**

B Exercises

1. Express Δu in polar coordinates and verify that $u(r,\theta)$ in Theorem 6.4a is harmonic for $r < 1$.
2. Verify that $u(x,y)$ in Theorem 6.4b solves the Neumann problem.

6.5 Fourier Series

The subject of Fourier series is historically connected with analytic and harmonic functions and we will use our knowledge of the Dirichlet problem to formulate an introduction to the subject. It is concerned with periodic functions on the real numbers $\mathbb{R}$. Let us recall that $g(x)$ is periodic with period 2π on $\mathbb{R}$ if $g(x) = g(x+2\pi)$. For example, if a function is defined on a circle, then it is a periodic function of the central angle. We will restrict our attention to functions with period 2π, and *when we say a function is periodic we mean it has period 2π*. Thus, if $f(z) = u(z) + v(z)$ is analytic for $|z| \le 1$, then $f(1,\theta)$ is periodic, as are $u(1,\theta)$ and $v(1,\theta)$. Furthermore, if the Taylor series of $f(z)$ at the origin is

$$f(z) = \sum_0^\infty c_n z^n$$

then

$$f(1,\theta) = \sum_0^\infty c_n e^{in\theta}$$

and so $f(1,\theta)$ is expressible as a series of the complex exponentials $e^{in\theta}$, $n = 0$ to ∞. Also

$$\bar{f}(1,\theta) = \sum_0^\infty \bar{c}_n e^{-in\theta}$$

so

$$u(1,\theta) = \frac{1}{2}\left[f(1,\theta) + \bar{f}(1,\theta) \right]$$

$$= \frac{1}{2} \sum_0^\infty c_n e^{in\theta} + \frac{1}{2} \sum_0^\infty \bar{c}_n e^{-in\theta}$$

$$= \sum_{-\infty}^\infty \gamma_n e^{in\theta}$$

where

$$\gamma_n = \begin{cases} \dfrac{1}{2}\, c_n, & n > 0 \\[2mm] \dfrac{1}{2}\, (c_0 + \bar{c}_0), & n = 0 \\[2mm] \dfrac{1}{2}\, \bar{c}_n, & n < 0. \end{cases}$$

These calculations raise the question as to what conditions on a periodic function $g(\theta)$ would permit it to be written as

$$g(\theta) = \sum_{-\infty}^{\infty} \gamma_n e^{in\theta}, \tag{1}$$

and, if (1) is possible, how to determine coefficients γ_n, $n = -\infty$ to $+\infty$. Or equivalently (see Exercise A1) when can $g(\theta)$ be written as

$$g(\theta) = \frac{1}{2}\, a_0 + \sum_{1}^{\infty} (a_n \cos n\theta + b_n \sin n\theta)? \tag{2}$$

Suppose $g(\theta)$ has a representation in the form (1), and suppose termwise integration is permissible. Then if we multiply both sides of (1) by $e^{-ik\theta}$ and integrate from 0 to 2π, we get

$$\int_0^{2\pi} g(\theta) e^{-ik\theta} d\theta = \int_0^{2\pi} \left(\sum_{-\infty}^{\infty} \gamma_n e^{in\theta} \right) e^{-ik\theta} d\theta$$
$$= \sum_{-\infty}^{\infty} \gamma_n \int_0^{2\pi} e^{in\theta} e^{-ik\theta} d\theta \tag{3}$$

But if $n \neq k$

$$\int_0^{2\pi} e^{in\theta} e^{-ik\theta} d\theta = \int_0^{2\pi} e^{i(n-k)\theta} d\theta$$
$$= \frac{1}{i(n-k)}\, e^{i(n-k)\theta} \Big|_0^{2\pi} = \frac{1}{i(n-k)}\, [1-1] = 0.$$

and if $n = k$

$$\int_0^{2\pi} e^{in\theta} e^{-ik\theta} = \int_0^{2\pi} e^{ik\theta} e^{-ik\theta} d\theta$$
$$= \int_0^{2\pi} d\theta = 2\pi.$$

Thus (3) reduces to

$$\int_0^{2\pi} g(\theta)e^{-ik\theta}\,d\theta = 2\pi\gamma_k,$$

which evaluates the coefficients γ_k:

$$\gamma_k = \frac{1}{2\pi}\int_0^{2\pi} g(\theta)e^{-ik\theta}\,d\theta. \tag{4}$$

Incidentally we have proved and used

$$\frac{1}{2\pi}\int_0^{2\pi} e^{in\theta}\cdot e^{-ik\theta}\,d\theta = \begin{cases} 1, & n = k \\ 0, & n \neq k. \end{cases} \tag{5}$$

This equation (5) is called the **orthogonality condition** (see Exercise A2).

A natural question then is: Under what conditions on $g(\theta)$ does equation (1) hold true when the coefficients are given by (4)? Clearly this is closely related to the problem of convergence of a power series *on* the circle of convergence.

If $g(\theta)$ is periodic on the reals $\mathbb{R}$, and is integrable on $[0, 2\pi]$, then the series (1) with the coefficients given by (4) is called the **Fourier series of** g, and the numbers γ_n are called the **Fourier coefficients** of g. First we connect these notions to the Dirichlet problem.

6.5a Theorem. *If $g(\theta)$ is piecewise continuous and periodic, then the solution $u(r, \theta)$ to the Dirichlet problem*

$$\Delta u = 0, \quad |z| = r < 1$$

$$u(1, \theta) = g(\theta), \quad r = 1$$

is given by

$$u(r, \theta) = \sum_{-\infty}^{\infty} \gamma_n r^{|n|} e^{in\theta}, \quad r < 1$$

where γ_n are the Fourier coefficients of $g(\theta)$:

$$\gamma_n = \frac{1}{2\pi}\int_{-\infty}^{\infty} g(t)e^{-int}\,dt.$$

Proof. We calculate

$$\frac{1}{2\pi}\sum_{-\infty}^{\infty} r^{|n|}e^{ik\phi} = \frac{1}{\pi}\left[\sum_{0}^{\infty} r^n \cos n\phi - \frac{1}{2}\right]$$

$$= \frac{1}{\pi}\left[\text{Re}\sum_{0}^{\infty} r^n e^{in\phi} - \frac{1}{2}\right]$$

$$= \frac{1}{\pi}\text{Re}\frac{1}{1 - re^{i\phi}} - \frac{1}{2\pi}$$

$$= \frac{1}{\pi}\frac{1 - r\cos\phi}{1 - 2r\cos\phi + r^2} - \frac{1}{2\pi}$$

$$= \frac{1}{2\pi}\frac{1 - r^2}{1 - 2r\cos\phi + r^2}$$

If we set $\phi = \theta - t$, then the last formula we calculated is $P(1, t; r, \theta)$:

$$\frac{1}{2\pi}\sum_{-\infty}^{\infty} r^{|n|}e^{in\theta}\cdot e^{-int} = P(1, t; r, \theta).$$

We can multiply by $g(t)$ and integrate from $t = 0$ to $t = 2\pi$. On the right we get $u(r, \theta)$; so

$$\begin{cases} u(r, \theta) = \dfrac{1}{2\pi}\displaystyle\int_0^{2\pi}\sum_{-\infty}^{\infty} r^{|n|}e^{in\theta}e^{-int}g(t)dt \\[2mm] \qquad = \dfrac{1}{2\pi}\displaystyle\sum_{-\infty}^{\infty} r^{|n|}e^{in\theta}\int_0^{2\pi} g(t)e^{-int}dt \qquad\qquad (6) \\[2mm] \qquad = \displaystyle\sum_{-\infty}^{\infty}\gamma_n r^{|n|}e^{in\theta}. \end{cases}$$

(See Exercise A3.) □

It seems we are close to what we want—for if we take $r = 1$ on the right and have $u(1, \theta) = g(\theta)$, then we have our Fourier series being equal to our boundary function. But in essence we are discussing convergence of a power series (of $f(z) = u(z) + v(z)$) *on the circle of convergence.* This is a very difficult question in general. But we do get something out of this idea, given by the following lemma.

6.5b Lemma. *If $g(\theta)$ is integrable and if $\displaystyle\sum_{-\infty}^{\infty}|\gamma_n|$ converges, then*

$$g(\theta) = \sum_{-\infty}^{\infty}\gamma_n e^{in\theta},$$

where the convergence is uniform, and so $g(\theta)$ is itself continuous and periodic.

Proof. By the Weierstrass M-test with $M_n = |\gamma_n|$ we have the result that

$$\sum_{-\infty}^{\infty} \gamma_n r^{|n|} e^{in\theta} \ll \sum_{-\infty}^{\infty} |\gamma_n|, \quad 0 \leq r \leq 1$$

so the series on the left converges uniformly for $0 \leq r \leq 1$, and hence represents a continuous function in the closed disk $0 \leq r \leq 1$. For $0 \leq r < 1$ its sum is

$$u(r, \theta) = \int_0^{2\pi} P(1, t; r, \theta) g(t) dt, \quad 0 \leq r < 1$$

by Theorem 6.5a. *And by Theorem 6.3c its value on the circle $r = 1$ is $g(\theta)$. This means, setting $r = 1$,*

$$g(\theta) = \sum_{-\infty}^{\infty} \gamma_n e^{in\theta}. \quad \square$$

We now give conditions on g which guarantee that $\sum_{-\infty}^{\infty} |\gamma_n|$ converges.

6.5c Theorem. *If $g(\theta)$ is periodic on $\mathbb{R}$, $g'(\theta)$ is continuous, and $g''(\theta)$ is piecewise continuous, then $\sum_{-\infty}^{\infty} |\gamma_n|$ converges, so that $g(\theta) = \sum_{-\infty}^{\infty} \gamma_n e^{in\theta}$.*

Proof. For $n \neq 0$ we have

$$
\begin{aligned}
\gamma_n &= \frac{1}{2\pi} \int_0^{2\pi} g(\theta) e^{-in\theta} d\theta \\
&= \frac{1}{2\pi} \left[\frac{g(\theta) e^{-in\theta}}{-in} \bigg|_0^{2\pi} + \frac{1}{in} \int_0^{2\pi} g'(\theta) e^{-in\theta} d\theta \right] \\
&= \frac{1}{2\pi in} \int_0^{2\pi} g'(\theta) e^{-in\theta} d\theta \\
&= \frac{1}{-2\pi n^2} \int_0^{2\pi} g''(\theta) e^{-in\theta} d\theta
\end{aligned}
$$

so that

$$|\gamma_n| < \frac{1}{n^2} \cdot \frac{1}{2\pi} \int_0^{2\pi} |g''(\theta)| d\theta = \frac{K}{n^2}$$

and so $\sum_{1}^{\infty} |\gamma_n|$ and $\sum_{-\infty}^{-1} |\gamma_n|$ each converges by comparison with $K \sum_{1}^{\infty} 1/n^2$. $\square$

There exist much stronger convergence theorems than 6.5c. Rather than exploring these theoretical questions we turn to some applications. We have seen (Theorem 6.5a) that Fourier series can give us the solution to the Dirichlet problem for Laplace's equation in a disk. As a further example we consider a problem in heat conduction.

If we have an insulated one-dimensional heat conductor (a wire, say), the one (space) dimensional heat equation is

$$u_{xx} = u_t \tag{7}$$

where subscripts indicate partial derivatives. (Compare Part A, Section 2.2.)

We assume we have a segment of wire of length π (chosen for computational simplicity) and that we know the initial temperature distribution in the wire. If the temperatures of the ends are controlled, so that we know what heat flows in and out of our segment, we should be able to determine the temperature within the segment for all later times. The case we consider is that in which the end temperatures are held to zero. This leads to the following problem for the heat equation.

For $0 \le x \le \pi$, $t \ge 0$ we seek $u(x,t)$ for which

$$\begin{cases} u_{xx} = u_t, & 0 < x < \pi, \ t > 0; \\ u(x,0) = f(x), & 0 \le x \le \pi; \\ u(0,t) = u(\pi,t) = 0, & t \ge 0. \end{cases}$$

In fact, we're going to lengthen the wire: We would like $f(x)$ defined on $-\pi \le x \le \pi$ to be an odd function so that the thermal effects will cancel out and give $u(0,t) = 0$, and then extended by periodicity to all x. Thus we end up with an infinite wire with an odd periodic function defined on it. Note also that for compatibility of the initial and boundary conditions we need $f(0) = f(\pi) = 0$.

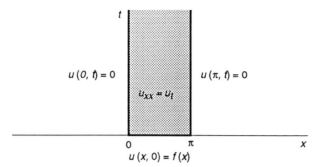

The Dirichlet Problem for $u_{xx} = u_t$

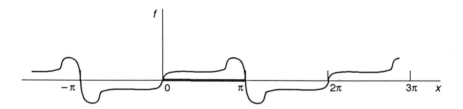

The initial values on $[0, \pi]$ continued as an odd periodic function

We will seek a solution to our problem (which is a special case of the **Dirichlet problem for the heat equation**) in the form of a Fourier series in x for each t. Thus we examine

$$u(x,t) = \sum_{-\infty}^{\infty} \beta_n(t)e^{inx}, \tag{8}$$

and try to determine the β's so as to get a solution. We will proceed to calculate, keeping track of what we must eventually justify to validate our calculations.

(A) We differentiate termwise and substitute into equation (7):

$$u_{xx} = \sum_{-\infty}^{\infty} -n^2\beta_n(t)e^{inx} = \sum_{-\infty}^{\infty} \beta_n'(t)e^{inx} = u_t.$$

(B) We multiply this equation by e^{-ikx} and integrate termwise from $-\pi$ to π to get

$$\beta_k' = -k^2\beta_k(t).$$

From this we get (how?)

$$\beta_k(t) = \gamma_k e^{-k^2 t}$$

where γ_k are constants.

Substituting these β's into (8) we get

$$u(x,t) = \sum_{-\infty}^{\infty} \gamma_n e^{-n^2 t} e^{inx}. \tag{9}$$

(C) Then taking termwise limits as $t \to 0$

$$u(x,0) = \sum_{-\infty}^{\infty} \gamma_n e^{inx} \overset{?}{=} f(x).$$

Clearly then, if we choose the γ_n's as the Fourier coefficients of $f(x)$, where f is continuous, $f(0) = f(\pi) = 0$, and f'' is piecewise continuous, we have that the series in (9) converges uniformly (how?) for all x and all $t \geq 0$. This then justifies both (B) and (C) and removes the question about equality to $f(x)$.

Furthermore, since $e^{-n^2 a}$ is such a strong convergence factor we have that the differentiated series for u_x, u_{xx} (and hence u_t) converge uniformly for $t \geq a > 0$ (how?). Hence, since $a > 0$ is arbitrary, the differentiations are justified for all $t > 0$ and all x. Thus we have shown that our proposed solution (9) satisfies all except the boundary conditions.

To see that it also satisfies those we examine

$$u(0,t) = \sum_{-\infty}^{\infty} \gamma_n e^{-n^2 t}$$

$$= \gamma_0 + \sum_{1}^{\infty} (\gamma_n + \gamma_{-n}) e^{-n^2 t}.$$

But $f(x)$ is odd so

$$\gamma_0 = \frac{1}{2\pi} \int_{-\pi}^{\pi} f(x)dx = 0, \tag{10}$$

and also (Exercise B2)

$$\gamma_n + \gamma_{-n} = 0. \tag{11}$$

Thus $u(0,t) = 0$ for $t > 0$. Similarly (Exercise B3) $u(\pi, t) = 0$ for $t > 0$.

We proceed to rewrite our solution (9)

$$
\left\{
\begin{aligned}
u(x,t) &= \sum_{-\infty}^{\infty} \gamma_n e^{-n^2 t} e^{inx} \\
&= \sum_{1}^{\infty} \gamma_n e^{-n^2 t} [e^{inx} - e^{-inx}] \text{ (how?)} \\
&= \sum_{1}^{\infty} (2i\gamma_n) e^{-n^2 t} \sin nx \\
&= \sum_{1}^{\infty} b_n e^{-n^2 t} \sin nx
\end{aligned}
\right.
\tag{12}
$$

where $b_n = 2i\gamma_n$.

We compute b_n:

$$
\begin{aligned}
b_n = 2i\gamma_n &= \frac{2i}{2\pi} \int_{-\pi}^{\pi} f(x) e^{-inx}\, dx \\
&= \frac{i}{\pi} \left[\int_0^{\pi} f(x) e^{-inx}\, dx - \int_{-\pi}^0 f(x) e^{-inx}\, dx \right] \\
&= \frac{i}{\pi} \left[\int_0^{\pi} f(x) e^{-inx}\, dx - \int_0^{\pi} f(x) e^{inx}\, dx \right] \\
&= \frac{i}{\pi} \int_0^{\pi} f(x) [e^{-inx} - e^{inx}]\, dx \\
&= \frac{2}{\pi} \int_0^{\pi} f(x) \sin nx\, dx.
\end{aligned}
$$

Thus we come finally to our solution:

$$
u(x,t) = \sum_1^{\infty} b_n e^{-n^2 t} \sin nx
$$

where

$$
b_n = \frac{2}{\pi} \int_0^{\pi} f(x) \sin nx\, dx.
$$

We consider also a special case of the **Neumann problem for the heat equation**. This is to find $u(x,t)$ such that

$$
\begin{aligned}
u_{xx} &= u_t, & 0 &< x < \pi,\ t > 0; \\
u_x(0,t) &= u_x(\pi,t) = 0, & t &\geq 0; \\
u(x,0) &= f(x), & 0 &\leq x \leq \pi.
\end{aligned}
$$

We leave this as an exercise (B4). Give a physical argument as to why this time $f(x)$ should be continued as an *even periodic function* and why we need $f'(0) = f'(\pi) = 0$. Then proceed to an analogous solution. Your answer should be

$$u(x,t) = \frac{a_0}{2} + \sum_1^\infty a_n e^{-n^2 t} \cos nx$$

where

$$a_n = \frac{2}{\pi} \int_0^\pi f(x) \cos nx\,dx.$$

Significant Topics   **Page**

A Exercises

1. Show formally (i.e., without raising convergence questions) that equations (1) and (2) are equivalent. Show that

$$a_n = \frac{1}{\pi} \int_{-\pi}^\pi g(\theta) \cos n\theta d\theta, \quad n = 0, 1, 2, \ldots,$$

and

$$b_n = \frac{1}{\pi} \int_{-\pi}^\pi g(\theta) \sin n\theta d\theta, \quad n = 1, 2, \ldots, \ldots.$$

2. Prove equation (5).
3. How is termwise integration justified for equation (6)?
4. Find the Fourier series in exponential form for
   (a) $g(\theta) = |\theta|, -\pi \le \theta \le \pi$
   (b) $g(\theta) = \sin \theta$
   (c) $g(\theta) = (\pi - \theta)/2, 0 < \theta < 2\pi$.

5. Convert the series found in A4 to trigonometric form.
6. Show that if f is periodic (2π), f' is continuous, and f'' is piecewise continuous, then the series (9) converges uniformly for all x and all $t \geq 0$.
7. Show that (10) is correct.
8. Carefully discuss the first step in the calculation (12).
9. In the Neumann problem begun at the end of this section, give physical reasons why we would like $f(x)$ extended as an even periodic function.

B Exercises

1. Show that for $t \geq a > 0$ the series (9) can be differentiated termwise as many times as we please.
2. Show that equation (11) is correct.
3. Show that if $u(x,t)$ is given by (9), then $u(\pi, t) = 0$ for $t > 0$.
4. Complete the solution of the Neumann problem begun at the end of this section and validate the answer given there.
5. In B4 above, what replaces equation (11)?

C Exercises

1. Solve a wave equation problem
   (i) $u_{xx} = u_{tt}$, $0 < x < \pi$, $t > 0$
   (ii) $u(0,t) = u(\pi, t) = 0$, $t > 0$
   (iii) $u(x,0) = f(x)$, $u_t(x,0) = 0$, $0 \leq x \leq 1$.
2. Replace (iii) above by

$$u(x,0) = 0, \quad u_t(x,0) = g(x), \qquad 0 \leq x \leq 1$$

3. Consider the Neumann problem for Laplace's equation in polar coordinates in the disk. Seek $u(r, \theta)$ for $r < 1$ such that
   (i) $u_{rr} + u_r/r + u_{\theta\theta}/r^2 = 0$, $r < 1$
   (ii) $u_r(1, \theta) = g(\theta)$
   (iii) $\displaystyle\int_{-\pi}^{\pi} g(\theta)d\theta = 0.$

   If a solution exists, it is periodic in θ for each r so seek a solution

$$u(r, \theta) = \sum_{-\infty}^{\infty} R_n(r)e^{in\theta}.$$

4. Consider the Dirichlet problem

$$\Delta u = 0 \quad \text{in} \quad D : |z| < 1$$

$$u = (\pi - \theta)/2 \quad \text{on} \quad C : |z| = 1.$$

where $\theta = \arg z,\ 0 < \theta < 2\pi$.

(a) Use Theorem 6.5a to show the solution is

$$u(r, \theta) = \sum_1^\infty (r^n \sin n\theta)/n.$$

(b) Write $\sin n\theta = (e^{in\theta} - e^{-in\theta})/2i$ and show

$$u = (1/2i)[\text{Log}\,(1 - \bar{z}) - \text{Log}\,(1 - z)]$$
$$= \text{Arg}\,(1 - z)$$

(c) By drawing a figure show geometrically that $\text{Log}\,(1-z) = (\pi-\theta)/2$ when $z = e^{i\theta}$.

Suggestions for Further Study

Here are four classic texts and, though they are in many ways dated, they are still excellent.

1. Copson, E. T. 1935. *An Introduction to the Theory of Functions of a Complex Variable*, Oxford University Press.
2. Phillips, E. G. 1957. *Functions of a Complex Variable*, 8th ed., Oliver and Boyd.
3. Titchmarsh, E. C. 1939. *The Theory of Functions*, Oxford University Press.
4. Whittaker, E. T. and Watson, G. N. 1927. *A Course in Modern Analysis*, Cambridge University Press.

The following two collections of problems are an excellent source of exercises from elementary to quite advanced levels.

5. Knopp, K. 1948, 1952. *Problem Book in the Theory of Functions*, Vol. I, Vol. II, Dover.
6. Polya, G. and Szegö, G. 1989, 1990. *Problems and Theorems in Analysis*, Vol. I, Vol. II, Springer–Verlag.

Here is an excellent and extensive list of conformal maps with figures.

7. Kober, H. 1957. *Dictionary of Conformal Representations*, 2nd ed., Dover.

These next three books are very good to excellent intermediate level expositions.

8. Boas, R. P. 1987. *Invitation to Complex Analysis*, McGraw–Hill.

9. Levinson, N., and Redheffer, R. M. 1970. *Complex Variables*, Holden–Day.

10. Marsden, J. E. 1973. *Basic Complex Analysis*, W. H. Freeman.

Here are two excellent advanced level texts. The first has essentially defined the graduate course in the subject for the last few decades.

11. Ahlfors, L. 1979. *Complex Analysis*, 3rd ed., McGraw–Hill.

12. Hille, E. 1962. *Analytic Function Theory*, Vol. II, Blaisdell.

The following are texts and reference books which make extensive use of complex variables theory.

13. Baggett, L. and Fulks, W. 1979. *Fourier Analysis*, Anjou Press.

14. Erdélyi, A. 1956. *Asymptotic Expansions*, Dover.

15. Erdélyi, A. et al. 1953, 1953, 1955. *Higher Transcendental Functions*, Vol. I, II, III, McGraw–Hill.

16. Kantorovich, L. V. and Krilov, V. I. 1958. *Approximate Methods of Higher Analysis*, Interscience.

17. Watson, G. N. 1941. *A Tretaise on the Theory of Bessel Functions*, Cambridge University Press.

Answers, Hints, and Solutions

Section 1.1

A Exercises

1. (a) 3;   (b) 4;   (c) 6;   (d) -5
3. $(11 - 10i)(\frac{1}{2} + 3i) = 35\frac{1}{2} + 28i$; $(3 + 2i)(12\frac{1}{2} + i) = 35\frac{1}{2} + 28i$
5. (a) $-10 - 4i$;   (b) $20(1 + i)$
7. (a) $x^2 - y^2$;   (b) $x^4 - 6x^2y^2 + y^4$;
   (g) $-y/(x^2 + y^2)$;   (h) $-2xy/(x^2 + y^2)^2$

B Exercises

1. $iz = -y + ix$
3. Apply Cramer's rule; then compute $\bar{\alpha}\beta/|\alpha|^2$ in terms of components.
5. As in elementary algebra.
7. $|\alpha/\bar{\alpha}| = |\alpha|/|\bar{\alpha}| = |\alpha|/|\alpha| = 1$
9. (a) By B8(c),(d) and (j);   (b) By B8(f);
   (c) Same as a_k is real if and only if $a_k = \bar{a}_k$.
11. As in elementary algebra

$$\frac{\alpha^{2k+1} + \beta^{2k+1}}{\alpha + \beta} = \sum_{j=0}^{2k}(-1)^j \alpha^{2k-j}\beta^j.$$

13. By division of polynomials as in elementary algebra.

C Exercises

1. $P(z) - P(\alpha) = \displaystyle\sum_{k=1}^{n} a_k(z^k - \alpha^k)$

3. By induction.

Section 1.2

A Exercises

1. By (6), p. 16:
   (a) $E(0), \quad E(2\pi/3), \quad E(4\pi/3)$
      $1, \quad \cos 2\pi/3 + i\sin 2\pi/3; \quad \cos 4\pi/3 + i\sin 4\pi/3$
      $1, \quad -\frac{1}{2} + i\frac{1}{2}\sqrt{3}, \quad -\frac{1}{2} - i\frac{1}{2}\sqrt{3}$
   (c) $3E(\pi/3), \quad 3E(\pi), \quad 3E(5\pi/3)$
      $\frac{3}{2}(1 + i\sqrt{3}), \quad -3, \quad \frac{3}{2}(1 - i\sqrt{3})$

3. (a) Equation: $x = y$. It is the perpendicular bisector of the segment
       between 1 and i.
   (b) $x > y$. Those points closer to 1 than to i. The half-plane below
       and to the right of $x = y$.
   (c) $x^2 + y^2 - 4y + 2x + 1 = 0$. Circle centered at $(-1, 2)$ with radius
       2.
   (h) Ellipse
   (i) Degenerate ellipse: the segment between -1 and 1.
   (j) Equation $x = y$, $x > 0$, $y > 0$. This is a ray from the origin at
       $45°$.

5. $\cos 3\theta + i\sin 3\theta = (\cos\theta + i\sin\theta)^3$
   $$= \cos^3\theta - 3\cos\theta\sin^2\theta + i(3\cos^2\theta\sin\theta - \sin^3\theta)$$

7. $|z^2 + 1| = \mathrm{dist}\,(z^2, -1)$. For $|z| \le 1$, $|z^2|$ is also ≤ 1, and distance is
   greatest when $z^2 = 1$, i.e., $z = \pm 1$ which gives a maximum of 2.

9. By Corollary 1.2c.

11. (a) On the same ray as z, but further out.
    (b) As in (a) but closer in.
    (c) In opposite directions from the origin.

(d) On the circle, center 0, radius $|z|$.

(e) As in (d), an angle φ from z.

(f) Angle φ from z, but arbitrary distance out from origin.

13. Equate absolute values, then divide by r.

B Exercises

1. $|\alpha| = |(\alpha + \beta) - \beta| \leq |\alpha + \beta| + |\beta|$

   $|\beta| = |(\alpha + \beta) - \alpha| \leq |\alpha + \beta| + |\alpha|$

3. $\arg(z/\zeta) = \arg[z \cdot (1/\zeta)] = \arg z + \arg(1/\zeta) = \arg z - \arg \zeta$.

5. If $\operatorname{Arg}\alpha = \operatorname{Arg}\beta$, then $\operatorname{Arg}(\alpha + \beta) = \operatorname{Arg}\alpha$, so $|\alpha + \beta + \gamma| = |\alpha + \beta| + |\gamma| = |\alpha| + |\beta| + |\gamma|$ if and only if $\operatorname{Arg}\alpha = \operatorname{Arg}\beta = \operatorname{Arg}\gamma$. Now use induction.

7. Set $z = \cos\theta + i\sin\theta$, so that $\dfrac{1}{z} = \cos\theta - i\sin\theta = \cos(-\theta) + i\sin(-\theta)$.
   Then apply de Moivre's theorem for $n > 0$.

9. Write $(\alpha - \beta)(\bar{\alpha} - \bar{\beta}) + (\alpha + \beta)(\bar{\alpha} + \bar{\beta})$ and multiply out.

11. Multiply by $1 - z$ to verify. Then by de Moivre's theorem on each term with $z = \cos\theta + i\sin\theta$,

$$\operatorname{Re}\left(\sum_0^n z^k\right) = \sum_0^n \cos k\theta = \operatorname{Re}\frac{1 - \cos(n+1)\theta - i\sin(n+1)\theta}{1 - \cos\theta - i\sin\theta}$$

and simplify. For part (b)

$$\sum_0^{n-1} \omega^k = (1 - \omega^n)/(1 - \omega)$$

and use B6.

C Exercises

1. The sum is the circumference of a regular polygon of n sides inscribed in the circle $|z| = 1$, and hence is $< 2\pi$. Each side subtends a central angle of $2\pi/n$, and hence is $> \sin(2\pi/n)$.

3. Just follow the angles around.

5. If $p(z, \zeta) = \sum_{n=0}^{p} \sum_{m=0}^{q} a_{nm} z^n \zeta^m$ with each $a_{nm} \geq 0$, then

$$|p(z, \zeta)| \leq \sum_{n=0}^{p} \sum_{m=0}^{q} a_{nm} |z|^n |\zeta|^m = p(|z|, |\zeta|).$$

Section 1.3

A Exercises

1. Set $w = u + iv$, $z = x + iy$ and yet

$$u = 2 - 3y, \qquad v = 3x,$$
$$x = v/3, \qquad y = (2 - u)/3,$$

and then substitute.

3. (a) 1;   (b) i;   (c) 0;   (d) 0;   (e) $e + i/e$;   (f) i.

B Exercises

1, 2, and 3. Just as in calculus, so see your calculus text.

5. If $p(z)$ is the polynomial and has degree > 1, then $|p(z) - p(\alpha)| = |z - \alpha| \cdot |q(z, \alpha)|$ where $q(z)$ is a polynomial of degree ≥ 1 and so can be made large by taking z and α large.

7. Similar to the proof of Theorem 1.3a.

9. $0 \leq \arg z \leq 3\varphi$, etc.

11. $\left| \sum_{1}^{n} z^k \right| = |(z - z^{n+1})/(1 - z)| \leq 2/|1 - z|$.

C Exercises

1. By substitution of

$$u = x/(x^2 + y^2), \quad v = -y/(x^2 + y^2),$$
$$x = u/(u^2 + v^2), \quad y = -v/(u^2 + v^2).$$

3. By the substitutions in C1, we get

$$a\left(\frac{u^2}{(u^2+v^2)^2} + \frac{v^2}{(u^2+v^2)^2}\right) + \frac{bu}{u^2+v^2} - \frac{cv}{u^2+v^2} + d = 0$$

which simplifies to

$$d(u^2 + v^2) + bu - cv + a = 0.$$

5. Verify for $\alpha = 0$, $\beta = 2$, $\gamma = 1 + i\sqrt{3}$. Then observe that each side of the equation is invariant under translation, rotation and dilation, that is, under any linear transformation. From this the necessity and sufficiency follow easily.

Section 1.4

A Exercises

1. $2 + \frac{1}{2} i$ 3. $\frac{1}{3} - \frac{1}{4} i$ 5. $25 + 295i$

7. $\ln(27/4) + i(e^3 - e^2)$

9. $(\cos\theta + i\sin\theta)' = -\sin\theta + i\cos\theta = i(\cos\theta + i\sin\theta)$

11. (a) $\varphi' = E(\pi/4)$ which is continuous and $\neq 0$.

(b) $\varphi' = 2tE(\pi/4)$ which is continuous but is 0 at $t = 0$.

13. $z = 2E(t)$ $dz = 2iE(t)dt$

(a) $\int_0^\pi (2\cos t)2i(\cos t + i\sin t)dt = 4i\int_0^\pi \cos^2 t \, dt - 4\int_0^\pi \cos t \sin t \, dt = 2\pi i$

(b) , (c) similar.

15. (a) $\left|\int_C (2 + 3x)dz\right| \leq ML = 8 \cdot 4\pi = 32\pi.$

(b) $\left|\int_{C_R}\right| + \left|\int_{C_L}\right| = 8 \cdot 2\pi + 2 \cdot 2\pi = 20\pi.$

B Exercises

1. Similar to the proof of Theorem 1.3a.

3. Derivative of a sum is the sum of the derivatives. $\varphi'(t) \neq 0$ same as $|\varphi'(t)| > 0$.

5. (a) By taking absolute values. (b) By periodicity of $E(\theta)$. (c) $k(t)$ is continuous and integer valued, and so is a constant.

7. (a) $|12 + 5z| \geq 12 - 5|z| = 12 - 5 = 7$.

(b) On right $|12 + 5z| \geq 13$ so

$$\left|\int_R\right| + \left|\int_L\right| \leq \frac{1}{13} \cdot \pi + \frac{1}{7} \cdot \pi = \frac{20}{91}\pi.$$

C Exercises

1. Apply the chain law to $r = \sqrt{\xi^2 + \eta^2}$.

Section 1.5

A Exercises

1. (a) 0;   (b) 0;   (c) $\pi i/2$
3. (a) 1;   (b) 0;   (c) πi
5. (a) $-\pi i$;   (b) $-\pi i/3$;   (c) 0;   (d) $-2\pi i/3$

B Exercises

1. $\pi/2$ 3. $\pi/6$

5. $\displaystyle\int_C (z - \alpha)^k dz = \begin{cases} 0, & k \neq -1 \\ 2\pi i, & k = -1. \end{cases}$

7. Substitute $\theta = \varphi + \pi/2$.

C Exercises

1. (b) Set $az^2 + bz + c = a_1(z - \alpha)^2 + b_1(z - \alpha) + c_1$ and compute $a_1 = a$, $b_1 = 2a\alpha + b$, $c_1 = a\alpha^2 + b\alpha + c$. Then the integral is

$$\frac{1}{2\pi i}\left[\int_C a_z da + \int_C \frac{b_1 dz}{z - \alpha} + \int_c \frac{c_1 dz}{(c - \alpha)^2}\right] = b_1 = 2a\alpha + b$$

3. Set $e^{i\theta} = z$ and the integral becomes

$$I = \int_C \frac{1}{2^{2n}} \left(z + \frac{1}{z}\right)^{2n} \frac{dz}{iz} = \frac{1}{i4^n} \int_C \sum_{k=0}^{2n} \binom{2n}{k} z^k \frac{1}{z^{2n-k}} \frac{dz}{z}$$

Each term integrates to 0 except when $k = n$, so

$$I = \frac{1}{i4^n} \binom{2n}{n} \int_C \frac{dz}{z} = \binom{2n}{n} \frac{2\pi}{4^n} = \frac{(2n)!}{(n!)^2} \frac{2\pi}{4^n}$$

Section 1.6

A Exercises

1. If $z \in N(z_0, r)$, then

$$|z - \alpha| = |(z - z_0) + (z_0 - \alpha)| \le |z - z_0| + |z_0 - \alpha|$$
$$\le r + (a - 2r) = a - r < a,$$

and so $z \in N(\alpha, a)$.

3. If β and γ are two points in $N(\alpha, a)$, then the radii from α to them form a connecting polygonal line.

5. $\bar{R} = \{z : 4x^2 + 9y^2 \le 36\}$; $\tilde{R} = \{z : 4x^2 + 9y^2 \ge 36\}$.

7. Choose a sequence of neighborhoods $N(\alpha, 1/k) \equiv N_k$. Then for each k, there is a $z_k \in N_k \cap R$ with $z_k \ne \alpha$.

9. ∂S itself.

11. Unbounded since $n \in R$ for each $n > 0$. $\partial R = \{z : \operatorname{Re} z = 0\}$.

13. One center: 0.

B Exercises

1. $\partial S = \partial \tilde{S}$.

C Exercise

1. If $R = I(C)$, and Γ is a simple closed curve in R, then if $I(\Gamma) \not\subset R$, it must contain points of $\tilde{R}$, and so must cross C. This is a contradiction.

Section 2.1

A Exercises

1. (a) See your elementary calculus text; the calculations here are the
   same.

   (b) By part (a).

3. If $f'(\alpha)$ exists, then

$$[f(z) - f(\alpha)]/(z - \alpha) = f'(\alpha) + r(z)$$

where $r(z) \to 0$ as $z \to \alpha$. Then

$$f(z) - f(\alpha) = f'(\alpha)(z - \alpha) + r(z)(z - \alpha)$$

$$\to 0 \quad \text{as} \quad z \to \alpha.$$

5. By the third formula of A1(a).

7. To avoid zero in the denominator in the integrations done there.

B Exercises

1. If $u = c$, then $u_x = u_y = 0$ and by the Cauchy–Riemann equations
   $v_x = v_y = 0$, so both u and v are constant and hence f is constant.
   Similar argument if $v = c$. If $|f| = c$, then $u^2 + v^2 = c^2$. Then we can
   differentiate with respect to x and y, and use the Cauchy–Riemann
   equations to get $u_x = u_y = v_x = v_y = 0$ and so u and v are constant.

3. All of u_x, u_y, v_x, v_y are continuous and satisfy the Cauchy–Riemann
   equations.

5. (a) $[f(z) - f(\alpha)]/(z - \alpha) = (\bar{z} - \bar{\alpha})/(z - \alpha)$. Suppose $z = \alpha + h$ where
   h is real, then

$$\frac{\Delta f}{\Delta z} = 1 \to 1 \quad \text{as} \quad z \to \alpha.$$

Suppose $z = \alpha + ik$ where k is real, then

$$\frac{\Delta f}{\Delta z} = -1 \to -1 \quad \text{as} \quad z \to \alpha.$$

Thus the complex limit does not exist.

7. See your elementary calculus book.

9. $\dfrac{2}{\zeta^2} + \dfrac{3}{\zeta} + 5$ is not analytic at $\zeta = 0$.

11.

$$h^{(n)}(\alpha) = n!a_n = 0, \qquad\qquad\qquad \Rightarrow a_n = 0,$$

$$h^{(n-1)}(\alpha) = n!a_n\alpha + (n-1)!a_{n-1} = 0 \Rightarrow a_{n-1} = 0.$$

By iteration to $a_0 = 0$.

13. (a) If $w = uv$, compute w_{xx}, and w_{yy}, and use Cauchy–Riemann to
show $w_{xx} + w_{yy} = 0$.

   (b) $\operatorname{Im}[f(z)]^2 = 2uv$.

C Exercises

1. Set $f(z) = u + iw$. Then $v = \ln|f(z)| = \frac{1}{2}\ln(u^2 + w^2)$. Differentiate
and use the Cauchy–Riemann equations.

3. $p(1/\zeta) = \displaystyle\sum_0^n a_k/\zeta^k$ is not analytic at $\zeta = 0$.

Section 2.2

A Exercises

1. Easy calculations.

3. $F = \mu \left[\dfrac{1}{v} - \dfrac{1}{b+a}\right] \to \mu/b$ as $a \to \infty$.

B Exercises

1. $z^2 = [rE(\theta)]^2 = r^2 E(2\theta) = r^2\cos 2\theta + ir^2\sin 2\theta$, so $\operatorname{Re} z^2 = r^2\cos 2\theta$
is harmonic and also $\operatorname{Im} z^2$.

3. Set $u(x,y) = \sin x \sinh y$. Then $u_{xx} = -\sin x \sinh y$ and
$u_{yy} = \sin x \sinh y$, so $u_{xx} + u_{yy} = 0$. And $u(0,y) = 0$, $u(\pi,y) = 0$,
$u(x,0) = 0$.

5. $\mathbf{F} = \mu \int_0^\infty \dfrac{(-x\mathbf{i} - y\mathbf{j} + z\mathbf{k})dz}{(x^2 + y^2 + z^2)^{3/2}}$, and use

$$\int \frac{d\tau}{(1 + \tau^2)^{3/2}} = \frac{\tau}{(1 + \tau^2)^{1/2}} + c$$

Section 2.3

A Exercises

1. Clear.

3. For $\alpha_n = 1/n$, $\sqrt[n]{|\alpha_n|} = \sqrt[n]{1/n} = 1/\sqrt[n]{n} \to 1$. For $\alpha_n = 1/n^2$, $\sqrt[n]{|\alpha_n|} = \sqrt[n]{1/n^2} = 1/\sqrt[n]{n^2} = 1/(\sqrt[n]{n})^2 \to 1$.

5. (a) con;   (b) div;   (c) div.

7. M-test with $M_n = 2/n^2$.

B Exercises

1. It is sufficient to show that $n^2 r^n \to 0$. Set $r = 1/\rho$ and consider, as $x \to \infty$.

$$\frac{x^2}{\rho^x} \sim \frac{2x}{\rho^x \ln \rho} \sim \frac{2}{\rho^x (\ln \rho)^2} \to 0$$

since $\rho > 1$.

3. There is an N for which $|\alpha_n| < 1$ for $n > N$. Then $|\alpha_n|^2 < |\alpha_n|$ so $\sum |\alpha_n|^2$ converges by comparison with $\sum |\alpha_n|$, and $\sum \alpha_n^2$ converges since absolute convergence implies convergence

C Exercise.

1. Set $|\alpha_n| = a_n$, $|\beta_n| = b_n$. Then $(a_n - b_n)^2 \geq 0$ from which $\frac{1}{2}(a_n^2 + b_n^2) \geq a_n b_n$. Thus $\sum a_n b_n$ converges by comparison with $\sum \frac{1}{2}(a_n^2 + b_n^2)$. Then $\sum \alpha_n \beta_n$ converges since absolute convergence implies convergence.

Section 2.4

A Exercises

1. (a) 1;   (b) 0;   (c) ∞;   (d) ∞;   (e) 0;
   (f) 1/3;   (g) 1/a;   (h) 1;   (i) 1.

3. Because their partial sums are S_n and zS_n, respectively. The z does not influence the convergence behavior.

5. (a) 1;   (b) 1;   (c) ∞.

B Exercises

1. $m^{1/n} \leq \sqrt[n]{|a_n|} \leq M^{1/n}$, and each end $\to 1$ as $n \to \infty$.

3. (a) 1;   (b) 0;   (c) ∞;   (d) 1;   (e) 1.

5. Set $a = (\rho + r)/2$. Then $\Sigma |c_k a^k|$ converges and serves as an M-test comparison for $\Sigma c_k z^k$, $|z| \leq r$

7. (a) $\rho \geq \min(\rho_1, \rho_2)$;   (b) $\rho \geq \rho_1 \rho_2$;   (c) nil.

C Exercises

1. Set $q = \overline{\lim} \sqrt[n]{|a_n|}$. Then $\overline{\lim} \sqrt[n]{|a_n z^n|} = |z|q$ so that we have convergence for $|z|q < 1$:
   (a) all z ($\rho = \infty$) if $q = 0$
   (b) only $z = 0$ ($\rho = 0$) if $q = \infty$
   (c) for $|z| < 1/q$ if $0 < q < \infty$.

Section 2.5

A Exercises

1. $\rho = 1$ by the ratio test. Compute

$$f'(z) = \sum_1^\infty \frac{\alpha(\alpha - 1)\cdots(\alpha - n + 1)}{(n - 1)} z^{n-1} \qquad (*)$$

Now (i) multiply $(*)$ by z; (ii) change indices in $(*)$; (iii) add (i) and
(ii) termwise.

6. By Lemma 2.5a.

7. $w = \sum_0^\infty z^{2n}/n!$

8. (a) $w = \sum_0^\infty z^{2n}/(2n)!$

   (b) Exercise A2.

   (c) $w = \sum_{k=0}^\infty z^{kn}/(kn)!$

Section 2.6

A Exercises

1. Clear.
3. $\sinh z = \sinh x \cos y + i \cosh x \sin y$

   $\cosh z = \cosh x \cos y + i \sinh x \sin y$

5. $\dfrac{\sin z}{e^y} \to \dfrac{i}{2} e^{-ix}$ as $y \to \infty$.

7.
$$\dfrac{\cos z}{\sin z} = \left[\sum_0^\infty (-1)^n z^{2n}/(2n)! \right] \Big/ \left[z \sum_0^\infty (-1)^n z^{2n}/(2n+1)! \right]$$
$$= \dfrac{1}{z}\left[1 + \sum_1^\infty c_k z^k \right] = \dfrac{1}{z} + g(z)$$

   where $g(z) = \sum_1^\infty c_k z^{k-1}$. $g(z)$ has radius of convergence π where $\sin z$
   is 0.

9. Circles, radius a, centers 0 and α.
11. As in elementary calculus.
13. (a) $2\pi i$;   (b) πi.
15. $|e^z| = |e^x||e^{iy}| = e^x \le e^{|x|} \le e^{|z|}$.

B Exercises

1. (b) $\cos z = \cos x \cosh y - i \sin x \sinh y$

$$
\begin{aligned}
|\cos z|^2 &= \cos^2 x \cosh^2 y + \sin^2 x \sinh^2 y \\
&= \cos^2 x(1 + \sinh^2 y) + \sin^2 x \sinh^2 y \\
&= \cos^2 x + \sinh^2 y \\
&= \cos^2 x \cosh^2 y + \sin^2 x(\cosh^2 y - 1) \\
&= \cosh^2 y - \sin^2 x \\
&= \frac{1}{2}(1 + \cosh 2y) - \frac{1}{2}(1 - \cos 2x) \\
&= \frac{1}{2}(\cosh 2y + \cos 2x).
\end{aligned}
$$

3. $\tan z = \sin z \overline{\cos z}/|\cos z|^2$. Multiply out the numerator and use 1(b).

5. (a) by (9), Section 2.3.

   (b) Take real and imaginary parts of $\displaystyle\sum_0^m e^{ik\theta}$.

   (c) Take real parts.

C Exercise

1. (a) $Ae^z + be^{z(-1+i\sqrt{3})/2} + Ce^{z(-1-i\sqrt{3})/2}$

   (b) $A = B = C = 1/3$ (c) same as (b).

Section 2.7

A Exercises

1. (a) $2k\pi i$;   (b) $(\pi i/2) + 2k\pi i$;   (c) $\frac{1}{2}\ln 2 + i[(\pi/4) + 2k\pi]$;

   (d) $\ln 2 + i[(\pi/3) + 2k\pi]$.

3. (a) $e^{2k\pi^2 i}$;   (b) $e^{-2k\pi^2}$;   (c) $e^{-\pi/2} \cdot e^{-2k\pi}$

   (d) $2^{1/2} e^{-\pi/4 + 2k\pi} \cdot e^{-i\pi/4 + 2k\pi i - i\ln\sqrt{2}}$

   (e) $e^{-\pi/4 - 2k\pi} \cdot e^{(i/2)\ln 2}$.

5. $\mathrm{Log}\, 1 = 0$.

B Exercises

1. By integration.

3. Similar to Example 4.

5. $\log \dfrac{1}{\zeta} = \log \left(\dfrac{1}{\rho} e^{-i\varphi} \right) = \log \dfrac{1}{\rho} - i\varphi$

   $\qquad = -\log \rho - i\varphi = -\log \zeta.$

9. $\dfrac{p}{q} \log z = \dfrac{p}{q} \operatorname{Log} z + i(2p\pi)k/q$ gives q distinct values $k = 0, 1, \ldots, q-1$.

   Then repeats.

11. The cut keeps $\theta = \operatorname{Arg} z$ so we get

$$\int_1^\alpha z^{1/2} dz = (2/3)\alpha^{3/2} - 2/3$$

   where $\alpha^{2/3}$ is the principal value.

15. Similar to Example 4 and B3.

C Exercises

1. Since e^z is never zero, there are no branch points of $\sqrt{e^z}$.

3. Since $f(z)$ is never zero, the integral is independent of the path so the only ambiguity in

$$\log f(z) = \int_\alpha^z f'(\zeta)/f(\zeta)d\zeta + \log f(\alpha)$$

   is in the choice of the branch of $\log f(\alpha)$.

5. There is a $c > 0$ for which $|e^z| \geq 2c$ for $|z| \leq a$, and an N for which $|e^z - P_n(z)| \leq c$ for $|z| \leq a$ and $n \geq N$.

7. By C3 Re $\log f(z) = $ Re $\log g(z)$ from which $\log f(z) = \log g(z) + i\alpha$, where α is real.

9. $\Delta \ln |f(z)| = \Delta[\ln h(x) + \ln g(y)] = \left(\dfrac{h'(x)}{h(x)} \right)' + \left(\dfrac{g'(y)}{g(y)} \right)' = 0, \left(\dfrac{h'}{h} \right)' = $

   $2a, \left(\dfrac{g'}{g} \right)' = -2a$

Section 3.1

A Exercises

1. The integrand is odd.

3. $\int_C x\,dx = \int_C y\,dy = 0;\ \int_C y\,dx = A,\ \int_C x\,dy = -A.$

B Exercises

1. (a) $2\pi i e^{\alpha}$;   (b) $2\pi i e^{\alpha}$;   (c) $2\pi i e^{\alpha}/k!$

3. All integrals are zero.

5. $\int_0^{\infty} \cos r^2\,dr = \int_0^{\infty} \sin r^2\,dr = \sqrt{\pi}/2\sqrt{2}.$ Then set $r^2 = t.$

C Exercises

1. (a) By Theorem 3.1d.

   (b) $2\pi i.$

   (c) 2π and 0, respectively.

Section 3.2

A Exercises

1. 0

3. (a) $(e-1)2\pi i$;   (b) $(e+1)2\pi i$;   (c) $-2\pi i.$

5. $\dfrac{2\pi i}{3!}\left(\dfrac{d}{dz}\right)^3 \left(\dfrac{z}{1-z}\right)\bigg|_{-2} = 2\pi i/81$

7. (a) 0;   (b) $\pi i/12$

9. (a) Use $\dfrac{1}{z^2+1} = \dfrac{1}{2i}\left[\dfrac{1}{z-i} - \dfrac{1}{z+i}\right]$

   (b) Use $\dfrac{z}{z^2+1} = \dfrac{1}{2}\left[\dfrac{1}{z-i} + \dfrac{1}{z+i}\right]$

11. Because $f(z)$ and its derivatives are continuous on $C.$

B Exercises

3. $\rho < \min[|\alpha - \beta|, \ \mathrm{dist}\,(\alpha, C), \ \mathrm{dist}\,(\beta, C)]$

5. $|e^{f(z)}| = e^{\mathrm{Re}\,f(z)} \leq e^{M}$.

7. $\pi e/4$.

9. $f(z)$ is bounded.

C Exercises

1. If $|z| < 1$, then $|1/\bar{z}| > 1$ so

$$f(z) = \frac{1}{2\pi i} \int_{\zeta=1} \frac{f(\zeta)d\zeta}{\zeta - z}; \quad 0 = \frac{1}{2\pi i} \int \frac{f(\zeta)d\zeta}{\zeta - (1/\bar{z})}.$$

In the second integral multiply numerator and denominator by $\bar{z}$, then subtract the second integral from the first.

3. The proof of Theorem 3.2d by induction works here. The proof by integration by parts does not.

Section 3.3

A Exercises

1. (a) $\displaystyle\sum_0^\infty (-1)^n (z-1)^n$;   (b) $\displaystyle\sum_0^\infty (-1)^n (z-2)^n / 2^{n-1}$;

(c) $\displaystyle\sum_0^\infty \frac{(-1)^{n+1}(z - \pi/2)^{2n+1}}{(2n+1)!}$;   (d) $\displaystyle\sum_0^\infty \frac{(-1)^{n+1}(z - \pi)^{2n+1}}{(2n+1)!}$;

(e) $\displaystyle\sum_0^\infty (-1)^{n+1} z^{2n+1}/(2n+3)!$;   (f) $\displaystyle\sum_0^\infty \frac{(-1)^n z^{2n+1}}{(2n+1)[(2n+1)!]}$;

(g) $\displaystyle\sum_0^\infty (-1)^{n+1} z^n / n$.

3. (a) $\displaystyle -\sum_0^\infty 1/z^{n+1}$;   (b) $\displaystyle\sum_0^\infty \frac{(-1)^n}{z^{2n}(2n)!}$;   (c) $\displaystyle\sum_0^\infty z^{n-1}$;

(d) $\displaystyle -\sum_0^\infty 1/z^{n+2}$;   (e) $\displaystyle\sum_0^\infty (-1)(z-1)^{n-1}$;   (f) $\displaystyle\sum_0^\infty (-1)^{n+1}(z-1)^{n+2}$

(g) $1 + \dfrac{1}{z}$;   (h) $\dfrac{1}{z}$;   (i) $\displaystyle\sum_{0}^{\infty} \dfrac{(-1)^n z^{2n-2}}{(2n+1)!}$;   (j) $\displaystyle\sum_{0}^{\infty} \dfrac{(-1)^n z^{2n-3}}{(2n)!}$

5. Each of these functions has no term of negative index in the Laurent expansion. Thus each can be considered analytic at $z = 0$.

B Exercises

1. Each end is a power series in $(z - a)$ or $1/(z - a)$, and so is analytic. Termwise integration verifies that the c_n's are the Laurent coefficients.

3. Integrate around $|z| = r$. Then

$$|c_n| \leq \frac{A + Br^k}{r^n} \to 0 \text{ as } r \to \infty \text{ for } n > k.$$

5. (a) Each branch is single valued in the indicated rings by arguments like those in Example 3, Section 2.7.

    (b) $|z - 1| > 2; \ 2 < |z + 1| < 3$ and $|z + 1| > 4$.

7. Set $f(z) = \displaystyle\sum_{0}^{\infty} c_n(z - a)^n$. Let $z - a = re^{i\theta}$ and take real parts.

C Exercises

1. Set $e^{i\theta} = z$, then the integral becomes

$$\int_{|z|=1} e^{z + \frac{1}{z}} \frac{dz}{iz} = \int_{|z|=1} e^z \cdot e^{1/z} \frac{dz}{iz}.$$

Expand e^z in a Taylor series, and $e^{1/z}$ in a Laurent series.

5. $|c_n| \leq Mr^{k-n} \to 0$ as $r \to \infty$ if $n > k$

            $\to 0$ as $r \to 0$ if $n < k$.

  so $c_n = 0$ for all n.

7. (a) The exponential function is analytic for all finite $z \neq 0$.

    (b) In the integral for the Laurent coefficient $J_n(w)$, set $z = e^{i\theta}$ and simplify.

    (c) In the equation

$$\exp\left\{ \frac{w}{2}\left(z - \frac{1}{z}\right) \right\} = \sum_{-\infty}^{\infty} J_n(w)z^n$$

replace z by $-\dfrac{1}{z}$.

(d) The constant term in the product of the series for $e^{\frac{wz}{2}}$ and $e^{-\frac{w}{2z}}$.

(e) From (d) and C1.

Section 3.4

A Exercises

1. (a) 1;   (b) 2;   (c) 3;   (d) 1.
3. $1/f(z)$ is analytic in R.
5. (a) Easy.   (b) Cauchy's theorem.   (c) Cauchy's theorem.

B Exercises

1. Much like A5 above.
3. u is either constant, or cannot achieve its maximum and minimum in R, but must achieve it in $\bar{R}$.
5. If $f(z)g(z) \equiv 0$, then either (a) both are $\equiv 0$, or (b) one, say $f(z)$, is $\not\equiv 0$. Then in case (b) there is a neighborhood in which $f(z)$ is never zero, and so $g(z) \equiv 0$ there, and by the identity theorem $g(z) \equiv 0$ in R. So at least one of them is $\equiv 0$.

C Exercises

1. Set $g(z) = f(z)/z^n$. Then $|g(z)| \leq 1$ for $0 < |z| < a$, and for each $r, 0 < r < a$, there is a z_0 with $|z_0| = r$ for which $|g(z_0)| = 1$.
3. $f(z)/P(z)$ has a removable singularity at each α_j and is never zero.
5. $F(re^{i\theta}) \to 0$ uniformly as $r \to 1-$, $0 \leq \theta \leq 2\pi$. Thus $\max\limits_{|z| \leq 1} |F(z)| = 0$.

Section 3.5

A Exercises

1. (a) $|1/(n+z) - 0| \leq 1/|n+z| \leq \dfrac{1}{n-1} \to 0$ as $n \to \infty$.

   (b) $|e^{-nz}/n - 0| \leq e^{-nx}/n \leq 1/n \to 0$.

   (f) $\left| \dfrac{z^{2n}}{1+z^{2n}} - 1 \right| = \left| \dfrac{1}{1+z^{2n}} \right| \leq \dfrac{1}{|z|^{2n}-1} \leq \dfrac{1}{b^{2n}-1} \to 0$.

3. Apply Theorem 3.5c.

5. (a) Theorem 3.5d.

   (b) Integrate $f(z)$ by parts.

7. Apply Theorem 3.5i with z replaced by $z + n$.

B Exercises

1. Choose an arbitrary $\alpha \neq 1, 2, 3, \ldots$, and let $f(z)$ denote the series. Let $2\rho = \min_{n>1} |\alpha - n|$ and $R = \{z : |z - \alpha| < \rho\}$, and choose $N > 2(|\alpha| + \rho)$. Write $f(z) = \sum_1^N + \sum_{N+1}^\infty \equiv \sigma_1 + \sigma_2$, respectively. Then σ_1 is analytic at α since it is a finite sum of analytic functions. In σ_2 we have $|z| \leq |\alpha| + \rho < n/2$ so

$$\frac{1}{n(n-1z)} \leq \frac{1}{n} \frac{1}{n-|z|} \leq \frac{2}{n^2}.$$

   Thus σ_2 converges uniformly in R and so is analytic at α.

3. The Taylor series for $\sin(z/n^2)$ at $z = 0$ is

$$\sin \frac{z}{n^2} = \frac{z}{n^2}\left[1 + \sum_{k=1}^\infty \left(\frac{z}{n^2}\right)^{2k} \frac{(-1)^k}{(2k+1)!}\right]$$

so, for $|z| < a$ we have

$$\left| \sin \frac{z}{n^2} \right| \leq \frac{a}{n^2}\left[1 + \sum_{k=1}^\infty \left(\frac{a}{n^2}\right)^{2k} \frac{1}{(2k+1)!}\right],$$

and there is an N for which the bracketed terms are < 2 for $n > N$. Thus

$$\sum_0^\infty \sin \frac{z}{n^2} = \sum_0^N \sin \frac{z}{n^2} + \sum_{N+1}^\infty \sin \frac{z}{n^2}$$

The first sum is analytic in $\mathbb{C}$, and

$$\left| \sum_{N+1}^\infty \sin \frac{z}{n^2} \right| \le \sum_{N+1}^\infty \frac{2a}{n^2}$$

and so is analytic for $|z| < a$. Thus the original series is analytic for $|z| < a$. But a is arbitrary, so the series is an entire function.

5. $\left| \int_0^\infty f(z,t)dt - \int_0^n f(z,t)dt \right| \le \int_n^\infty M(t)dt \to 0.$

7. $a/(z^2 - a^2); \ -2az/(z^2 - a^2)^2; \ z/(z^2 - a^2); \ (z^2 + a^2)/(z^2 - a^2).$

9. Set $e^{-t} = \tau$ in the defining integral.

C Exercises

1. Let $R_a = \{z : \operatorname{Re} z = x > a > 0\}$. Then for $0 < t \le 1$

$$\left| \frac{t^{z-1}}{t(t+1)} \right| \le \frac{t^x}{t(t+1)} \le \frac{t^{a-1}}{t+1} \le \frac{1}{2} t^{a-1}$$

and $\int_0^1 t^{a-1} dt$ converges (and $= 1/a$—but it's the convergence that's important). Thus $\int_0^1 \frac{t^{z-1}}{1+t} dt$ is analytic for $\operatorname{Re} z > a$. But if $\operatorname{Re} z = b > 0$, we can take $a = b/2$ so the integral is analytic for $\operatorname{Re} z > 0$. Similarly $\int_1^\infty \frac{t^{z-1} dt}{1+t}$ is analytic for $\operatorname{Re} z < 1$.

3. Mimic the proof of Theorem 3.5d.

Section 3.6

B Exercises

1. If $f(z) = \sum c_k (z - \alpha)^k$, then $\overline{f(\bar z)} = \sum \bar c_k (z - \bar\alpha)$.

3. $\rho = 1$ and $\sum \dfrac{1}{2^k+1}$ converges. It has a natural boundary at $|z| = 1$ since $f'(z)$ does.

5. $\Gamma(z+1) = z\Gamma(z)$. Hence $\Gamma(z) = \Gamma(z+1)/z$, where the right side is analytic for $z > -1$ except at $z = 0$. $\Gamma(z+2) = (z+1)\Gamma(z)$ so $\Gamma(z) = \Gamma(z+2)/[z(z+1)]$ and the right side is analytic except at $z = 0, -1$. by induction we can continue to any z except $z = -n$.

C Exercises

1. $f(z) = 2f\left(\frac{1}{2}\,z\right)f'\left(\frac{1}{2}\,z\right)$ and the right side is analytic for $|z| < 2a$. Iterate: $f(z)$ is then analytic for $|z| < 2^n a$ for arbitrary n.

3. From C2 as B3 followed from Example 5.

5. (a) Set $t = (1 - \tau)$ in the defining integral.

   (b) Because $\displaystyle\int_0^1 t^{a-1}\,dt$ converges for $a > 0$.

   (c) Direct substitution.

   (d) $B(z, 1-z) = \displaystyle\int_0^\infty \dfrac{t^{z-1}\,dt}{1+t} = \pi\cos\pi z.$

   (e) Direct substitution.

Section 4.1

B Exercises

1. $f'(z) \neq 0$ since the map is conformal, and $z'(t) \neq 0$ since C is smooth. Thus
$$\frac{dw}{dt} = \frac{d}{dt}\,f(z(t)) = f'(z(t))z'(t) \neq 0.$$
And the length of Γ is
$$\int_I \left|\frac{dw}{dt}\right|\,dt = \int_I |f'(z(t))||z'(t)|\,dt,$$

3. By Liouville theorem $f(z)$ is a constant and so cannot map onto the unit circle.

C Exercises

1. If $\beta = f(\alpha)$, then $\alpha = f^{-1}(\beta)$:

$$\frac{f^{-1}(w) - f^{-1}(\beta)}{w - \beta} = \frac{z - \alpha}{f(z) - f(\alpha)} \rightarrow \frac{1}{f'(\alpha)}.$$

So $[f^{-1}(\beta)]' = 1/f'(\alpha)$.

Section 4.2

A Exercises

1. Each z maps onto a unique $w = \alpha z + \beta$, and each w onto a unique $z = (w - \beta)/\alpha$, and w and z are both large when either is.

3. This is the condition that $(\alpha z + \beta) \equiv c(\gamma z + \delta)$ for some constant c.

5. Each z gives one w, and conversely each w gives one z.

7. $x = a$ gives $a(u^2 + v^2) = u$, then complete the square on the u terms.

B Exercises

1. Substitute from equations (7).

3. Substitute and simplify.

5. If ∞ is a fixed point, the transformation is linear and apply B2. If all three are finite points, we have three distinct solutions of $\gamma z^2 + (\delta - \alpha)z - \beta = 0$. Let these be z_1, z_2, z_3. Then

$$\gamma z_1^2 + (\delta - \alpha)z_1 - \beta = 0$$
$$\gamma z_2^2 + (\delta - \alpha)z_2 - \beta = 0.$$

Thus

$$\gamma(z_1^2 - z_2^2) + (\delta - \alpha)(z_1 - z_2) = 0,$$

so

$$\gamma(z_1 + z_2) + (\delta - \alpha) = 0.$$

Similarly

$$\gamma(z_1 + z_3) + (\delta - \alpha) = 0$$

so

$$\gamma(z_1 - z_3) = 0.$$

Thus $\gamma = 0 \Rightarrow \alpha = \delta \Rightarrow \beta = 0.$

Section 4.3

A Exercises

1. $\alpha = i(5e^{i\pi/3} - 2).$ 3. $|w| < 1.$

5. $w = [(3 - i)z - 2]/(z - i).$

7. $w = (z\bar{\alpha} - \alpha)/(z - 1).$

9. $w = \dfrac{ze^{3\pi i/4} - (1/\sqrt{2})i - \alpha}{ze^{3\pi i/4} - (1/\sqrt{2})i - \bar{\alpha}}$ where $\operatorname{Im}\alpha > 0.$

B Exercises

1. (a) Rotate the z plane in Example 1.

   (b), (c) See **Remarks** after Example 1.

3. For $a < b$, $w = (z - a)/(b - a).$

C Exercises

1. $w = (z - 1)/(z - 3)$ maps C_1 onto L_1, and C_2 onto L_2 where each L is a line through $w = 0$. By conformality the slope of L_1 is the value of $y' = dy/dx$ at $(1,0)$ from C_1, and similarly for L_2 and C_2. Thus the slope of L_1 is 1 and of l_2 is $1/2$. Thus $\varphi = \pi/4 - \tan^{-1}\frac{1}{2}$. Let $\theta = \tan^{-1}\frac{1}{2}$. Then $w = e^{-i\theta}(z - 1)/(z - 3)$ is the desired mapping.

3. Let C consist of the segments $u \leq 0$ and $u \geq 1$ of the u-axis ($v = 0$), and the arc of $(u - \frac{1}{2})^2 + (v + \frac{1}{2})^2 = \frac{1}{2}$ which lies above the segment $0 < u < 1$. Then the image region is that part of the w-plane on and above C.

Section 4.4

B Exercises

1. (a) Inside the parabola $u = 1 - v^2/4$ $(u < 1 - v^2/4)$ slit along the negative u-axis.

   (b) $u > v^2/4 - 1$, slit along the positive u-axis.

   (c) $\{v^2/4b^2 - b^2 < u < a^2 - v^2/4a^2,\ v > 0\}$.

3. If $w = \rho e^{i\varphi}$, then $\{e^a < \rho < e^b,\ c < \varphi < d\}$.

5. (a) Angles from the origin $a < \theta < b$.

   (b) $w = e^{iz}$ maps the strip onto the upper half-plane. Then use Example 1, Section 4.3.

7. Translate: $\zeta = z - \pi$; then the ζ strip maps onto the negative of the sine mapping discussed in the text.

9. $\{2n\pi < x < (2n + \frac{1}{2})\pi,\ y > 0\}$ for all n.

11. Take w from C1, Section 4.3, and set $\zeta = (w)^{\pi/\theta}$. Apply Example 1, Section 4.3.

Section 4.5

A Exercises

1. See any trigonometry book.

3. An angle of 2π spans a full circle.

B Exercises

1. (a) $\cos x \sinh y$ is zero on the boundary.

   (b) Both y and xy are zero when $y = 0$.

3. Similar to Example 4 with $\theta = \arg(w - 1)$.

5. $\dfrac{1}{2}\dfrac{y}{x^2 + y^2} - 1$.

7. L maps onto $u > 0$. Then take $h(w) = (1/\pi)\cot^{-1}[(u - 1)/v]$ and transplant back to the z-plane.

9. We have $a_1 = 1$, $a_2 = 1$, and $a_3 = \infty$ with $\beta_1 = 2/3$, $\beta_2 = 2/3$ so β_3 must also be $2/3$. Thus the triangle is equilateral.

11. $b = \dfrac{1}{2} B\left(\dfrac{1}{4}, \dfrac{1}{4}\right) = \Gamma^2\left(\dfrac{1}{4}\right)/2\sqrt{\pi}$,

$b' = B\left(\dfrac{1}{2}, \dfrac{1}{4}\right).$

Section 5.1

A Exercises

$(P, n) = $ pole of order n; $R = $ removable; $E = $ essential

1. (a) 0, R;   (b) 0, R;   (c) 0, (P, n);
   (d) 0, $(P, 1)$; $2n\pi$, $(P, 2)$;   (e) 0, R; $2n\pi$, $(P, 2)$;
   (f) 0, R; $2n\pi$, $(P, 2)$;   (g) $n\pi$, $(P, 1)$
   (h) $(n + \frac{1}{2})\pi$, $(P, 2)$;   (i) $2n\pi i$, $(P, 1)$.

3. $P = \displaystyle\sum_{k=1}^{n} \dfrac{a_{-k}}{(z - \alpha)^k} = \dfrac{1}{(z - \alpha)^n} \displaystyle\sum_{k=1}^{n} a_{-k}(z - \alpha)^{n-k}$. Then set $Q(z)$ equal to the last sum: $Q(\alpha) = a_{-n}$.

B Exercises

1. If the radius is r, then $e^{1/z} = \alpha$ for $|z| < r$ is equivalent to $e^\varsigma = \alpha$ for $|\varsigma| > 1/r$. Then periodicity applies.
3. If $f(z)$ has a pole at α, then $1/f(z)$ has a zero there, and zeros are isolated.
5. Apply equation (4) to each of g and h.

C Exercise

1. Use the argument of Theorem 5.1a.

Section 5.2

A Exercises

1. (a), (b) No poles;   (c) $R = \frac{1}{2}$ at 0

    (d) $R = 2$ at $2n\pi$, $n = 0, \pm 1, \pm 2, \ldots$;

    (e) $R = 4n\pi$ at $2n\pi$, $n = \pm 1, \pm 2, \ldots$;

    (f) $R = 24n^2\pi^2$ at $2n\pi$, $n = \pm 1, \pm 2, \ldots$;

    (g) $R = 1$ at $(n + \frac{1}{2})\pi$, $n = 0, \pm 1, \pm 2, \ldots$;

    (h) $R = 0$ at $n\pi$, $n = 0, \pm 1, \pm 2, \ldots$;

    (i) $R = 1$ at $2n\pi i$, $n = 0, \pm 1, \pm 2, \ldots$.

3. By the formula for $\Gamma(z)$ in Example 4, Section 3.6.

B Exercises

1. (a) $R = 1$;   (d) $R = 2n\pi i$ when $\log z = \operatorname{Log} z + 2n\pi i$

3. See the arguments in the proofs of Theorem 5.3a and 5.3c in the next Section.

5. By B4 above.

7. Similar to B6: The residue at α_k is $n_k g(\alpha_k)$ and at β_j is $m_j g(\beta_j)$.

C Exercises

1. $\dfrac{1}{2\pi i} \displaystyle\int_C f'(z)/f(z)\,dz = \dfrac{1}{2\pi i}\,\Delta_C(\log f(z))$

$$= \frac{1}{2\pi i}[\Delta_C |f(z)| + i\Delta_C \arg f(z)] = \frac{1}{2\pi}\Delta_C \arg f(z).$$

3. By arguments similar to the proof of 5.3a, we get $|g(z)| < |f(z)|$ on $|z| = R$ when R is large.

Section 5.3

A Exercises

1. $1/x^2$ is not integrable $\int_0^1 1/x^2\,dx$ does not exist. See A5(a) below.

5. (a) $\int_a^2 \dfrac{dx}{x^2} = -\dfrac{1}{x}\Big]_a^2 = \dfrac{1}{a} - \dfrac{1}{2} \to \infty$ as $a \to 0$.

   (c) $\int_0^a \sqrt{|x|} = \int_0^a x^{1/2}\,dx = 2x^{3/2}/3\Big]_0^a = 2a^{3/2}/3 \to \infty$ as $a \to \infty$.

B Exercises

1. π/a

2. Use
$$\frac{1}{(x^2 + a^2)(x^2 + b^2)} = \frac{1}{b^2 - a^2}\left[\frac{1}{x^2 + a^2} - \frac{1}{x^2 + b^2}\right]$$

   to get
$$\frac{1}{b^2 - a^2}\left[\frac{\pi}{a} - \frac{\pi}{b}\right]$$

3. $\pi/2a^3$.

5. From B2, B3 above.

Section 5.4

A Exercises

2. $\pi/(2e^a)$ 4. $\pi/(4e)$ 6. $\pi/(2e)$ 8. 0

B Exercises

2. $\dfrac{\pi}{2(b^2 - a^2)}\left[\dfrac{1}{a}e^{-a} - \dfrac{1}{b}e^{-b}\right]$ 4. $\pi e^{-1}\sin 1$

6. $\pi/6$ 8. $2\pi/(3\sqrt{3}\,a^2)$

Section 5.5

B Exercises

6. (b) $\pi(\ln a - 1)/2a^3$

Section 5.6

A Exercises

2. Set $t - \tau = s$ in the integral.

3. If $|f(t)| \le Me^{at}$ and $|g(t)| \le Ne^{at}$, take $c = \max(a, b)$ and get
$|f * g(t)| \le MNte^{ct} \le (MN/e)e^{(c+1)t}$.

5. (a) $-\frac{1}{3} t - \frac{1}{9} + \frac{1}{2} e^{3t}$;   (b) te^{-3t};
    (c) $\frac{1}{2} - e^{-t} + \frac{1}{2} e^{-2t}$ (d) $\frac{1}{2} - e^t + \frac{1}{2} e^{2t}$;
    (e) $\frac{1}{6} e^{-t} - \frac{1}{2} e^t + \frac{1}{3} e^{2t}$.

B Exercises

1. $\mathcal{L}(f^{(n)}) = z^n \mathcal{L}(f) - \sum_{k=0}^{n-1} z^{n-1-k} f^{(k)}(0)$.

Section 6.1

A Exercises

3. $p(t; x, y) = \dfrac{1}{\pi} \operatorname{Im} \dfrac{1}{t - z}$

7. Theorem 6.1a for $x \ge a$ becomes: Suppose
    (i) $f(z)$ is analytic for $x \ge a$,
    (ii) $f(z) \to 0$ uniformly as $z \to \infty$ for $x \ge a$,
    (iii) $\displaystyle\int_{-\infty}^{\infty} \dfrac{|f(a + i\tau)| d\tau}{1 + |\tau|}$ converges. Then

$$-\frac{1}{2\pi} \int_{-\infty}^{\infty} \frac{f(a + i\tau) d\tau}{a + i\tau - z} = \begin{cases} f(z) & x > a, \\ 0 & x < a \end{cases}$$

B Exercises

1. Jordan's Lemma applies.

C Exercises

1. We only need show that the integral over the semi-circle of the usual contour goes to zero as the radius $\to \infty$.

$$\left| \int_S \frac{f(\zeta)d\zeta}{\zeta - z} \right| \leq \int_0^\pi \frac{|f(Re^{i\theta})|Rd\theta}{R - |z|}.$$

Take $R > 2|z|$, then the integral is $\leq$

$$2 \int_0^\pi |f(Re^{i\theta})|d\theta = 2 \left(\int_0^a + \int_a^{\pi-a} + \int_{\pi-a}^\pi \right) |f(Re^{i\theta})|d\theta.$$

The two end integrals are small when a is small; the middle one is small when R is large.

Section 6.2

A Exercises

1. (a) $\theta = \operatorname{Im} \operatorname{Log}(z - c)$
3. $u = (y + 1)/[x^2 + (y + 1)^2]$

B Exercises

1. (a) h_n is continuous and the $\int_{-\infty}^\infty |h_n(t)|/(1 + |t|)dt$ converges.

(b)
$$\frac{1}{\pi} \operatorname{Im} \{z[\operatorname{Log}(z + n - 1) - \operatorname{Log}(z + n)]$$
$$+ \operatorname{Log}(z + n - 1) - \operatorname{Log}(z - n + 1)$$
$$+ z[\operatorname{Log}(z - n) - \operatorname{Log}(z - n + 1)]\}$$

C Exercises

1. (a) Take $|t| \geq 2|x|$, and $n \geq 1$ and $\geq \dfrac{1}{2|y|}$. Then

$$(x-t)^2 + y^2 \geq \frac{1}{4} t^2 + y^2 \geq \frac{1}{4n^2} t^2 + \frac{1}{4n^2} = \frac{1}{4n^2} (t^2 + 1).$$

Thus $\displaystyle\int_{2|x|}^{\infty} pg\,dt$ and $\displaystyle\int_{\infty}^{-2|x|} pg\,dt$ converge by comparison, and $\displaystyle\int_{-2|x|}^{2|x|} pg\,dt$
exists by piecewise continuity.

(b) Use A3 of Section 6.1.

Section 6.3

B Exercises

3. $u(r,\theta) = \displaystyle\int_{0}^{2\pi} P(a,t; \frac{1}{r}, -\theta) g(-t)\,dt$

C Exercises

1. Let α be an arbitrary point in R and $N(\alpha, a)$ be a neighborhood whose
   closure is in R. Then translate the origin to α (see B4 above) and write

$$u_n(r,\theta) = \int_{0}^{2\pi} P(a,t; r,\theta) u_n(a,t)\,dt, \quad 0 \leq r < a.$$

Then by uniform convergence on $\partial N(\alpha, a)$, we get

$$u(r,\theta) = \int_{0}^{2\pi} P(a,t; r,\theta) u(a,t)\,dt$$

and so u is harmonic in $N(\alpha, a)$. Since α is arbitrary, u is harmonic in
R.

Section 6.4

B Exercises

1. $\Delta u = u_{rr} + \dfrac{1}{r} u_r + \dfrac{1}{r^2} u_{\theta\theta}$

2. $\dfrac{\partial u}{\partial y} = \displaystyle\int_{-\infty}^{\infty} p(t; x, y)g(t)dt.$

Section 6.5

A Exercises

5. (a) $\dfrac{\pi}{2} - \dfrac{4}{\pi} \displaystyle\sum_{1}^{\infty} \dfrac{\cos(2k + 1)\theta}{(2k + 1)^2}$

(b) $\sin\theta$

(c) $\displaystyle\sum_{1}^{\infty} \dfrac{1}{k} \sin k\theta$

7. Because $f(x)$ is extended to $-\pi \le x \le \pi$ as an odd function.

9. To flatten the graph at 0 and π so that the derivative is zero there.

B Exercises

1. Because the differentiated series are uniformly convergent.

3. By (10), $\gamma_0' = 0$. Then from the sine series

$$u(\pi, t) = \sum_{1}^{\infty} b_n e^{n^2 t} \sin n\pi = 0.$$

5. $\gamma_n = \gamma_{-n}.$

C Exercises

1. $u = \displaystyle\sum_{1}^{\infty} b_n \sin nx \cos nt; \quad b_n = \dfrac{2}{\pi} \displaystyle\int_{0}^{\pi} f(x) \sin nx\, dV$

3. $u = \displaystyle\sum_{1}^{\infty} r^n (\gamma_n e^{in\theta} + \gamma_{-n} e^{-in\theta})/n;$

$\gamma_n = \dfrac{1}{2\pi} \displaystyle\int_{-\pi}^{\pi} g(\theta) e^{-in\theta}\, d\theta,\ n = \pm 1, \pm 2, \ldots.$

Index